AF845557

Water Management in the Premodern Middle East

Middle East Environmental Histories

Middle East Environmental Histories seeks to publish the best work on the environmental history of the Middle East from late antiquity to the present. People in the Middle East, as everywhere, have lived in, with, and against the natural world. The multiple engagements of these peoples with environments over time helped to produce ideas, states, cultures, literatures, families, and societies. This series aims to further elaborate these histories. It will examine the Middle East in its global context while always keeping the particularities of the Middle East and its various environments in focus—the region's historic role as the crossroads of trade between the Mediterranean and Asia, its cultural and religious diversity, the place of Islam, the balance between settled and sown, and the role of oil, to name a few. Although a series about the history of the Middle East, it will lean on interdisciplinary influences from other fields. Overall, it desires to mark the best of this growing field and help shape its future.

Series Editor
Alan Mikhail, Yale University

Series Board
Maaike van Berkel, Radboud University
Amina Elbendary, American University of Cairo
Faisal Husain, Pennsylvania State University
Onur Inal, University of Vienna
Michael Christopher Low, University of Utah
Zozan Pehlivan, University of Minnesota
Caterina Scaramelli, Boston University
Sarra Tlili, University of Florida

Other titles in this series:

Yonca Köksal Özyaşar and Can Nacar, *Anatolian Livestock Trade in the Late Ottoman Empire*, 2024

Richard W. Bulliet, *Exploring Animal Energy in the Arid Zone: More Camels, Fewer Wheels*, 2024

WATER MANAGEMENT IN THE PREMODERN MIDDLE EAST

Forces from "Above" and "Below"

Edited by

Josephine van den Bent, Maaike van Berkel,

and Edmund Hayes

Leiden University Press

This publication is part of the project "Source of Life: Water Management in the Premodern Middle East (7th-15th c.)" with file number VI.C.181.016 of the research programme VICI SGW which is financed by the Dutch Research Council (NWO).

Cover design: Andre Klijsen
Cover illustration: Image of the Nilometer from *'Ajā'ib al-makhlūqāt wa-gharā'ib al-mawjūdāt* by al-Qazwīnī (d. 682/1283). British Library: Oriental Manuscripts, Or 14140, folio 62v; via Qatar Digital Library.
Lay-out: Crius Group, Hulshout
Printer: Scanlaser bv, Zaandam

Every effort has been made to obtain permission to use all copyrighted illustrations reproduced in this book. Nonetheless, whosoever believes to have rights to this material is advised to contact the publisher.

ISBN 9789087284718
e-ISBN 9789400605084 (e-PDF)
e-ISBN 9789400605657 (e-PUB)
https://doi.org/10.24415/9789087284718
NUR 692

© 2026, Josephine van den Bent, Maaike van Berkel & Edmund Hayes / Leiden University Press, Rapenburg 73, 2311 GJ Leiden

All rights reserved. Without limiting the rights under copyright reserved above, no part of this book may be reproduced, stored in or introduced into a retrieval system, or transmitted, in any form or by any means (electronic, mechanical, photocopying, recording or otherwise) without the written permission of both the publisher and the editors of the book.

Creative Commons Licence CC BY-NC-ND (https://creativecommons.org/licenses/by-nc-nd/4.0/deed)
Contact EU General Product Safety Regulation (GPSR): productsafety@lup.nl

Printed and bound by CPI Group (UK) Ltd, Croydon, CR0 4YY

Table of Contents

Preface and Acknowledgements

This book is the collaborative outcome of the "Source of Life" project, an NWO (Dutch Research Council) VICI project at Radboud University in Nijmegen, the Netherlands. The idea of exploring urban life through the prism of water management was first developed during a sabbatical in Cairo in the tumultuous spring and summer of 2013, when I spent long hours in the Dār al-Wathāʾiq al-Qawmiyya (the National Archive) and the Wizārat al-Awqāf (Ministry of Endowments) to find out whether *waqf* documents could contribute to our understanding of the organisation of urban "public services". Water was omnipresent in the *waqf* documents and soon presented itself as the most precious commodity (urban) societies needed to manage. Water management therefore became key, for me, in thinking about the social institutions, practices, and people involved in organizing urban life in the premodern Middle East. *Waqf* documents were also instrumental in demonstrating that the decentralized organization of these cities, and the blurring of state and society, made it necessary to move beyond a clear-cut public-private divide in analysing water management, and instead consider not only state institutions, but also "everyday governance"—community-rooted forms of organization. The primary source of the resilience and endurance of water supply in Middle Eastern cities was not a strong centralized administration and water policy, but its organizational hybridity.

Despite the obvious importance of this vital resource in premodern urban society, there are hardly any historical sources which place water management centre stage. Therefore, studying water management, especially in different eras and cities, requires an interdisciplinary team working with diverse textual sources, as well as with archaeological evidence. In 2019 the Dutch Research Council gave me the opportunity to put together such a team of researchers and it is, first and foremost, thanks to them—Josephine van den Bent, Peter J. Brown, Edmund Hayes, and Angela Isoldi—that we can now present our collaborative results. For me, working with Josephine, Angela, Peter, and Ed, building a database, organizing workshops, discussing sources, and life, and above all, travelling together in Egypt, Jordan, Turkey, and Saudi Arabia, to explore pipes, canals, cisterns, and latrines, has been the greatest inspiration and joy of this project. I owe a lasting debt of gratitude to all four of them.

However, the outcomes of this project have been truly collaborative in many other respects as well. As a team we profited tremendously from a large group of scholars who were willing to explore the water history of the premodern Middle Eastern city with us. Their advice, questions, insights, and sharing of data, at various stages of the project, during seminars, workshops, conferences and informal meetings, have been a huge inspiration, and their support and enthusiasm have immensely contributed to the outcomes of this project. I am grateful to: Janet Alberda, Ajab Alotibi, Dina Ishak Bakhoum, Massimiliano Borroni, Janna Coomans, Jim Crow, Mark Driessen, Amina Elbendary, Roeland Emaus, Maurits Ertsen, Yehoshua Frenkel, Alison Gascoigne, Maartje van Gelder, Guy Geltner, Ingrid Hehmeyer, Carola Hein, Jasser Al-Herbish, Jaafar Jotheri, Sufyan Al Karaimeh, Hugh Kennedy, Derek Kennet, Majid Labbaf Khaneiki, Remke Kruk, Jen Lavris Makovics, Amalia Levanoni, Alan Mikhail, Marcus Milwright, Heba Mostafa, Hagit Nol, Alastair Northedge, Astrid van Oyen, Yossef Rapoport, Louise Rayne, Ieva Reklaityte, Tarek Sabraa, Henk van Schaik, Petra Sijpesteijn, Petra Smulders, Mohamed Soliman, Vanessa Van Renterghem, Jo Van Steenbergen, Mariëtte Verhoeven, Peter Verkinderen, the late Bert de Vries, Bethany Walker, Nicholas Warner, Peter Webb, Ghazwan Yaghi, and Abdullah al-Zahrani.

At Radboud University, our project found a warm home and a vibrant research environment among our colleagues from the chair group of Ancient and Medieval History. I want to say a special thanks to two colleagues with whom I share a love for watery things. Nathalie de Haan and Mariëtte Verhoeven both wholeheartedly contributed to our project in its various stages. The colleagues of the Humanities Lab built a database for us. Several students helped us in our research and with organizing events. I would like to thank Celis Tittse for his work on the database, Rick Schouwstra for editing the footnotes and bibliographies of this volume, and Rick van Brummelen, Marieke Ceelaert and Romeo Burger for helping us organize various meetings and workshops. Finally, I would like to thank the staff of the Netherlands-Flemish Institute in Cairo for their hospitality and support with organizing the conference of which this volume is the result. It is with great gratitude that I remember the late Rudolf de Jong as the host of our conference.

Maaike van Berkel
Nijmegen, March 2026

CHAPTER 1

Introduction
Water Management in the Premodern Middle East: Forces from "Above" and "Below"

Josephine van den Bent, Maaike van Berkel, Edmund Hayes

[A town] should be on a river,
or springs with plenty of fresh water should be facing it.
The existence of water near the place
simplifies the water problem for the inhabitants,
which is urgent ...[1]
Ibn Khaldūn, *Muqaddima*

When planning a city, the presence of water is crucial, says the historian Ibn Khaldūn (d. 808/1406). Unfortunately, he continues, the Arabs have not always taken his good advice into account. Ibn Khaldūn laments that "they [the Arabs] only see to it that they have pastures for their camels and they do not care whether the water is good or bad ..."[2] As a result, cities such as Basra "fell prey to ruin and disintegration."[3] Managing access to fresh water for large numbers of people, whether city dwellers or pilgrims and travellers, was without doubt one of the greatest challenges the Middle East and North Africa faced then, and still faces today. Nevertheless, in the premodern era, societies in the region were rather successful in providing water to highly populated cities such as Abbasid Baghdad and Mamluk Cairo, and along vital and busy roads such as the pilgrimage roads across the Arabian desert to the holy cities in the Hijaz. We still do not fully understand how this was achieved. Who was responsible, who financed, and who built water installations? Who were the people involved in the management and maintenance: Authorities, collectives, neighbourhood communities, elites, or private individuals? And for whom were the installations built: For households, broader communities or also for the common good?

[1] Ibn Khaldūn, *al-Muqaddima*, 2:184; Transl. Rosenthal, 2:246.
[2] Ibn Khaldūn, *al-Muqaddima*, 2:203; Transl. Rosenthal, 2:269.
[3] Ibn Khaldūn, *al-Muqaddima*, 2:204; Transl. Rosenthal, 2:270.

This volume brings together research on the theme of water and management covering a broad geographical and chronological scope, from Roman Egypt and Damascus and Medina in the first centuries of Islam to Baghdad under the Buyids and Seljuqs and Cairo under the Mamluks. Each chapter focuses on the management of fresh water and asks questions on the (sometimes fraught) relationships between forces from "above"—central authorities, formal institutions and elites—and "below"—more traditional and informal practices and arrangements organised and carried out by and within local communities. At the same time, we take into account diverse landscapes and their related sources of water, that in turn ask for diverse infrastructure for collecting, transporting, storage and distribution. We can broadly divide water infrastructure into sources (rivers, wells, rainwater); collection (e.g., the roof of a building);[4] storage (cisterns,[5] ponds, ceramic jars); transport (canals, pipes, waterskins); and provision at the end point (drinking fountains, water shops,[6] bathhouses and ablution facilities for washing). Furthermore, the provision of water implied safeguarding it: in particular by segregating clean water from wastewater and other sources of contamination through the creation and management of sewers, cesspits, and storage and disposal solutions for grey water and the assorted waste products that the pre-modern city generated.[7] In practice many of these aspects often overlapped: Thus, a canal that was designed to transport water to a specific place would also provide water for the communities along the way who helped themselves, as well as providing for the transport of other goods on boats.[8]

The volume as a whole is the result of a conference held in Cairo at the Netherlands-Flemish Institute in February 2024 and forms, together with the forthcoming monograph (Maaike van Berkel, *Urban Water Management in the Medieval Islamic World*, Bloomsbury), the culmination of the "Source of Life" project—an NWO (Dutch Research Council) VICI project at Radboud University in Nijmegen, the Netherlands (2019-2026). It brings together archaeologists and historians who explore a wide range of textual and material sources of evidence which relate to a multitude of institutions involved in the transport, storage and distribution of water. Together they explore the intersection between institutions and the physical reality of water use within cities (and their hinterlands) in the Mediterranean and the Middle East.

4 For example, the rainwater-collection network of the Amman Citadel, as the arid climate and its distance from stable sources of water made it particularly reliant on rainwater. As the archaeologist Ignacio Arce notes at information panels in situ: "Not a drop was misused or lost."

5 See the contribution of Bethany Walker in this volume.

6 See Ibn al-Ukhuwwa's instructions for a city's market inspector (*muḥtasib*) to regulate water shops, Ibn al-Ukhuwwa, *Maʿālim al-qurba*, 97.

7 Insufficient attention has been paid to waste, sanitation and sewage in the MENA region with the exception of the *Médiévales* special issue "Lieux d'hygiène," among which Reklaityte, "Les latrines en al-Andalus."

8 Van den Bent, "Caliphal Involvement," 14. See also the case of the rights-holders along the channel that supplied the Damascus mosque, mentioned in Hayes "Land, Law and Water Politics."

Water History

Since the mid-20th century, scholarship on the relation between water and power or administration has often built on Wittfogel's thesis of hydraulic societies.[9] Building on Weberian notions of a clear West-East divide and Weber's characterization of the Eastern (or sultan's) rule as despotic, Wittfogel argued that organizing water, especially in arid and semi-arid regions, required central bureaucratic and despotic rule. This prompted the emergence of the centralized state in, among others, ancient Egypt and ancient Mesopotamia, and in subsequent centuries under the Islamic rulers of the region. In his *Oriental Despotism* he emphasised the state's control of natural resources, including water, and the top-down involvement of authorities in the construction and maintenance of major water works. More recent studies have successfully refuted this overarching claim on the management of water and state formation, emphasizing the more egalitarian, society-based and bottom-up forms of organisation they encountered in their specific case studies.[10] Nevertheless, Wittfogel's thesis is still persistent and continues to appear in scholarly debates. It is only very recently that water historians have tried to formulate alternative pathways for analysing the relation between state formation and water management, informed by historical and archaeological research worldwide.[11]

In line with recent scholarship this volume moves beyond a top-down view on the relation between water and power. Instead, it focuses on the full range of institutions and stakeholders which played a role in the collection, transport, storage and distribution of water. As such it builds on insights from studies in political anthropology on the blurring of state and society and it tries to include all forms of agency. Scholarship in this field has reconceptualized non-Western states by avoiding a narrow interpretation of the state as main provisioner of basic (public) services and focused instead on the interplay of state and community-based mechanisms in the organisation of services, including the provision of water.[12] A good example of such an approach is Tessa Farmer's book on everyday water practices in present-day Cairo. It demonstrates the variety of interpersonal relationships which the residents of an informal neighbourhood obtain and maintain—both within the

[9] Wittfogel, *Oriental Despotism.*

[10] See, for example, Harrower, *Water Histories*; Wilkinson *et al.*, "Hydraulic Landscapes;" Wilkinson *et al.*, *Models of Mesopotamian Landscapes.*

[11] For example by Harrower, *Water Histories.*

[12] For political anthropological studies on state and society, see Gupta, "Blurred Boundaries;" Van Klinken and Barker, *State of Authority*; Migdal, *State in Society*. See also: Bayat, *Life as Politics*; Chatterjee, *Politics of the Governed*; Ismail, *Political Life in Cairo's New Quarters*; Simone, *City Life from Jakarta to Dakar*. For a water management focused study which does exactly this, see Farmer, *Well Connected.*

community and with authorities—to collaboratively organize clean water and wastewater infrastructure for their community.[13]

The water history of the medieval MENA region is still in its infancy and has so far mostly dealt with miscellaneous topics based on specific case studies. For example, archaeologists and architectural historians have often focused on the natural landscape, various types of water systems, knowledge of hydraulic technology and the ways in which water infrastructure was used. Especially fruitful for our understanding of water management have been studies that combined material and textual evidence. Good examples of such scholarship are the studies on Samarra by Alastair Northedge, Derek Kennett, and Peter J. Brown, on the water infrastructure along the Pilgrimage road from Iraq to Mecca, the Darb Zubayda, by Saad Al-Rashid, on early Islamic Egyptian Tinnīs by Alison Gascoigne, on Yemen by Ingrid Hehmeyer, on (mainly rural) sites in the Levant by Bethany Walker, and on Iraqi waterways by Peter Verkinderen.[14] In addition, art historians have analysed the use and diverse functions of water in court life, for example its presence in the lush gardens and palaces of al-Andalus.[15] Recently, some illuminating histories of some of the main rivers of the region have appeared, such as John Cooper's analysis of the navigational landscape of the medieval Nile,[16] Peter Verkinderen's study of the waterways of Iraq and Iran in the early Islamic era based on both satellite imagery and Arabic historical and geographical sources,[17] and, for a slightly later period, Faisal Husain's *Rivers of the Sultan* on the Tigris and Euphrates in the Ottoman era.[18] Historians working with *fiqh* and hadith texts have written on topics such as water rights, hygiene and moral obligations of rulers with regard to water management.[19] Then there are scholars interested in the history of science who dealt with knowledge transfer and the technicalities of water infrastructure.[20] Finally, some studies focused on water management in

[13] Farmer, *Well Connected.*

[14] See, for example, Northedge, *Historical Topography*; Northedge and Kennet, *Archaeological Atlas*; Brown, "Water in an Abbasid City;" Brown and Van Berkel, "Water Provision in Early Islamic Cities;" Van den Bent and Brown, "Constructing Hydraulic Infrastructure;" Al-Rashid, *Darb Zubaydah*; Gascoigne, "Water Supply of Tinnīs;" Hehmeyer, *History of Water Engineering*; Walker, "Liquid Landscapes;" Verkinderen, *Waterways*. On hydraulic techniques see also Zaqzouq, "Les norias;" Wilson, "Classical Water Technology."

[15] See, for example, Blessing, "Vessel as Garden;" Ruggles, *Islamic Gardens.*

[16] Cooper, *Medieval Nile*

[17] Verkinderen, *Waterways.*

[18] Husain, *Rivers of the Sultan.*

[19] See, for example, Van Berkel and Hayes, "Governing Water;" Salameh, "Customary Water Rights;" Wilkinson, "Muslim Land and Water Law."

[20] See, for example, Ruggles, *Islamicate Environments*; Hill, *A History of Engineering*; de Miranda, *Water Architecture*; Wilkinson and Rayne, "Hydraulic Systems in the Middle East;" Wilson, "Classical Water Technology;" Zaqzouq, "Les norias."

the countryside, exploring irrigation practices and regulations. Good examples of such studies are Yossef Rapoport's and Ido Shahar's analysis of rural society of the medieval Fayyūm, based on the $7^{th}/13^{th}$-century account of these villages by al-Nābulsī, and Ingrid Hehmeyer's *History of Water Management in Yemen.*[21] In addition, the "Source of Life" project focused more on urban water management, profiting from work by, among others, Amalia Levanoni who analysed aspects of water provision in Mamluk Cairo and Yehoshua Frenkel who explored the same for the city of Damascus under Mamluk rule.[22]

Such studies on specific cases are crucial to our understanding of water history. Indeed, Alan Mikhail has emphasised the gaping hole the MENA region forms in the global history of the environment in general and how global, large-scale, studies need to be fed by analysis of sources on regional situations.[23] This volume aims to contribute to the environmental history of the region by focusing on water history. It roots these detailed individual water studies in local experiences within a region-wide comparative perspective. A detailed region-wide comparative perspective on water management helps us to understand differences and similarities with regard to the environment and the local adaptations to it. Thus, this volume brings together studies dealing with related questions as to how local communities in the MENA region throughout the premodern period lived with their (changing) natural environments and adjusted these environments to access and wield one of its most crucial natural resources.

To date, we have a limited understanding of the actors involved and how communities at large managed to deal with the crucial task of water provision under challenging climatological circumstances. The environmental approach of the "Source of Life" project and the chapters of this volume not only illuminate the management of water, but also function as a microcosm for understanding broader processes and evolutions of urban governance. By studying the way in which societies dealt with a vital commodity such as water, and the challenges caused by organizing access to it and protection from it, we are able to refine our understanding of the ways in which societies were organized in a more general sense and which institutions and actors were involved in its administration. Although the project and the focus of this volume have been dedicated to urban governance, the environmental lens has encouraged us to look at urban administration beyond the

[21] Rapoport and Shahar, "Irrigation in the Medieval Islamic Fayyum;" Hehmeyer, *History of Water Engineering*. See also Haug, *Garden of Egypt*.

[22] Levanoni, "Urban Water Management;" Frenkel, "Management of Water in Fourteenth-Century Damascus."

[23] Mikhail, *Water on Sand*, 1-25. For another recent region-wide comparative publication in the field of environmental history, see Ruggles, *Islamicate Environments*.

borders of the better documented cities alone, and to analyse its management in close relation to the diverse landscapes and ecosystems of its hinterlands.

The importance of the local becomes evident when we take a closer look at the case studies presented in this volume. The Middle East is home to tremendously diverse climates and landscapes, from the forests of Anatolia to the arid deserts of the Arabian Peninsula. Some of its areas are traversed by major rivers like the Nile, the Euphrates and the Tigris, which facilitated both water supply and transport, but also caused risk of flooding. Other areas lacked perennial water sources and completely depended upon (limited) rainfall. The case studies in this volume demonstrate administrative diversity in the light of ecological differentiation as well as changes throughout the centuries. The water situation in Cairo, for example, was highly dependent on the specifics of the river Nile, with its rise in summer, its movement westward in the Mamluk era, and its fluctuating volume per year (see Isoldi's and Bakhoum's chapters). On the other hand, the small-scale management of household hydraulic systems in rural and urban Mamluk and Ottoman Bilād al-Shām, where communities lacked perennial water sources and had to collect and store rainwater, form a different, locally specific, way of dealing with water, which also changed across time (see Walker's chapter). The similarly arid area in and around Arabia's Medina furnishes us with yet other divergent examples: In the early Islamic era landowners of local estates invested in complex and large-scale systems such as dams, wells and irrigation canals, and even used the control of water as a means to oppose the highest authorities, the Umayyad caliphs (see Munt's chapter). The Abbasid investments in the Darb Zubayda, in turn, were not only dependent on specific geographical circumstances of the arid climate, but also related to specific seasonable differences (see Brown's chapter). In Baghdad, the Tigris and Euphrates, and the many man-made canals fed by those rivers—some of which connected the Euphrates with the Tigris—formed the main source of fresh water (see Van den Bent's and Van Renterghem's chapters).

Methodology and Sources

The chapters in this volume use a wide range of sources of evidence to explore diverse, sometimes competing, perspectives on water management. With the exception of some technical (e.g., al-Karajī's *Inbāṭ al-miyāh al-khafiyya* (The Extraction of Hidden Waters) and legal treatises (e.g., the Egyptian *Qānūn al-riyy*),[24] few contemporary texts are truly dedicated to water provision or the management thereof. However, given the importance of water as a life necessity and its omnipresence,

[24] On the *Qānūn al-riyy*, see Borsch and Sabraa, "Qānūn al-Riyy."

the topic rears its head in a wide variety of sources, albeit in some cases more systematically than in others. Within the broad category of textual sources, there are a great many different source types in which references to water management are found, each with their own focuses, benefits, and challenges. Examples range from documentary sources such as *waqfiyyas* (endowment charters of charities) or court records to instruction manuals, from chronicles to geographies, and from legal texts to ethnographic material. A wide range of archaeological sources also play a key role in researching water management in the premodern Middle East. They include, for instance, surviving structures, excavations of monumental hydraulic infrastructure such as aqueducts but also of private homes, as well as aerial photography and satellite imagery. This volume as a whole and several chapters within it scrupulously incorporate and combine textual and archaeological sources and methods because of their complementary character.

First, general textual analysis allows for the identification of various aspects of qualitative information on water management, while archaeological research offers further insight in more practical and quantitative respects. Qualitative information includes the cultural and legal structures that underlie water management as well as its various stakeholders, and the interplay between them. *Waqfiyyas*, for instance, may describe water provision to endowed buildings as well as the water's beneficiaries (see Isoldi's and Bakhoum's chapters). Administrative records mention the installation of and repairs to drains (see Milwright's chapter). And chronicles and geographies describe construction activity of major projects and the main actors within these projects, but can also reveal day-to-day tensions related to these hydraulic installations (see Van Renterghem's and Munt's chapters).

Texts, however, are sometimes silent on a number of other important aspects of water management. Archaeological sources provide insight into the material and technological design (see Brown's chapter), the geographic distribution and chronologies of water systems and water sources, on the building, maintenance and restoration of the installations, and on dealing with wastewater, grey water, drains, sewers, cesspits, and sumps. Where possible, therefore, combining textual and archaeological sources provides a clearer picture. For instance, Bethany Walker (chapter 9) combines textual records on the spread, use and filling systems of cisterns in Mamluk and Ottoman Jordan and Palestine—demonstrating their financial and cultural importance—with archaeological excavations of cisterns in Tall Ḥisbān and Khirbet Beit Mazmīl. The chapter shows how these cisterns were used, abandoned, restored, and reused over the course of centuries.

Second, cultural norms on water management are often more easily derived from textual sources (although certainly not absent in archaeological remains), while archaeological data at times can add details on the realities of everyday life. Sources such as legislation, tax-, or *ḥisba* manuals (advice treatises for the *muḥtasib*, the market inspector) discuss water in various ways. Islamic legal texts give instructions on hygiene and purity, but also on (equal) water distribution; *kharāj* (land-tax) manuals are very concerned with various aspects of irrigation and access to water; and some *ḥisba* manuals contain instructions ranging from the upkeep of water infrastructure to bathhouse hygiene and regulations for water carriers. A challenge in using these sources, however, is that they are inherently prescriptive and normative. They voice what religious authorities, government officials, or others saw as desirable, as ideal. Their information can consequently not be translated directly to an understanding of actual practice. Despite risking a fine of €120, approximately a quarter of Amsterdam cyclists ignores a red light on a regular basis. Medieval inhabitants of the Mediterranean and the Middle East are unlikely to have been a lot more norm- and/or law abiding: despite repeated injunctions against drains emptying into the street, many an innocent passer-by fell victim to a sudden onslaught of drain water, and in Rome illegal tapping of aqueduct water was a recurring issue even though it was a fineable offense (see De Haan's chapter). Consequently, these texts reveal a lot about these norms and laws, as well as about motivations behind them and what various types of authorities considered important.

The same applies to more descriptive texts—such as histories, geographies or biographies—albeit perhaps to a lesser extent. While on the one hand they offer glimpses into everyday life, these too may contain material that describes norms more than practices. They may give actual historical examples of good and bad governance, but may also emphasise anecdotes that drive home ideals rather than reflect actual events, or idealise "the good old days" in favour of a murky present. Descriptions of Baghdad, for instance, on closer scrutiny are often descriptions of an (idealised) view of the Abbasid capital as it was newly built by al-Manṣūr (r. 136–58/754–75) rather than the situation at the time of writing. Yet, while it is sometimes difficult to evaluate to what extent propagated norms were adhered to, fact is that the sources voicing these norms were driven by practice and daily experiences and in turn affected practices and experiences, influencing the ways in which people shaped their environment and habits (see Hayes' chapter).

Written norms will emphasise, for instance, the importance of proper wastewater management and the reasons why, but may not necessarily explain in detail how that is done. Archaeology at times gives more information about daily practices and overall practicalities of water management that may not be as visible in texts, such as the ways in which wastewater drains were connected and cesspools easily accessible for cleaning, or the ways in which aqueducts were designed to be easily repaired.[25] For some areas under study in this volume, archaeological material is the most important primary source material offering all sorts of insights

[25] Scanlon, "Preliminary Report," 16–18.

about daily practice, while for others it supplies relevant supporting evidence. This is not always possible, however. Precious little archaeological evidence for water management in early Islamic Medina has survived, for instance. Yet even then, remains discovered elsewhere can be used to compare and elucidate literary descriptions (see Munt's chapter).[26]

Archaeological sources are especially helpful to uncover the daily practices of non-elites. While textual sources tend to have a primary focus on elites, archaeological sources can offer access to a broader social context. In terms of water management, this includes for instance water provision for individual households, or smaller-scale infrastructure constructed and maintained by local communities. Archaeological evidence has revealed that in Samarra, for which the textual sources report that drinking water came primarily from the Tigris,[27] areas with higher population densities were further away from the river. That meant it was harder for them to access water, and these areas may well have been lower-status neighbourhoods than the bigger plots close to the river.[28] In her contribution to this volume, De Haan (chapter 2) points out that smaller-scale, locally-maintained infrastructure was probably the dominant form of water management in many areas of the Roman Empire, although it rarely appears in the written sources, usually only in case of conflicts. Even so, written sources too can be important in deducing and analysing non-elite circumstances. De Haan also describes an ostracon—at the intersection of archaeological and textual sources—from Egypt's Western Desert that reveals that high-ranking officials received three times as much water as the labourers who did the hard work, showing the struggles of the disadvantaged when it came to access to water. This information could not have been deduced from archaeological sources alone, as is the case for the key role played by water carriers in premodern Islamic cities, which again underlines the importance of combining various source types were possible.

Finally, archaeological and textual sources offer different temporal perspectives. Textual sources generally emphasise specific moments and events in history, offering key information on important moments and events and all their water-related consequences. Founding documents for instance reflect the (intended) situation at the time of a *waqf*'s establishment (see Isoldi's chapter), and chronicles recall weighty events, such as the foundation of a city and its water systems, or moments of crisis, such as dam breaks, riots, floods or extended droughts (see Van den Bent's, Van Renterghem's, and Munt's chapters). In some cases, this may mean

[26] See also, for instance, Northedge, *Historical Topography*; Van den Bent and Brown, "Constructing."

[27] al-Yaʿqūbī, *Kitāb al-buldān*, 263.

[28] Brown, "Water in an Abbasid City," 16.

that the sources show us *ad hoc* solutions in times of crisis rather than a reflection of how maintenance and management was normally carried out. At the same time, "normalcy" is also frequently reflected in the narrative sources, especially when they relate anecdotes or events that are not actually concerned with water management, but in which matters or people related to water figure in the background. For instance, caliphal injunctions in a campaign against the praising of Muʿāwiya ibn Abī Sufyān (r. 41–60/661–680) directed towards "the drinkers and water providers in the two Friday Mosques" alert us to their presence.[29] Archaeological sources, on the other hand, often provide testimony of the long-term developments, over centuries if not longer. That includes environmental aspects such as the changing courses of riverbeds and siltation, but also the human-made practicalities on how hydraulic infrastructure was built, maintained, and adapted. Walker's contribution to this volume about cisterns, for instance, underlines the long history of household cisterns, which were built and (re)used over centuries and some of which show traces of being modified to address changing needs more than once.

The universality of the need for water—whether for drinking, bathing, washing, industry, or any other purpose—means that the topic often emerges in the textual sources and is eminently visible in all sorts of archaeological remains. By integrating various source types, and especially by combining textual and archaeological evidence, this volume gathers as many jigsaw pieces as possible in order to forge a clearer image of the region's water history.

The Interplay of Tangible Infrastructures and Intangible Institutions

The current volume is primarily focused on cities, though it is acknowledged that when thinking about water provision, cities cannot be disconnected from their geological, ecological and political environments. When thinking about how an individual gets access to usable water in a given city at a given moment in time, we have to imagine a vast network of overlapping systems: human, non-human, tangible and intangible. The tangible infrastructure that provides water to its users is perhaps the most visible part of these systems: canals in the landscape, ditches in the archaeological record, wells and cisterns in surviving buildings and jars and ornamental fountains in museums, but such infrastructure requires social and political processes and institutions to ensure the provision of water at every step. The users *are* the system, not merely recipients of it.

Tangible infrastructures require constant human participation to create, maintain, and regulate them. Not only governments and elites, but people at all levels

[29] al-Ṭabarī, *Taʾrīkh*, 3.4:2165.

of society were engaged in the effort to ensure that enough water reached them. While large infrastructures such as the great canals of Cairo or Baghdad required significant investment that only the richest of elites, often linked with the state, could provide, many crucial parts of the system were created and maintained by individuals, families, and communities further down the social scale. While little information is often accessible about such low-level local initiatives, our sources sometimes give us hints, as in Ibn ʿAsākir's description of water infrastructure in Damascus. He notes that, in the Damascus of his day (6th/12th century):

> There are various channels[30] with defined endowments which are under the supervision of the overseer of endowments (*mutawallī al-awqāf*), [and these] are well-known and clear. Most of them are not endowed but are managed by the Muslims [i.e. the local citizens] and benefit is obtained by all of them, and the wells are kept fresh by the neighbouring [people]. I will mention them and write down the number of them so that whoever wishes to count them may know them.[31]

The way people created and managed infrastructure was sometimes *ad hoc*, but given the importance of the task at hand, such arrangements often crystallised into customs and institutions (both formal and informal) and these were sometimes recorded and codified. We see water infrastructure managed through the decree of rulers and their representatives at all levels, the normative judgements of lawyers, and the practical oversight of judicial authorities,[32] and the agreements and consent of specific user communities. Technical solutions to specific infrastructural needs (such as the creation of water-tight plaster for lining cisterns, or the casting of interlocking sections of ceramic pipes) were passed down by craftspeople embedded in particular social and economic contexts (see Milwright's chapter). Religious and cultural mores influenced both what kind of value was given to specific types of water provision (for example the provision of water for ritual ablution at mosques),[33] but also how tasks should be carried out, and by whom. Both the informal and the formal processes and institutions of managing water and its infrastructure could be highly regionalised, responding to local ecologies, economies,

[30] The Arabic word *qanāt* can have various meanings including aqueduct, underground conduit drawing water from the water table, and simple channel or canal.

[31] Ibn ʿAsākir, *Tārīkh Madīnat Dimashq*, 2:377–78. Ibn ʿAsākir seems to imagine his readers walking round the city and counting them off as they walk. For an in-depth study of this chapter, see Hayes, "Land, Law and Water Politics."

[32] See the arrangement for maintenance of drinking fountains in *waqf* documents studied by Isoldi in this volume. See also Isoldi, "Confluent Waters;" Van Berkel, "Waqf Documents."

[33] For a discussion of the hierarchy of purity of different types of water and their purposes, see Van Berkel and Hayes, "Governing Water."

and cultures, but were also transmissible between regions, as good ideas caught on and were carried around with migrants, merchants, scholars and rulers.[34]

While the division of the world into public and private spheres of enterprise has justly been problematised in its applicability to pre-modern contexts,[35] it is important to acknowledge a spectrum of users and user communities from purely private initiatives (like constructing a cistern or a cesspit to serve one's own domestic use); to infrastructures beyond the domestic space like groups of families sharing a cistern; to provision aimed at far greater communities like the provision of water at a large mosque; or a *waqf*-funded dispensing of water from a *sabīl* at particular times of day. Pre-modern Middle Eastern governments certainly did not come with the same expectations of provision of public services associated with modern states. Furthermore, our modern expectations of a divide between public and private duties must be adjusted to the blurring of public and private duties by the fact that major infrastructural works that benefitted the public could often be provided by actors from within the state elite acting as private individuals to gain thereby for themselves prestige, honour, and a reward in heaven. Successful governments therefore were associated with the provision of certain public amenities, without these amenities being provided directly by a state in a *de jure* sense. We do, it is true, occasionally see statements that suggest that providing clean water was one of the duties of ruler to ruled. But in general, the duties, theories and motivations for providing water in society were various: from the efforts of individuals and families to provide themselves with water, to a charitable gift from an elite person to the less advantaged, to the acts of rulers, to activities carried out within the framework of profit, such as the operation of a fee-paying bathhouse, or a water carrier paid to provision elite households, or an irrigation canal constructed as an investment by a landowner with the aim of increasing the value and productivity of the land. Such interlocking motivations gave different actors different scopes for intervening in the patchwork of structures and institutions, that, taken together, were able to sustain very large premodern cities in the Middle East.

34 The history of underground *qanāts* is an interesting though controversial case study in the question of the diffusion of technology. For this, and the wider debate about the transmission of agricultural crops, techniques and technologies, see Kirchner *et al.*, "Re-Thinking the 'Green Revolution'." See also al-Yaʿqūbī on hiring experts from different places for the construction of water infrastructure in Samarra. al-Yaʿqūbī, *Kitāb al-buldān*, 264.

35 See, Denoix, "Les notions de « privé » et de « public » ."

Outline of the Book: Water across Time and Space

We have chosen to structure this volume according to region, while sticking loosely to chronology, which allows us to emphasise the experience of people, including local communities, elites and rulers, and the constraints and possibilities of the diverse environmental conditions in specific localities. The four sections of the book cover four of the main localities in the region at large.

To emphasise the *longue durée* history of water institutions and traditions in the broader Mediterranean, the volume starts with an overview chapter on water management in the Roman world, including examples from the MENA region in pre-Islamic times. Nathalie de Haan's "Water, Wealth and Inequality in the Roman World: Forces from 'Above' and 'Below'" explores the relationship between power and water distribution in urban and rural contexts and analyses the stakeholders involved in regulating access to water. It challenges the widespread misconception that large quantities of water were available for everyone in the Roman Empire thanks to Roman technological and organisational qualities. The available amount of water from the impressive Roman aqueducts, for example, was unevenly divided, and as such De Haan's chapter demonstrates much more continuity with post-Roman practices than is commonly assumed.

The first regional section is on the Hijaz during the early Islamic era, the Umayyad and early Abbasid reigns. Both contributions within this section, Harry Munt's "Water Management in Early Islamic Medina: Infrastructure, Labour and Politics" and Peter J. Brown's, "An Environmental History of the Darb Zubayda: Towards a First Approximation", take a clear environmental approach to water management. While Munt links environmental questions to issues of power, Brown focuses on the particular challenges caused by the changing seasons of the hajj. Munt analyses the precarity of water provision in Medina in the early Islamic era, but also emphasises that it was plentiful enough to sustain cultivation of the area if the water infrastructure was carefully maintained and managed. On the basis of (principally local) histories, Munt argues that the complex mechanisms for collecting, transporting, storing and accessing water offered opportunities for the ruling elite, the Umayyads, to expand their influence in the region. Local representatives and relatives of the Umayyads were instrumental in this process, to the disadvantage of other rival families. Peter J. Brown links caliphal investment in the pilgrimage routes, the Darb Zubayda, to favourable (during winter) and unfavourable (during hot and dry summers) hajj seasons. Combining archaeological and textual evidence with details on regular meteorological cycles and fluctuating climatic conditions, Brown argues that investments in the Darb Zubayda, including the establishment of waystations and water facilities, were not only motivated by political and socio-economic conditions but probably also by environmental challenges. Especially during the years when the hajj took place in the hot summer months, the Abbasid caliphal family seems to have been more inclined to extensively invest in the water infrastructure along the pilgrimage roads to minimize the risks of water shortages.

The next chapter, "The Infrastructure of Purity: Plumbing Advice from Hadith and Early *Fiqh*", by Edmund Hayes, bridges the geographical foci of the first and second sections, discussing hadith texts attributed to scholars located both in the Hijaz and Iraq. The texts analysed relate to very specific pieces of infrastructure (drains, sewage ditches, wells), and the cases provide a level of granular detail that suggest they might emerge from concrete lived experiences. However, in addition to the level of local detail, they also engage with a broader normative concern which ultimately made them relevant to pious readers across the Islamic empire: the motivation to maintain pure water that would be suitable both for drinking and ritual ablution, in urban environments packed with the peril of potential pollutants. The Iraqi context, and more specifically the Baghdadi context, takes centre stage in Josephine van den Bent's chapter "Baghdad's System of Water Provision: Infrastructure, Maintenance, and Decline in the City's First Two Centuries". Van den Bent focuses on the development of the system of water provision in Baghdad over the first two centuries of the city's existence, paying special attention to the division of responsibilities in the maintenance of water infrastructure. She emphasises the multiple foundations of the city's system of water provision and how this diversity allowed the city to endure, especially in times of crisis. Vanessa Van Renterghem's chapter, "Fighting over Water: Everyday Water Issues in Late Buyid and Early Seljuq Baghdad ($5^{th}/11^{th}$ Century)" moves chronologically to Baghdad in the $5^{th}/11^{th}$ century—a period in which the city was the stage for recurring crises of diverse character, all of which affected inhabitants' access to water. The "forces from below" are particularly prominent in her chapter. By focusing on a number of case studies in which access to water caused conflict and tension, Van Renterghem not only analyses the role of rulers and their administration but also the influence of neighbourhood communities and individuals in the water provision of the population.

The third section is on the Levant and again moves on in time. It focuses on the physical evidence involved in the collection, storage and transportation of water. In his "Urban Crafts Associated with the Provision of Water in Islamic Syria", Marcus Milwright uses archaeological and textual sources alongside ethnographic observations to analyse artisans' practices in producing water-related infrastructure and its components. Ranging from the medieval period until the early 20th century, Milwright explores the socio-economic significance of these practices, as well as their continuity. Continuity and discontinuity of water infrastructure is also pivotal in Bethany Walker's "Traditional Water Harvesting in Mamluk and Early Ottoman

Bilād al-Shām: The Family Cistern". The local environment, especially the lack of perennial sources of water in large parts of Jordan and Palestine, urged local urban and rural communities to construct and maintain family and community cisterns. Basing herself on two case studies, Tall Ḥisbān in Jordan and Khirbet Beit Mazmīl in the hinterland of Jerusalem, Walker studies the long-term history of these water storage installations and analyses why these small-scale cisterns, which had hardly received any attention from the government throughout their existence, suddenly became of financial interest to the Mamluk and early Ottoman administrations. Similarly to Milwright, she combines textual and archaeological sources with ethnographic evidence.

The final section of this volume is on Egypt. In her chapter, "'From a Spring Where Allah's Servants Will Drink': Reassessing the Social Significance of *Sabīl*s in Mamluk-Era Cairo", Angela Isoldi argues that the water dispensers of the later medieval Cairene landscape, the *sabīl*s, were really not that important volume-wise in the provision of water for the city. Challenging the idea that *sabīl*s were crucial to urban water provision, Isoldi instead understands these facilities as part of a policy of undertaking charitable deeds on the part of the *ahl al-dawla*. Dina Ishak Bakhoum's chapter, "Cairo's *Sabīls* during Egypt's Modern Period", then discusses the preservation and afterlife of these monuments from the 19th century onwards. Exploring the reason why these facilities lost their original function in the course of the modern era, she focuses on various stakeholders involved in their preservation and particularly on the role of the Comité de conservation des monuments de l'art arabe.

The book concludes with an epilogue by Alan Mikhail in which he reflects on the various chapters in this volume and how they contribute to broader questions of the environmental history of the region in the premodern era. Thanks to these case studies from different MENA regions, this volume demonstrates the wide range of practices and arrangements related to water management in the region during that era, the many ways in which these were influenced by local constraints and possibilities, and the involvement of a huge variety of players, both from "above" and "below".

Bibliography

Primary Sources

Ibn ʿAsākir. *Tārīkh Madīnat Dimashq*. Edited by ʿUmar ibn Gharāma al-ʿUmrawī. 80 volumes. Dār al-Fikr, 1995.

Ibn Khaldūn. *al-Muqaddima*. Edited by Abdesselam Cheddadi. 5 vols. Bayt al-Funūn wa-l-Ādāb, 2005.

Ibn Khaldūn. *The Muqaddimah: An Introduction to History*. Translated by Franz Rosenthal. 3 vols. Princeton University Press, 1967.

Ibn al-Ukhuwwa. *Maʿālim al-qurba fī aḥkām al-ḥisba*. Edited and partially translated by Reuben Levy. Cambridge University Press, 1938.

al-Karajī, Muḥammad. *Inbāṭ al-miyāh al-khafiyya*. Edited by Baghdād ʿAbd al-Munʿīm. Maʿhad al-Makhṭūṭāt al-ʿArabiyya, 1997.

al-Ṭabarī. *Taʾrīkh al-rusūl wa-l-mulūk*. Edited by M.J. de Goeje. Vol. 3.4. E.J. Brill, 1890.

al-Yaʿqūbī. *Kitāb al-buldān*. Edited by M.J. de Goeje. E.J. Brill, 1892.

Secondary sources

Arce, Ignacio. "The Umayyad Hydraulic System at Amman Citadel: Collection, Storage, Distribution, Use and Sewage." In *Men of Dikes and Canals: The Archaeology of Water in the Middle East*, edited by Hans-Dieter Bienert and Jutta Häser, 243–60. Marie Leidorf, 2004.

Bayat, Asef. *Life as Politics: How Ordinary People Change the Middle East*. Amsterdam University Press, 2010.

Bent, Josephine van den. "Caliphal Involvement in Water Provision in the Cities of the Early Abbasid Period." *Journal of Abbasid Studies* 12, no. 1 (2025): 1–30.

Bent, Josephine van den, and Peter J. Brown. "Constructing Hydraulic Infrastructure in the ʿAbbāsid and Ṭūlūnid Capitals: Water Conduits in Baghdad, Samarra, and Cairo between the Eighth and Ninth Centuries." *Al-Masāq* 37, no. 1 (2025): 126–45.

Berkel, Maaike van. "Waqf Documents on the Provision of Water Management in Mamluk Egypt." In *Legal Documents as Sources for the History of Muslim Societies*, edited by Maaike van Berkel, Petra Sijpesteijn, and Léon Buskens, 217–32. Brill, 2017.

Berkel, Maaike van, and Edmund Hayes. "Governing Water and Public Health in a Muslim City: Cairo in the Thirteenth and Fourteenth Century CE." In *Public Health in the Premodern World: Dynamic Balances*, edited by Guy Geltner, Janna Coomans and Ronit Youli-Tlalim, 134–55. Oxford University Press, 2026.

Blessing, Patricia. "The Vessel as Garden: The 'Alhambra Vases' and Sensory Perception in Nasrid Architecture." In *Sensory Reflections: Traces of Experience in Medieval Artefacts*, edited by Fiona Griffiths and Kathryn Starkey, 116–41. De Gruyter, 2018.

Blyn, George. "The Green Revolution Revisited." *Economic Development and Cultural Change* 31, no. 4 (1983): 705–25.

Borsch, Stuart and Tarek Sabraa. "Qānūn al-Riyy: The Water Law of Egypt." *Sophia Journal of Asian, African, and Middle Eastern Studies* 35 (2017): 87–124.

Brown, Peter J. "Water in an Abbasid City: The Sources, Distribution, and Significance of Water at Samarra." *Journal of Abbasid Studies* 12, no. 1 (2025): 36–60.

Brown, Peter J. and Maaike van Berkel. "Water Provision in Early Islamic Cities: A Multidisciplinary Approach to Urban Water Governance." In *City, Citizen, Citizenship 400–1500: A Comparative Approach*, edited by Els Rose, Robert Flierman, and Merel de Bruin-van de Beek, 75–103. Palgrave MacMillan, 2024.

Chatterjee, Partha. *The Politics of the Governed: Reflections on Popular Politics in Most of the World.* Columbia University Press, 2004.

Cooper, John P. *The Medieval Nile: Route, Navigation, and Landscape in Islamic Egypt.* American University in Cairo Press, 2014.

Cressier, Patrice, Sophie Gilotte, and Marie-Odile Rousset, eds. "Lieux d'hygiène et lieux d'aisance en terre d'Islam (VIIe-XVe siècle)." Special issue, *Médiévales* 70 (2016).

Denoix, Sylvie. "Les notions de «privé» et de «public» dans le monde musulman sunnite médiéval." In *Public et privé en Islam*, edited by Mohamed Kerrou. Institut de recherche sur le Maghreb contemporain, 2002.

Farmer, Tessa. *Well Connected: Everyday Water Practices in Cairo.* John Hopkins University Press, 2023.

Frenkel, Yehoshua. "The Management of Water in Fourteenth-Century Damascus." In *Egypt and Syria under Mamluk Rule: Political, Social and Cultural Aspects*, edited by Amalia Levanoni, 235–54. Brill, 2022.

Gascoigne, Alison L. "The Water Supply of Tinnīs: Public Amenities and Private Investments." In *Cities in the Pre-Modern Islamic World: The Urban Impact of Religion*, edited by Amira K. Bennison and Alison L. Gascoigne, 161–76. Routledge, 2007.

Gupta, Akhil. "Blurred Boundaries: The Discourse of Corruption, the Culture of Politics, and the Imagined State." *American Ethnologist* 22, no. 2 (1995): 375–402.

Harrower, Michael J. *Water Histories and Spatial Archaeology. Ancient Yemen and the American West.* Cambridge University Press, 2016.

Haug, Brendan. *Garden of Egypt: Irrigation, Society, and the State in the Premodern Fayyūm.* University of Michigan Press, 2024.

Hayes, Edmund. "Land, Law and Water Politics: Ibn ʿAsākir's Chapter on Umayyad Canal Building in the Hinterland of Damascus." *Al-Masāq* 38, no. 1 (2025): 1–17.

Hehmeyer, Ingrid. *A History of Water Engineering and Management in Yemen: Material Remains and Textual Foundations.* Brill, 2019.

Hill, Donald. *A History of Engineering in Classical and Medieval Times.* Croom Helm, 1984.

Husain, Faisal H. *Rivers of the Sultan: The Tigris and Euphrates in the Ottoman Empire.* Oxford University Press, 2021.

Ismail, Salwa. *Political Life in Cairo's New Quarters: Encountering the Everyday State.* University of Minnesota Press, 2006

Isoldi, Angela. "Confluent Waters, Fragmented Powers: Social Geographies of Water in Late Medieval Cairo." PhD diss., Radboud University Nijmegen, 2026.

Kirchner, Helena, Guillermo García-Contreras, Corisande Fenwick, and Aleks Pluskowski. "Re-Thinking the 'Green Revolution' in the Mediterranean World." *Antiquity* 97, no. 394 (2023): 964–74.

Klinken van, Gerry, and Joshua Barker, eds. *State of Authority: State in Society in Indonesia.* Cornell University Press, 2009.

Levanoni, Amalia. "Urban Water Management in the Medieval Middle East: The Case of Mamluk Cairo." In *Islamic History and Civilization*, edited by Amalia Levanoni 255–76. Brill, 2022

Migdal, Joel S. *State in Society: Studying how States and Societies Transform and Constitute One Another.* Cambridge University Press, 2001.

Mikhail, Alan, ed. *Water on Sand. Environmental Histories of the Middle East and North Africa.* Oxford University Press, 2013.

de Miranda, Adriana. *Water Architecture in Lands of Syria: The Water-Wheels*. L'Erma di Bretschneider, 2007.

Northedge, Alastair. *The Historical Topography of Samarra* [Samarra Studies 1]. The British School of Archaeology in Iraq, 2007.

Northedge, Alastair, and Derek Kennet. *Archaeological Atlas of Samarra* [Samarra Studies 2]. British School of Archaeology in Iraq, 2015.

Rapoport, Yossef, and Ido Shahar. "Irrigation in the Medieval Islamic Fayyum: Local Control in a Large-Scale Hydraulic System." *Journal of the Economic and Social History of the Orient* 55 (2012): 1–31.

Al-Rashid, Saad. *Darb Zubaydah. The Pilgrim Road from Kufa to Mecca*. Riyadh University Libraries, 1980.

Reklaityte, Ieva. "Les latrines en al-Andalus: leurs principales caractéristiques et les conditions sanitaires urbaines." *Médiévales* 70 (2016): 59–77.

Ruggles, D. Fairchild. *Islamic Gardens and Landscapes*. University of Pennsylvania Press, 2008.

Ruggles, D. Fairchild. *Islamicate Environments: Water, Land, Plants, and Society*. Cambridge University Press, 2025.

Salameh, Nadim H. "Customary Water Rights in Medieval Wādī Zabīd: Some Legal Cases on *al-ʿādil bi'l-qanāʿah*." In *Proceedings of the Seminar for Arabian Studies: Papers from the Thirty-Second Meeting of the Seminar for Arabian Studies held in London, July 16–18, 1998* 29 (1999): 137–42.

Scanlon, George T. "Preliminary Report: Excavations at Fustat, 1964." *Journal of the American Research Center in Egypt* 4 (1965): 6–30.

Simone, AbdouMaliq. *City Life from Jakarta to Dakar: Movements at the Crossroads*. Routledge, 2010.

Verkinderen, Peter. *The Waterways of Iraq and Iran in the Early Islamic Period: Changing Rivers and Landscapes of the Mesopotamian Plain*. I.B. Taurus, 2015.

Walker, Bethany. "The 'Liquid Landscapes' of the Late Mamluk Mediterranean: Rural Perspectives on the Ever-Changing Sultanate." In *The Mamluk Sultanate from the Perspective of Regional and World History: Economic, Social and Cultural Development in an Era of Increasing International Interaction and Competition*, edited by Reuven Amitai and Stephan Conermann, 261–80. University of Bonn Press, 2019.

Wilkinson, John C. "Muslim Land and Water Law." *Journal of Islamic Studies* 1 (1990): 54–72.

Wilkinson, Tony James, and Louise Rayne. "Hydraulic Systems in the Middle East." In *Encyclopaedia of the History of Science, Technology, and Medicine in Non-Western Cultures*, edited by Helaine Selin. Springer, 2014.

Wilkinson, Tony James, McGuire Gibson, and Magnus Widell. *Models of Mesopotamian Landscapes. How Small-Scale Processes Contributed to the Growth of Early Civilizations*. Archaeopress, 2013.

Wilson, Andrew. "Classical Water Technology in the Early Islamic World." In *Technology, Ideology, Water: From Frontinus to the Renaissance and Beyond* [Acta Instituti Romani Finlandiae, 31], edited by Christer Bruun and Ari Saastamoinen. Finnish Institute in Rome, 2004, 115–41.

Wittfogel, Karl. *Oriental Despotism: A Comparative Study of Total Power*. Yale University Press, 1957.

Zaqzouq, Abdul Razzak. "Les norias: anciens moyens d'irrigation les plus importants dans la région de Hama." In *Techniques et pratiques hydro-agricoles traditionnelles en domaine irrigué* [Actes de Colloque de Damas, 27 juin-1 juillet 1987], edited by Bernard Geyer, 2 vols., 337–65. Librairie Orientaliste Paul Geunther, 1990.

CHAPTER 2

Water, Wealth, and Inequality in the Roman World: Forces from "Above" and "Below"*

Nathalie de Haan

Abstract

This chapter discusses access to water for both city dwellers and people living in rural areas in parts of the Roman world. The aim is to unravel, with the help of both texts and archaeological evidence, some of the practices and patterns of water distribution. Even though many Roman cities seem to have had enough water at their disposal to meet the needs of their populations, both direct access to water and the amount of it available per capita were unevenly distributed. Consequently, urban water supply systems can be regarded as expressions of power relations. In the countryside, conversely, the distribution of water to various consumers (plots of land), seems to have been the responsibility of local irrigation communities and organised from below, especially where water was a scarce commodity.

Keywords: Frontinus; urban distribution; rural distribution; irrigation; water laws; commons; access to water

Ruling the Roman Empire: From Above and from Below

The Graeco-Roman world was a world of cities, as Moses Finley famously put it.[1] The over 2,000 more or less autonomous cities in the Roman empire, big or small, managed their local affairs according to their own local laws (*suis legibus*) under the umbrella of the provincial and central administrations that had constituted the organizational and legal framework of the empire ever since it had begun its expansion in the mid-3rd century BCE. In other words, "[t]he Roman Empire relied on a dual-governance system, engaging both levels—the cities and the Roman state—in an exchange of demands and responses, in an empire where intermediary institutions,

* My warm thanks go to the team members of the "Source of Life" project who organised the inspiring workshop in Cairo, and to the workshop's participants, the peer-reviewers and the editors of this volume for their useful comments. Unless otherwise stated, translations of ancient sources are my own.

1 Finley, "The Ancient City," 305.

such as the regional assemblies of cities, had few real powers."[2] Rome was certainly not exceptional in this: polycentric governance was a characteristic of the many successive empires in the Middle East and Europe from ancient times onward. The term "polycentric political system" was introduced in the early 1960s by political scientists Vincent Ostrom, Charles Tiebout and Robert Warren, who defined the concept as follows: "'Polycentric' connotes many centres of decision-making which are formally independent of each other. Whether they actually function independently, or instead constitute an interdependent system of relations, is an empirical question in particular cases."[3] Moreover, they acknowledged that polycentric political systems are not limited to the field of metropolitan government: "The concept is equally applicable to regional administration of water resources, regional administration of international affairs, and to a variety of other situations."[4]

These autonomous cities, however, depended directly on their hinterlands in many ways, since the society of which they were part was essentially an agricultural society. When it came to the daily management of their city, as with water provision, local authorities had to deal with landowners (including the Roman state), tenants and the inhabitants of surrounding villages in their territories; likewise with tax collection. Moreover, whilst the natural environment provided opportunities, e.g. for local sale or even export of crops or other agricultural products, it could also impose limits on the growth of a city. Carrying capacity could differ dramatically from region to region, especially when it came to water supply. It is good to keep in mind here that the Mediterranean and Middle East consisted (and still consist) of a myriad of microregions that even within short distances can differ greatly when it comes to annual rainfall, winds and average temperatures, in part because of features such as the presence or absence of nearby rivers and springs, the sea, or mountains and hills.[5]

The aim of this contribution is, firstly, to show how the multilayered system of polycentric governance in the Roman empire affected water management, in both towns and rural areas, and both from above and from below. At the local level, regulations and cultural practices often emerged from below, but were sometimes eventually cast into laws from above, by provincial authorities (under the Roman Republic) or lawyers in the service of emperors (during the imperial period). Consequently, water supply and distribution, the maintenance of hydraulic infrastructure, both outside and inside cities, and conflict management were the result

2 Pont, "The Dynamics of Dual-level Governance in the Roman Empire," 22.

3 Ostrom, Tiebout, and Warren, "The Organization of Government," 831.

4 Ostrom, Tiebout, and Warren, "The Organization of Government," 831n2; Brélaz and Lau, "Introduction," 1–18; Part I of this work is entitled "Rome and Her Cities—The Polycentric Empire". See also Isoldi's contribution in the present volume.

5 Horden and Purcell, *The Corrupting Sea*, 1–5, 51–122.

of complex interactions between different stakeholders. Moreover, arrangements were often dynamic in nature and had to be negotiated and renegotiated when changes in the supply or demand of water, or shifts in power relations occurred.

The second goal is to scrutinize widespread yet largely incorrect ideas about the abundance of water in the Roman world. It is a tenacious misconception indeed, that large quantities of water were available for everyone, thanks to the knowledge and skills of Roman engineers and investments in large-scale infrastructure by the central authorities. Instead, access to water for agricultural needs and to (drinking) water for individuals resulted directly from social hierarchies and power relations. A focus on unequal power relations and access to resources is, of course, not novel. Its consequences, however, for the availability and distribution of water, need further exploration, if only to contradict the rather triumphalist myth of Roman technical and organisational skills which resulted in a presumed empire-wide abundance of water for all its inhabitants. By zooming in on examples from Roman Italy, I hope to show how varied the palette of possibilities and responses to the natural environment was, often requiring adapted and highly localised solutions, and always depending on power relations as well. First of all, I turn to the origins of the 'myth of water galore' and the biases in our sources.

Image and Self-Image

Several factors can be identified as contributing to this widespread yet largely incorrect image of Roman water management. The blurred picture is first of all the result of the impressive remains of hydraulic infrastructure across the Mediterranean. The water channels built on high arches, aqueduct bridges, siphons and major dams, however, represent only a small part of the water supply systems built and used across the empire. Numerous smaller features, offering small-scale solutions adjusted to local circumstances, such as water pipes of wood and small ditches or dams made of loam or mud, did not survive or are simply overlooked. The large-scale hydraulic structures, however, nicely fit into the picture of Roman genius, an image already promoted by ancient authors themselves. Dionysius of Halicarnassus, for instance, a Greek-speaking Roman citizen writing in the late 1st century BCE, states in his *Roman Antiquities*: "Indeed, in my opinion the three most magnificent works of Rome, in which the greatness of her empire is best seen, are the aqueducts, the paved roads and the construction of the sewers."[6]

[6] Dionysius of Halicarnassus, *Roman Antiquities* 3.67.5 (LCL vol. 246–47; transl. E. Cary). Cf. Frontinus (after a short description of the nine aqueducts that supplied Rome in his time): "With such an array of indispensable structures carrying so many waters, compare, if you will, the idle Pyramids,

But such large-scale, expensive projects, commissioned and financed by central authorities ("from above") seem to have been the exception rather than the rule. Just like many other autocratic rulers in the Mediterranean before and after them, Roman emperors invested in such huge projects to enhance their own (or their family's) prestige and to underline their presence in both a symbolic ("I take care of you") and a tangible way by ensuring the daily deliverance of drinking water.[7] The ruler's power was even more emphasised in the aqueduct bridges that served their own palaces, such as the branch of the Aqua Claudia in Rome that catered to the water needs of the imperial palaces, the so-called Arcus Neroniani built under Nero (54–68 CE), on the Palatine Hill for many centuries.[8]

The writings of Roman authors of the imperial period contain other biases. Their repeated suggestions that Roman emperors took actively care of the well-being of their subjects should be regarded as the result of their rhetorical (flattering) style, echoing in part the self-fashioning of the emperor Augustus and his successors.[9] In reality, Roman rulers lacked the expertise (and presumably interest) for developing extensive and long term health plans for the common people, although minimal provisions of water and food had to be made to guarantee social stability in major cities. To this end, regulations existed and water management was organised, but not necessarily to the equal benefit of all. The metropolis of Rome is a case in point, as will be demonstrated in the next section.

Frontinus and the City of Rome

We are much better informed about water management in the capital than we are for any other city in the Roman world. The remains of aqueducts in the countryside are the silent witnesses of a supply system that provided the city and its inhabitants

or the useless, though famous, works of the Greeks!", Frontinus, *Aqueducts* 16 (LCL vol. 174: 356–59; transl. C. Bennett).

[7] Horden and Purcell, *The Corrupting Sea*, 247–50, with examples and critical assessment.

[8] See Bruun, "Water Supply, Drainage and Watermills," 305. See also the contribution of Van den Bent in this volume on the water supply of the Abbasid palace in Baghdad.

[9] Frontinus, *Aqueducts* 1 (LCL vol. 174: 330–32; transl. C. Bennett): "and inasmuch as Nerva Augustus, an emperor of whom I am at a loss to say whether he devotes more industry or love to the State, has laid upon me the duties of water commissioner, an office which concerns not merely the convenience but also the health and even the safety of the City...". See also Frontinus, *Aqueducts* 88 (LCL vol. 174: 416–17; transl. C. Bennett): "this care of our most dutiful emperor Nerva". Cf. also the anecdote told in Suetonius' biography of Augustus, where Augustus is said to have stated, when people in Rome complained about the price of wine, that his son-in-law Agrippa had ensured that no one would go thirsty by building several aqueducts: Suetonius, *Lives of the Caesars: The Deified Augustus*, 42 (LCL vol. 31: 214–15).

with water of good quality. The traces of the inner urban system for the transport of water to the city's inhabitants are less visible, even if stamped lead pipes (*fistulae*) with the names of private water consumers and/or plumber's workshops are known.[10] What is more, we have a detailed overview of nine of the eventually eleven aqueducts and their daily deliveries thanks to the treatise *The Aqueducts of Rome* (*De Aquis Urbis Romae*) written by Sextus Julius Frontinus around 100 CE. Frontinus was appointed water commissioner (*curator aquarum*) under emperor Nerva (r. 96–98 CE) and wrote his treatise about the supply and water management of the capital when he entered office. He presents his text as a sort of a handbook, a blueprint for himself and his successors. But first and foremost, he shows himself as a reliable and diligent official in the service of the new emperor Nerva and of the people of Rome.[11] He starts his work with a brief historical overview, followed by a description of the courses and flow of the nine aqueducts built to supply Rome since the late 4th century BCE.[12]

The building of the first four aqueducts supplying Rome, the Aqua Appia (312 BCE), the Aqua Anio Vetus (273 BCE), the Aqua Marcia (144 BCE) and the Aqua Tepula (125 BCE) coincided with rapid population growth, thanks to Rome's successes in warfare and the expansion that followed. We possess other indications of the increase in population, for example the growing number of multi-storeyed apartment buildings. The continuous rise in the number of inhabitants and the need for more running water resulting from the boom in the building of bathhouses led both to repairs and to the construction of five new aqueducts during the early principate.[13] As a result, in Frontinus' day large quantities of running water flowed to the city via underground channels and pipelines, crossing valleys by aqueduct bridges or (inverted) siphons, and transported on arches in the immediate vicinity of Rome, in order to enter the city at height (and thus being easier to distribute). Within the city area the water was no longer freely flowing but transported under pressure in lead pipes (*fistulae*) to a variety of consumers, including public fountains.

Frontinus explains at length the benefits of the major reforms under the first emperor Augustus, who had the existing aqueducts repaired and built three

10 Bruun, *The Water Supply of Ancient Rome*; De Kleijn, *The Water Supply of Ancient Rome*.

11 Peachin, *Frontinus and the* curae *of the* curator aquarum.

12 After Frontinus' lifetime two more aqueducts were completed, the Aqua Traiana (initiated under Domitian) in the early 2nd century, and the early-3rd-century Aqua Alexandrina (named after Alexander Severus and supplying the rebuilt and enlarged Baths of Nero under the same emperor's rule). The Baths of Caracalla (also early 3rd century) were fed by the Aqua Antoniniana, a new branch of the Aqua Marcia.

13 For the relationship between the building of baths and aqueducts see e.g. Hodge, *Roman Aqueducts*, 5–6; Fagan, *Bathing in Public*, 69–74, 106.

new ones. Augustus' close friend and son-in-law Agrippa had been in charge of reorganizing Rome's water management, resulting in the *cura aquarum*, a body organised in a similar way to the already existing bodies for the inspection of roads (*cura viarum*) and the distribution of grain (the *cura frumenti*).[14] From the Augustan period onward all aqueducts supplying water to Rome were commissioned by the emperors and no longer by city magistrates. In the same vein, the emperor was directly responsible for the building of major aqueducts elsewhere in the empire as well, or at least gave his consent, as inscriptions and literary sources testify.[15]

Since Frontinus presents water as a common good, he is keen on combatting fraud. Illegal tapping from aqueducts, both outside and inside the city should be punished and supervision reinstalled, a practice that, according to him, had largely lapsed over time. Especially in the years prior to his acceptance of the office, if we are to believe Frontinus' words, corruption was rampant; furthermore, his predecessors "through laziness and inertia had neglected their duties."[16] Moreover, the tapping of water by private individuals should be monitored more closely: "no one shall draw public water without a permit and no one shall draw more than has been granted."[17]

Water could be granted upon request, a procedure described by Frontinus. Individuals who received a permit to have a private connection belonged to the elite. The grant of water was a personal privilege awarded by the emperor himself (the so-called *beneficium Caesaris*). This system of personal grants, initiated under Augustus, was part of the balance of power between the emperor and the senators, and to the benefit of the elite, a small group of people who laid claim to large quantities of water.

Before discussing the distribution of water elsewhere, it is important to address, again, some of the biases in our source material with regards to Rome. It is true that large amounts of water arrived at the city of Rome, day after day. The most cautious calculations, based on the information in Frontinus' treatise and the archaeological remains of the aqueducts in Rome's hinterland, lead to rough estimates of between 520,000 and 635,000 m^3 delivered on a daily basis in the late

[14] Frontinus, *Aqueducts* 98–101 (LCL vol. 174: 426–33).

[15] Inscriptions, some examples: EDR 094030 (Ostia, Vespasian); CIL 3, 2909 / 9983 (Iader / Zadar, Croatia, Trajan); CIL 7, 142 (Segontium / Caernarfon, Britain, Septimius Severus, Caracalla and Geta); CIL 3, 709 (Dyrrachium / Durrës, Albania, Alexander Severus; repair of an aqueduct financed by Hadrian); CIL 3, 549 (Athens, a new urban aqueduct begun under Hadrian and inaugurated by his successor Antoninus Pius in 139 or 140 CE). For the emperor's permission necessary for the construction of aqueducts and public baths, see e.g. Pliny the Younger, *Letters Book 10*, 37 and 38; 39 and 40; 90 and 91 (LCL vol 59: 211–17, 278–81).

[16] Frontinus, *Aqueducts* 101 (LCL vol. 174: 430–31; transl. C. Bennett).

[17] Frontinus, *Aqueducts* 103 (LCL vol. 174: 434–35; transl. C. Bennett).

1st century CE.[18] Basing ourselves on estimates of the number of Rome's inhabitants in this period as ranging from 800,000 to 1,000,000, the average quantity of water delivered would have been somewhere between 52 and 79 litres per capita per 24 hours. Undeniably, this is a large amount of water for any premodern society (and even modern societies), but one has to take into account the fact that more than half the daily water delivered was distributed to buildings selected by the emperor (e.g. the large imperial baths), as well as a significant number of private individuals, most of them belonging to the senatorial elite.[19] As a consequence, in reality the available amounts for most of Rome's inhabitants would have been much lower, somewhere between 23 and 35 litres a day.[20] Still, these numbers reflect access to water in an abstract way only. The majority of people, without a private connection to the urban system,[21] would have depended for their water needs on public fountains and the ornamental fountain basins that formed the endpoint of the aqueducts. Frontinus mentions thirty-nine ornamental fountains (*munera*), possibly including the terminal basins of the nine aqueducts of his time, as well as 591 public fountains with a basin or small reservoir (*lacus*) for which he himself and his team were responsible.[22] Together with an assistant (*procurator*) and with the help of an administrative staff the *curator aquarum* was at the head of more than 700 persons, for a large part enslaved people. Regular inspections, the prevention of illegal tapping, the cleaning of the channels and minor and major repairs to all aqueducts outside and inside the walls of Rome were their daily business. This kind of maintenance was needed everywhere aqueducts were built, but given the scale of Rome's demand for water and the number of people depending on its continuous supply, these tasks were centralized and put under direct control of the emperors and their entourage.

Still, part of Agrippa's masterplan for Rome consisted in delegating the responsibility for supervision over each neighbourhood (*vicus*) to two persons, living in that neighbourhood or owning property there, "under whose discretion water

18 Bruun, "Water Supply, Drainage and Watermills", 306–307.

19 Frontinus, *Aqueducts* 78 (LCL vol. 174: 408–409); See Bruun, "Water Supply, Drainage and Watermills", 307.

20 To clarify this point, based on Frontinus' text we may estimate the amount of water available for public use (that is freely accessible) as 44% of the total (see Bruun, "Water Supply, Drainage and Watermills", 307).

21 About 300 names of water rights owners are documented, most of them (but not exclusively) members of the senatorial elite. For documentation and important analyses see Bruun, *The Water Supply of Ancient Rome* and De Kleijn, *The Water Supply of Ancient Rome*.

22 Frontinus, *Aqueducts* 78 (LCL vol. 174: 408–409); Frontinus' contemporary Pliny the Elder mentions 500 fountains (*salientes*) as the result of Agrippa's activity alone, plus 700 *lacus* (basins) and 130 *castella* (reservoirs for inner urban regional distribution): see Pliny, *Natural History* 36.121 (LCL vol. 419: 96–97).

in the public space would flow", as Frontinus puts it, citing the jurist and former *curator aquarum* Ateius Capito.[23] It is an interesting but hitherto less discussed feature that during the first centuries CE, even in the metropolis, water management seems to have been in part organised at the microlevel of neighbourhoods. We lack information on the exact duties and competences of these neighbourhood supervisors, except for the fact that they were appointed directly by the *aediles*, magistrates who were, among other duties in the local government, responsible for the maintenance of public buildings.

The highly rhetorical style of writers such as Pliny the Elder and Frontinus suggest that water in Rome was first of all a public good, distributed to the benefit of all. The rough calculations above, based on close reading of some passages, already show a different picture. Moreover, the access to and use of water was bound by rules and restrictions. Frontinus lists a number of senatorial decrees from the late republic and reforms under Augustus, but it should again be kept in mind that these decrees and laws concerned Rome, and that the capital was in many ways the exception, not the rule.[24]

What we can also learn from Frontinus' treatise is how the balance was kept between the needs of Rome's urban population on the one hand, and the immediate hinterland on the other. The latter was crucial for the supply of foodstuffs, and used, at least in part, the same water sources that supplied the city. No wonder then that the *curator aquarum* carefully administered the amount of water that was distributed to estates and farms from the various aqueducts on their way to Rome. Moreover, illegal tapping seems to have been a recurring problem.[25]

Rural Areas: Water Laws and Common-Pool Resources

For rural areas we depend on the information provided by legal sources dealing with water use, mostly for irrigation. Roman jurists tried to achieve a balance, with water laws serving the needs of as many (if not all) stakeholders as possible, starting from the idea that water was in essence a common good. The sea, some

[23] Frontinus, *Aqueducts* 97 (LCL vol. 174: 426–27). C. Ateius Capito (c. 30 BCE–22 CE) was appointed *curator aquarum* in 13 CE.

[24] Laws and senatorial decrees: Frontinus, *Aqueducts* 94–111 and 125–29 (LCL vol. 174:422–43, 454–65). For Roman water legislation, see Bruun, "Water Legislation in the Ancient World," 575-604; Bannon, "A Short Introduction," 1–18.

[25] Water distributed outside Rome (*extra urbem*), in quantities per aqueduct: Frontinus, *Aqueducts* 78–86; On the problem of illegal tapping: Frontinus, *Aqueducts* 9, 65, 67, 69, 72 (LCL vol. 174: 348–51, 388–401). On corruption and bribery of *aquarii*: Frontinus, *Aqueducts* 74-76 (LCL vol. 174: 402–407); on other forms of fraud by consumers and/or *aquarii*: Frontinus, *Aqueducts* 112-115 (LCL vol. 174: 442–47).

lakes and most rivers were considered to be public goods. Some scholars maintain that in practice *all* flowing water was public property.[26] Public water resources could in principle be used by any Roman citizen, with the restriction, however, that no harm should be inflicted on neighbours.[27] Some evidence suggests that water consumers had to pay a fee or tax (*vectigal*) for the water they drew from rivers (the commons) and used to irrigate their gardens, orchards or seasonal pasture lands. Cicero, for example, states that he had to pay for the water he took from a nearby brook, the Crabra, to irrigate his estate in Tusculum near Rome.[28]

Some rivers, though, were regarded private property, as explained in two passages by the early 3rd-century jurist Ulpian:

> Some rivers are public; others are not. Cassius defines a public river as a perennial river. This definition, approved by Celsus, appears correct.[29]
> The Praetor's edict, above set forth, concerns public rivers. If private, the interdict does not apply; for a private river differs in no sense from any other private property.[30]

If water from a public river was used to supply a settlement or city nearby, the owners of the land along the river presumably had to allow the construction of a canal or conduit for the transport of the water.[31] The alternative would have been that the city would have had to purchase the necessary stretches of land.

Access to public water sources for consumers at a distance could also be guaranteed by means of the legal instrument of the easement or servitude (*servitus*). This legal concept entails the right of a person other than the owner of the land to make use of that land.[32] Servitudes relied on cooperation and agreements between parties and were thus crucial to balancing the needs of the various water consumers. In other words, this legal instrument, enforced by law ("from above"), helped smaller

26 Eck, "Die Wasserversorgung im römischen Reich," 58.

27 Dig. 43.20.3.1 *Ex flumine aquam ducere plures possunt, ita tamen, ut vicinis non noceant, vel, si angustus amnis sit, etiam ei, qui in alia ripa sit.*

28 Cicero, *De Lege Agraria*, 3.9 (LCL vol. 240: 562–63; transl. Dyck): *ego Tusculanis pro aqua Crabra vectigal pendam* ("At my Tusculan estate, I will pay a tax for the use of the Aqua Crabra").

29 Dig. 43.12.1.3. *Fluminum quaedam publica sunt, quaedam non. Publicum flumen esse Cassius definit, quod perenne sit: haec sententia Cassii, quam et Celsus probat, videtur esse probabilis.* Transl. Bruun, "Water Legislation," 577. Cassius and Celsus, mentioned in this passage, are jurists from the 1st and early 2nd c. CE.

30 Dig. 43.12.1.4. *Hoc interdictum ad flumina publica pertinet: si autem flumen privatum sit, cessabit interdictum: nihil enim differt a ceteris locis privatis flumen privatum.* Transl. Bruun, "Water Legislation," 577. A praetorian interdict was a legal order issued by the praetor that authorized an action.

31 Bruun, "Water Legislation," 582.

32 Berger, *Encyclopedic Dictionary*, 702–703 s.v. *Servitus* (*servitutes*).

communities to manage the local water resources ("from below"). The *servitus aquae ductus* (or *aquae ducendae*), the right to conduct water, was a rural servitude, "consisting in the right of the owner of the dominant land to conduct water from, or across, another's land through pipe or canals."[33] Some restrictions, however, applied: the water had to be drawn from a perennial spring (a *fons perennis*) or at the intake (*caput*) of a continuously flowing (and hence public) river.

Springs (*fontes*) were sometimes in private possession, at least if they were located on private land. Private springs could be bought and sold, but also used by benefactors to supply, for instance, the public baths of a nearby city. A well-preserved inscription dating to the 1st century CE refers to the local magistrate (*duovir*) Publius Faianius Plebeius, who ensured the water supply for his hometown Forum Novum (Latium, located in today's province of Rieti):

> P. Faianius Plebeius, *duovir* for the second time, brought water from his own land into the town of Forum Novum at his own expense, had all the public water basins (*lacus*) built, and oversaw the discharge of the water into the *piscina* that is in the Campus; he also approved the work. Also, since the seller of the lot on which the bath stands offered the buyer no guarantee concerning the water supply, P. Faianus Plebeius, in order that water would flow in the bathhouse, gave his own water to that bathhouse, so that the citizens of the town might not be deprived of its convenience.[34]

We will address the distribution of water in cities and the role of benefactors further below, but first we will discuss distribution practices in rural areas.

The Distribution of Water and Irrigation Communities in Rural Areas

Most of the water infrastructure in the Roman world was located in rural areas. Most of these hydraulic structures were built, maintained and managed by local authorities or local groups of users. In some regions of the Empire—those with a temperate climate—rainfed agriculture was possible, but this was not the case in large parts of the Mediterranean and Middle East. In rural areas where dry farming was out of the question, either because of limits imposed by the natural environment, or because of the type of crops that were grown, artificial irrigation with was a necessity.

33 Berger, *Encyclopedic Dictionary*, 703 s.v. *Servitus aquaeductus* (*aquae ducendae*).

34 ILS 5767 = CIL 9, 4786, Forum Novum (Italia, Regio IV). Translation (slightly adapted) after Fagan, *Bathing in Public*, 309.

Rivers and springs were the main water sources and water was transported to the fields with the help of irrigation channels. Dams, *qanāt* systems and wadis, the latter two identified in some parts of the Roman Mediterranean only, continued earlier local practices, and were, again, the result of local circumstances.[35] In most cases, a large number of farmers or landowners would rely on the same water resources, and the available amounts had to be divided among the various consumers. The rules of conduct on how to avoid depletion of the common water sources or prevent disproportionate use by some consumers at the expense of others, would somehow have to be agreed upon, either orally or in writing. Such rules of conduct fit into the theoretical framework known as "common-pool resources theory" developed by political scientist Elinor Ostrom.[36] Water, a common resource (or "commons") had to be distributed in such a way that, ideally, the interests of all users ("appropriators") who depended on the same resources (i.e. water sources) were met. Moreover, in order to prevent a situation where the quantities taken by one user damaged the interests of others, rules of conduct had to be established. Ostrom showed in her research that tight-knit communities (villages, neighbourhoods) are able to collaborate and share their common-pool resources in a successful way if they manage to agree on rules that are acceptable for all. The collaboration needed for this strategy often entails a bottom-up, locally organised arrangement.

For the Roman world, several such rules of conduct are known from inscriptions and literary texts; these will be discussed below. This is not to deny that power dynamics in some form were almost always also in operation. If a water shortage existed, cooperation was much needed in order to cater for the needs of all, but competition and conflict over access and its durations and quantities could arise. In consequence, disputes were settled with laws and interdictions, thereby restoring equilibrium in the use of the common-pool resources. In this respect, Roman law was instrumental in ensuring the proper functioning of local agreements and maintaining good relations between the various stakeholders.

In a rescript dating to the mid-2nd century, the emperors Antoninus Pius and Lucius Verus stated that the amount of water allotted depended on the size of the fields.[37] Meanwhile, other legal arrangements, collected in the *Digesta* (a late antique corpus of earlier legal texts, compiled in the 6th century), reflect different types of distribution practices.

35 Oleson, "Irrigation," 183–215. On *qanāts* see Hodge, "Qanats," 35–38.

36 Ostrom, *Governing the Commons.*

37 Dig. 8.3.17. *Papirius Justus, On Constitutions, Book I*: The August Emperors Antoninus and Verus stated in a Rescript, that, "Where water is taken from a public river for the purpose of irrigating fields, it should be divided in proportion to the size of the same; unless someone can prove that, by virtue of a special privilege, he is entitled to more." They also stated in a Rescript that, "A party should only be permitted to conduct water where this can be done without injury to another."

According to these legal sources, access to water might be ordered by regulations allotting days or hours to individual consumers. Similar water-sharing regulations are documented for the medieval and modern Middle East as well.[38] The right to conduct water (via an irrigation channel) or draw water (from a spring or well) from the same source for a number of consumers was thus possible.[39] If the quantities of water were sufficient for a larger number of consumers at the same time, such regulations were not necessary, as is made explicit—perhaps an indication of the fact that such schedules were rather the rule than the exception.[40] Stipulations were concrete and casuistic in nature, for example: "If I have the right to lead water (to my fields) during the day or during the night, I cannot lead water during another hour than for which I have the right to do so."[41] Pomponius, a jurist from the first half of the 2nd century, expresses the opinion that this right (*ius*) not only pertained to irrigation of the land, but also to watering cattle or to "amenity" (*amoenitas*), probably referring to baths or garden fountains that were part of the villa of the landowner.[42] The right to draw water for irrigation purposes could be granted to estates (*praedia*) or to persons (*personae*). In the latter case, the water right would automatically expire if the holder of that right were to pass away. Still, if the heir or new owner of the land could demonstrate the need for water, he (or she) would most probably obtain the right to water as well, and, as the jurists add, "this is not a favour, but it would be a form of injustice if the water right were not granted."[43]

Sharing water sources with rules of conduct concerning when water may be drawn from the commons can be regarded as a form of common-pool resource

[38] Hayes, "Land, Law and Water Politics", 1–17.

[39] Dig. 8.3.2.1 Nerva 4 reg.: *Aquae ductus et haustus aquae per eundem locum ut ducatur, etiam pluribus concedi potest: potest etiam, ut diversis diebus vel horis ducatur.* Cf. Dig. 8.6.7 Paulus in libro 13 ad Plautium (on servitudes), where there is mention of the right to transport water over land owned by another farmer (a servitude) on a monthly or seasonal basis: "If the right to conduct water is granted in such a way that this can only be done during the summer, or for one month, (...).

[40] Dig. 8.3.2.2 Nerva 4 reg.: *Si aquae ductus vel haustus aquae sufficiens est, potest et pluribus per eundem locum concedi, ut et isdem diebus vel horis ducatur.*

[41] Dig. 43.20.2 *Pomponius libro 32 ad Sabinum*: *Si diurnarum aut nocturnarum horarum aquae ductum habeam, non possum alia hora ducere, quam qua ius habeam ducendi.* (transl. author) See for comparable stipulations in Islamic law the contribution of Hayes in this volume.

[42] Dig. 43.20.3 *Pomponius libro 34 ad Sabinum*: pr. *Hoc iure utimur, ut etiam non ad irrigandum, sed pecoris causa vel amoenitatis aqua duci possit.*

[43] Dig. 43.20.1.43: *Et datur interdum praediis, interdum personis. Quod praediis datur, persona extincta non extinguitur: quod datur personis, cum personis amittitur ideoque neque ad alium dominum praediorum neque ad heredem vel qualemcumque successorem transit. Plane ei, ad quem dominium transit, impetrabile est: nam si docuerit praediis suis aquam debitam, etsi nomine eius fluxisse, a quo dominium ad se transiit, indubitate impetrat ius aquae ducendae, nec est hoc beneficium, sed iniuria, si quis forte non impetraverit.*

management. Still, these legal texts seem to reflect first and foremost the interests of rich landowners. It remains unclear how such rules would have affected the interests of small farmers or tenants. They are likely to have been the outcome of conflict management between the owners of larger estates and written down to avoid future frictions and similar disputes elsewhere.

As is the case with most of the passages in the *Digesta*, all phrased in very general terms and no further details are given. Therefore, we lack, for instance, information about the regions where these rules were applied. On the other hand, we do have information from both literary sources and inscriptions that the distribution of water based on hours or days was common in various rural areas of the Roman Empire, including the hinterlands of Rome, Roman Spain and North Africa.

The Aqua Crabra near Tusculum, mentioned above, is also briefly discussed by Frontinus in his account of the Aqua Julia, an aqueduct commissioned by Agrippa. He provides some interesting details that show the potential tensions between cities and their hinterlands with regards to water resources:

> Past the intake of Julia flows a brook, which is called Crabra. Agrippa refrained from drawing from this brook either because he had condemned it, or because he thought it ought to be left to the proprietors at Tusculum, for this is the water which all the estates of that district receive in turn, dealt out to them on regular days and in regular quantities.[44]

Frontinus also claims to have shut off the Crabra brook and to have restored its use entirely to the landowners nearby, after a period in which the water was used to increase the volume flowing in the Aqua Julia, an aqueduct that supplied Rome. The villa estates in Tusculum, owned by wealthy senators such as Cicero, possessed gardens and orchards with a high demand for water. Such estates, producing cash crops for the markets in Rome, were located in the immediate hinterland. After all, perishable products such as vegetables and fruits could only be transported over short distances. Frontinus offers information here about the interdependency of cities and hinterlands. Moreover, his remark that all estates received in turn regular quantities of water on regular days reflects one of the water distribution practices attested by the legal sources discussed above. Frontinus is keen to list for every aqueduct supplying Rome the quantities of water that were allotted to farms and estates outside the city (*extra urbem*). These quantities were considerable. He does, however, not give details of distribution practices or how control was organised, but does mention fraud, i.e. illegal tapping, and the bribing of *aquarii*.[45]

44 Frontinus, *Aqueducts* 9 (LCL vol. 174:348–49; transl. C. Bennett, slightly adapted).

45 See above footnote 25.

More evidence for hourly regulations for the distribution of water near Rome comes from an inscription found in Tibur (modern Tivoli).[46] The stone is heavily damaged, but the names of some of the villa owners are still readable, as are indications for the openings (*foramina*) in the aqueduct conduit for the tapping of water, and the hours allotted to the various estates.[47] Time was measured with the help of sundials (during daytime) and/or water clocks. Yet another inscription, perhaps also from the area of Tibur, but in any case from Rome's hinterlands, shows similar arrangements, listing the names of owners, and the number of (lead?) pipes (*aquae*) that transported water from an aqueduct to their land, according to fixed hours, e.g. "from the second till the sixth hour" (*ab hora secunda ad horam sextam*).[48] Arrangements like these also point at the necessity of dividing the available water in a reasonable way between cities and their hinterlands, where they drew water from the same sources. Rome is no exception here.[49]

It is hardly surprising that the distribution of water for irrigation was a practice common in arid zones as well. Pliny the Elder mentions such an arrangement for the city of Tacape, nowadays Gabès in Tunisia:

> There is a city-state of Africa called Tacape, in the middle of the desert on the route to the Syrtes Gulfs and Leptis Magna, which has the exceptionally marvellous blessing of a well-watered soil. There is a spring that distributes water over a space of about three miles in every direction, giving a generous supply, but nevertheless it is distributed among the population only at special fixed periods of the day.[50]

Pliny does not explain who decided the hours and amounts of water allotted to the inhabitants and landowners who needed water to irrigate their olive groves or date palms. All we can assume is that the city council must have had a role in the decision-making, since Tacape was a Roman *colonia*. We can also safely assume that the wealthier landowners, who were members of the local council, had a strong voice in the decision-making process and the control of the arrangements made.

The same goes for a well-known inscription from Lamasba in Roman Numidia (Henchir Merouana, Algeria), detailing the distribution of water from a perennial

46 CIL 14, 3676; Eck, "Die Verteilung des Mangels," 227–38.

47 The names of two villa owners: M. Sallvius (*sic*) Domitianus, L. Primus Sosianus; the number and size (measured in *digiti*, 'fingers') of the openings (*foramina*) and hours are indicated.

48 CIL 6, 1261. See for a discussion Eck, "Die Verteilung des Mangels," 228–29; Wilson, "Villas, Horticulture and Irrigation Infrastructure," 752–54. The stone, once visible in the garden of the church S. Maria del Priorato on the Aventine Hill in Rome, is now lost.

49 For the case of Tibur, supplied by Rome's aqueducts, see also Evans, "*In Tiburtinum usum*," 447–55.

50 Pliny, *Natural History* 18.188 (LCL 371: 306–307; transl. H. Rackham, slightly adapted).

spring to the fields of the land owners listed.[51] The inscription, dating to the early 3rd century CE, seems to fix a new arrangement after disagreement had arisen over an earlier one, as is mentioned in the exordium of the surviving text. The new arrangement, set in stone and visible for everyone, records agreements as to the duration of the water flow to the respective fields in accordance with their size. All fields were located on the sloping terraces of a hill. The agreement was reached by order of the council and the *coloni* (landowning citizens) of Lamasba, who had delegated the negotiations to Valentinus and a colleague whose name remains obscure, both presumably provincial officials. Still, the initiative had been a local one in order to settle the dispute over the use of the commons.

Pre-Roman Irrigation Communities

We even have some evidence for local, pre-Roman practices and types of arrangements for which normally no provincial or central authorities would be involved. Two inscriptions from Roman Spain offer glimpses of the agreements between members of local irrigation communities. The oldest one, a bronze tablet known as the *Tabula Contrebiensis*, was found in Contrebia Belaisca (modern Botorrita, Aragon) and dates to the year 87 BCE.[52] It settles a dispute between communities over the purchase of land and the right to construct a canal for irrigation (*rivi faciendi aquaive ducendae*) over both the recently bought land and "public land" (*ager publicus*, a typically Roman notion). Here again it is clear that the initiative was taken by the communities themselves. Moreover, these originally Celtiberian communities with Latinized names existed already, before Roman-style legal decisions were taken and inscribed in bronze, placed in a public space and made visible for all members of the community.[53] The autonomy of these communities in the late 2nd and early 1st centuries BCE was apparently significant. Roman provincial authorities and legislation were called upon to continue and formalise existing practices.

The second inscription, the *Lex Rivi Hiberiensis* or Bronze of Agón, comes from the same semi-arid region, the Middle Ebro Valley in NE Spain. The inscription, which dates from the early 2nd century CE, was found in the 1990s. It is a long text, consisting of 152 lines set in three columns, inscribed on a bronze table. The text settles a dispute between two local irrigation communities, situated on the right

51 CIL 8, 4440. For a detailed analysis see Shaw, "Lamasba: An Ancient Irrigation Community," 61–103.

52 Richardson, "The *Tabula Contrebiensis*," 33–41; Birks, Rodger, and Richardson, "Further Aspects," 45–73.

53 Richardson, "The *Tabula Contrebiensis*", 35.

bank of the Ebro River and using the same irrigation canal (the *rivus hiberiensis*), which drew its water from the river. The text has not survived in its entirety, but the surviving parts provide us with interesting details about a conflict that had arisen between two groups of farmers or landowners over water.[54]

The conflict itself remains to a large extent unclear, due to the fact that the text here is missing or inscribed on an extremely worn part of the bronze sheet. Most likely, the downstream party suffered from the behaviour of the upstream community.[55] The dispute was complicated by the fact that the two parties belonged to the hinterlands of two different Roman cities: the downstream irrigation community, the *pagus Gallorum*, belonged to the Roman colony Caesaraugusta (Zaragoza), whereas the *Belsinonensis* district, the area of the other community, upstream the Ebro river, was juridically speaking part of the *municipium* (city) of Cascantum (Cascante).[56] Therefore, the dispute not only regarded the two irrigation communities in this rural area but potentially had consequences for the two cities involved as well, relying on the supply of foodstuffs from their respective hinterlands. This must have been the reason why the dispute was eventually brought to the provincial court and had to be settled by the governor of the Roman province Hispania Citerior.

The governor's judgement was clearly formulated in a *lex*. Importantly, this new law issued by the governor merged the two former irrigation communities into a single community governed by a single set of regulations for its proper functioning.[57]

These regulations entail first and foremost the maintenance of the irrigation canal and sluices in common use, in stipulations enjoining cleaning and repairs over a fixed period in the early summer. Each member of the community had to contribute their fair share to the work, according to their water rights (*ius aquae*) and water shares (*aquatio*). Fines and terms were established for those members of the community who failed to fulfil these obligations, or who took more water than they were entitled to. Fines were also due for those members who threw waste or manure (*stercus*) into the canal (*rivus*).[58]

The *aquatio*, that is the amount of water, was established based on the size of the plot of land that had to be irrigated. The main body of the community was the assembly (*concilium*). All its members had voting right and all decisions were taken by the majority of votes of the members (*pagani*). Voting, however, occurred in proportion to the *ius aquae* of each member. Consequently, members with bigger

[54] Editio princeps: Beltrán Lloris, "An Irrigation Decree," 147–97; Crawford and Beltrán Lloris, "The *Lex Rivi Hiberiensis*," 233; Einheuser, *Studien zur lex rivi Hiberiensis*.

[55] Beltrán Lloris, "An Irrigation Decree", 165.

[56] Beltrán Lloris, "An Irrigation Decree", 160–62.

[57] Beltrán Lloris, "An Irrigation Decree", 164, 186–88.

[58] Beltrán Lloris, "An Irrigation Decree", 170–72.

plots of land had more voting rights in the assembly. On the other hand, their duties for maintenance work and levies were heavier because of their larger plots.[59]

Two pairs of *magistri pagi* were elected by the assembly and served for one year; they fulfilled a central role and were possibly assisted by an inspector in charge of the records (*libertus tabu[larius]*?). The *magistri pagi* were elected in early June. This was a consequence of the agricultural calendar and the need for the irrigation of different crops at different times of year. The assembly decided where and when to start the irrigation circle (taking turns) after the completion of maintenance work and the resumption of water flow. This was closely connected to priorities for certain crops and their need for water.

On the one hand, then, we see that the provincial authorities did intervene, and on the other, a presumably older tradition of local arrangements. Moreover, the stipulations formulated by the provincial governor seem to express the idea that future disagreements within the community could be solved internally, since the rules for this newly established irrigation community were now fixed, inscribed in bronze, placed in a public space and thus, again, visible to all. This way, all the parties involved were co-responsible and accountable, including the locally elected *magistri pagi*. It is striking, how strong the position of the individual members of the community who elected the *magistri pagi* and could held them accountable was—obviously a safety mechanism to avoid corruption.[60] Finally, the agreement served the interests of two nearby cities as well, both dependent on the agricultural activities of their hinterlands.

Access to Water and Inner Urban Distribution in Cities and Settlements: Some Examples

"For four hundred and forty-one years from the foundation of the City, the Romans were satisfied with the use of such waters as they drew from the Tiber, from wells, or from springs."[61] This passing comment, from the opening chapters of Frontinus' text on the situation in Rome before the building of the first aqueduct, reveals what must have been common practice in many small or medium sized cities and settlements within and outside Roman Italy. Springs, however, were not available everywhere and digging wells was often a difficult operation, especially when groundwater levels were low and the soil was either very loose or rocky.[62] The

[59] Beltrán Lloris, "An Irrigation Decree", 171–73, 177; Bannon, "A Short Introduction," 73–77.

[60] Beltrán Lloris, "An Irrigation Decree", 182–83, 190–91.

[61] Frontinus, *Aqueducts* 4 (LCL vol. 174: 338–39; transl. C. Bennett).

[62] For an excellent discussion of digging and using wells see: Gerrard, "The Social Lives of Wells," 77–104. For wells in Pompeii: Jansen, *Water in de Romeinse stad*, 20–22 with further references.

storage of rainwater in cisterns was another important method for securing access to water. In fact, the roofs of larger townhouses in Roman Italy and elsewhere in the Mediterranean were designed to collect rainwater and to lead the water via spouts or drainpipes and gutters to underground cisterns or basins. Research has shown that the cisterns of many houses in Pompeii, located in the central halls (*atria*) or garden areas (*peristylia*) had capacities varying from 12 to c. 75 cubic metres.[63] Most houses had several cisterns, and these cisterns were still in use when an urban water supply system was introduced in Pompeii in the early 1st century BCE, itself significantly improved in the Augustan period around 30 BCE. Some houses, notably the grand houses owned by the Pompeian upper class, had a private connection to the urban system that was built in the Augustan period, as part of a regional water supply system commissioned by Augustus' confidant and son-in-law Agrippa, known as the Aqua Augusta or Serino Aqueduct.[64] The cisterns in these houses stored both rain water and the water flowing from the fountains with basins in the *atria*, courtyards and peristyle gardens. This overflow water could be used for various purposes, such as cleaning and watering the garden afterwards.[65]

Most inhabitants of Pompeii, however, depended on the water they fetched from public fountains, fed by the urban water supply.[66] In the last phase of Pompeii, forty-three public fountains situated on pavements and at crossroads provided people with good quality (potable) water.[67] The actual number must have been higher since around one quarter of the city's intramural area remains unexcavated. Scaling up the number to take this into account, we may expect that there would have been fifty-five public fountains freely accessible for everyone. As a result, most people in Pompeii could draw water for free from a fountain nearby.

Epigraphic evidence offers some glimpses of water grants awarded to citizens by local councils. A clear example is mentioned in an inscription from Suessa Aurunca (Italy), dating to the late 2nd century CE. A local citizen, C. Titius Chresimus, was honoured with a water connection to his house. The diameter of the lead pipe transporting the water was specified: one finger wide.[68] This phrasing suggests that the urban distribution of water in Suessa Aurunca was also achieved with

63 Jansen, *Water in de Romeinse stad*, 22–26.

64 See for a brief overview with references Piras, "Cura Aquarum in Campania," 87–91.

65 Jansen, *Water in de Romeinse stad*, 48–58.

66 At least for drinking water, for which Romans preferred spring (mineral) water, as written sources reveal, e.g. Vitruvius, *On Architecture* Book 8 (dedicated to water) (LCL 280: 132–93).

67 Jansen, *Water in de Romeinse stad*, 45 (street fountains + three tap water points without a basin), Olsson, "Water Quantities" provides a new method for calculating the amounts; recent excavations revealed another water tower with a public fountain in regio V.

68 CIL 10, 4760 (= 6296): *ut aquae digitus in domo eius flueret*.

a network of lead pipes that transported water to the various consumers with a private connection, similar to the system in Pompeii.

A few other grants for private connections and even the privilege of water for free (so without paying a *vectigal*) are known from Cales (Italy, early 1st century CE), Ipolcobulcula (Spain, late 2nd century CE), from an unknown site in southern France and perhaps also from Thysdrus (Tunisia).[69] The phrasing in the Thysdrus inscription (late 3rd century CE) is somewhat vague but suggests that some houseowners enjoyed the privilege of a private connection, 'under certain conditions', while also mentioning public fountains with basins (*lacus*) placed on the main streets (*plateas*). A recently found edict from Laodicea, in southwest Turkey, confirms that private parties had to pay for the water directed to their homes or workshops.[70] The fee would be established on the basis of the diameter of the lead pipe that led the water to the private properties, as was normal practice elsewhere as well. The size of the pipes, therefore, had to be calibrated and approved with the use of an authorized standard pipe or bronze gauge (a *chalkeion metron* is mentioned in the text, line 15).[71] The text explicitly states that granting water for free was not possible: "In like manner no one should be allowed to take the city's water for free, on any pretext, nor a city magistrate (officer) decide this kind of exemption (concession)".[72] Moreover, the edict mentions fines for illegally tapping of water, damaging the aqueduct channel, or polluting the water. It was stipulated that the proconsul would appoint two citizens every year who acted as *curatores*.[73] Moreover, they were held accountable should illegal tapping of water occur. The edict was issued by the proconsul (governor) of Asia in 114 or 115 CE, who resided in the provincial capital Ephesus. The provisions promulgated in the detailed text of the edict seem to respond to an earlier petition written by the assembly of Laodicea. Thus, a local

[69] Cales: CIL 10, 4654 = ILS 5779: *in eius* [sc. Q. Paconii Leptae] *domum aqua pura duceretur*; Ipolcobulcula: CIL 2, 1643: *C. Annius Praesius Ipolcobulcu/le(n)sis Apueaclesis incola / ob honorem seviratus / et gratuitum aquae / usum*; From un unknown site in Gallia Narbonensis (southern France), a undated fragment of a honorific decree: CIL 12, 5413: *ipse ea domo utatur aquam g[ratuitam?]*; Thysdrus: ILS 5777: *aqua adducta...per plateas lacubus* [sic] *inpertita domibus etiam certa condicione concessa.*

[70] For an edition, translation and compelling interpretations of this inscription see Guizzi, "An Edict of a Proconsul of Asia," 143–64; Guizzi, "Proteggere l'acquedotto," 151–72; Guizzi, "Public Texts," 7–15. Now also Gautier di Confiengo and Santucci, "Water Control in Phrygia," 317–18.

[71] Guizzi, "Public Texts," 13.

[72] Guizzi, "Public Texts," 8 (Greek text, line 11) and 9 (translation).

[73] Guizzi, "Public Texts," 8 (Greek text, lines 20–22) and 9 (translation): "Considering the continuous safety of the water, I also thought necessary to define this also: that by the city be chosen during electoral assemblies, from among the most distinguished men, two superintendents every year, to take care of the superintendence in an uncorrupted (pure) and zealous (loving of honors) way, looking only for the common utility."

initiative to protect the supply and distribution of water, in order to safeguard the accessibility of the commons to all, was approved off by the provincial authorities.

In a similar manner, at a level intermediate between 'from below' and 'from above', wealthy citizens or patrons would sometimes invest in the hydraulic infrastructure of their cities, as epigraphic evidence testifies. For Italy, a corpus of fifty-three inscriptions relates to the building or restoration of aqueducts by public or private parties.[74] Interestingly enough, these projects seem to have above all served to boost the prominence and visibility of the benefactor. The high costs involved are mentioned explicitly in most cases.[75] Moreover, the aqueducts mentioned are all located in (micro-)regions of Italy where water shortage was not at stake as such. Often, (co-)financing of an aqueduct by a local benefactor went hand in hand with the building of public baths.[76] The boom in public baths and the growing bathing habit from the 1st century BCE onward were in many cases the driving forces behind the construction of urban aqueducts. Finally, as mentioned already, both inscriptions and literary sources reveal that emperors invested large amounts of money in water projects outside Rome as well. Their main objective was comparable to the aims of local or provincial benefactors: visibility and "presence". But behind all the rhetoric involved and the major constructions that caught the eye, this does not mean that access to sufficient (drinking) water was a given for any inhabitant of the Roman world. Quite on the contrary: the lived reality for many was far less glorious, as can be seen, for instance, in evidence from an imperial stone quarry in the Egyptian desert.

Access to Water and Inequality

An ostracon (inscribed potsherd) from the quarry at Mons Claudianus in the Eastern Desert of Egypt, speaks volumes.[77] The quarry, famous for its grey granite and hence in imperial hands since the mid-1st century, is located in the Red Sea

[74] Tamburrino, Riera, and Zanovello, "Looking at Water Shortage," 388–391. A total number of 66 inscriptions are given here, varying from the 2nd century BCE–5th century CE, including 13 inscriptions from the city of Rome.

[75] E.g.: the Aqua Augusta built in the 2nd century CE in Pula (Croatia), by *patronus coloniae* L. Menacius Priscus, who also secured 400,000 sesterces for maintenance (CIL 5, 47): *L(ucius) Menacius L(uci) f(ilius) Vel(ina) / Priscus / equo pub(lico) praef(ectus) fabrum aed(ilis) / IIvir quinq(uennalis) trib(unus) mil(itum) / flamen Augustor(um) patron(us) colon(iae) / aquam Aug(ustam) in superiorem partem coloniae et in inferiorem / i<m=N>pensa sua perduxit et in tutelam / eius dedit HS CCCC(milibus).*

[76] For some examples see Fagan, *Bathing in Public*, 309–312.

[77] O. Claud. inv. 1538 + 2921. The sherd (two pieces), part of the neck of an amphora, measures 26 x 19,5 cm; published by Cuvigny, "L'organigramme du personnel," 309–53.

Mountains. The settlement at the quarry, inhabited by around 1,000 persons, consisted of a small number of military men and the quarry workers, local people (*paganoi*) and men hired from other parts of the empire, both enslaved and freed or freeborn persons (*phameliarioi*). Possibly, some *paganoi* lived at Mons Claudianus with their families.[78]

The daily rations of water on a specific day in November for all inhabitants were assigned according to a strict hierarchy: the highest officers and engineers received triple the amount of water given to the unskilled labourers, who did the hardest work.[79] The list of water rations confirms that members of the workforce did not receive more than two to three litres per day, an insufficient amount for an adult performing such physically demanding work under the extreme climatic conditions of the Eastern Desert. The rationing of water was a clear expression of power relations here, not necessarily corresponding to the needs of individuals. The text is interesting for yet another reason: the job of *akouarioi* (*aquarii*), water carriers, is mentioned, namely the people who transported the water in waterskins (*askoforia*) with the help of camels.[80] Water was collected from springs and wells throughout the Eastern Desert. The people at Mons Claudianus had to be supplied with c. 2250 litres per day, for which the *akouarioi* were responsible.[81] Moreover, the animals (donkeys, mules, and camels) needed drinking water as well. Last but not least, the three blacksmiths mentioned in the list needed water for the constant sharpening of the tools used by the stone masons and quarry workers.

This one ostracon offers a unique insight into the daily lives of underprivileged people, a harsh reality rarely reflected in other sources. Notes like these, recording the daily rations of water that were or had to be handed out, were simply not saved. But this one ostracon in our corpus of sources, we may assume, stands for many more such notes, written on cheap, perishable materials and thrown away afterwards. This ostracon is only one tiny piece in the puzzle, but we may expect that this little piece of evidence on the stratified and low rates of daily water rations reflects the lived reality of many more people.

[78] On Mons Claudianus see Murray, "The Roman Roads and Stations," 138–50; Hirt, *Imperial Mines and Quarries*; Gates-Foster et al., "The Early Imperial Fortress of Berkou," 30–74; Breeze, *Frontiers of the Roman Empire*," 49–57.

[79] The distribution list dates to c. 110 CE: Cuvigny, "L'organigramme du personnel," 310, 320–21.

[80] Cuvigny, "L'organigramme du personnel", 332; Hirt, *Imperial Mines and Quarries*, 211–12.

[81] Hirt, *Imperial Mines and Quarries*, 211.

Conclusion

Despite the geographical and chronological differences in our evidence from both rural and urban contexts, there is some consistency in the picture that emerges. First and foremost, access to water and access to power were inextricably connected in the Roman world. The terms set for the accessibility of water for groups or individuals, and the quantities available, can be regarded as a direct expression of power relations. In rural areas, shares in water corresponded to the size of land. Land ownership not only defined the social status of Roman citizens, but also their possibilities for political careers. As members of local councils or even as magistrates in their hometowns, the local ruling classes commissioned, managed and sometimes financed water works for both the city and its hinterland. The central level, that is, the emperor and his entourage, would step in if the costs involved exceeded the financial possibilities of a city, and sometimes in the case of financial overstretch. None of this should come as a surprise, but rather fits with a wider body of evidence and ideas about how the Roman empire took shape.

There are, nonetheless, biases in our sources, both in the textual and material evidence. The extensive system of water licenses in the city of Rome, depending on the goodwill of the emperor, for instance, should be regarded as an exception. This system of privileges granted by the ruler, who used the water grants as an instrument to maintain good relations with the elite, would only later be parallelled in Constantinople, the new capital and main residence of the emperors from the 4th century onward. The few inscriptions we have that mention the privilege of a private water connection outside of Rome, as a way of honouring a person, should not be read as contradicting the normal state of affairs where only a handful of town dwellers had a connection to the urban running water system and still fewer would have received the connection as a grant. Even where such an urban system was present, the vast majority of inhabitants would have had to fetch water from public fountains or distribution points. As for the evidence in rural contexts, legal and literary sources and inscriptions alike all seem to refer to local arrangements made by and for local landowners or reflect the settlement of disputes that had arisen.

It is striking that many of those major water management projects in the Roman world initiated and financed from above were primarily realized in regions where water was abundant, or at least sufficient. These were prestige projects commissioned and paid for by the emperor or by local or provincial benefactors striving for a career in the highest echelons of the empire.

It was in regions with water shortages where small(er)-scale infrastructure was built and maintained by local communities who were also in charge of water distribution.[82] Only in rare cases, usually when conflict had arisen, do we read about these self-organised irrigation communities. Still, the phenomenon of local water communities in arid zones may have been the rule rather than the exception. After all, the need for cooperation was crucial for survival, and hence in the collective interest. As is often the case for the ancient world, the most common cultural practices are hardly visible in our source material, or not at all. Still, scrutinizing the snippets of evidence we have, water management, both from below and from above, turns out to be a relevant prism through which to shed more light on living conditions and inequality in the Roman world.

Bibliography

Abbreviations

CIL *Corpus Inscriptionum Latinarum* (1863–)

Dig. Alan Watson. *The Digest of Justinian*, vols. 1–4, rev. ed. University of Pennsylvania Press, 1998.

EDR Epigraphic Database Roma (online: http://www.edr-edr.it/it/present_it.php), accessed March 2026.

ILS *Inscriptiones Latinae Selectae*, edited by H. Dessau. Weidmann, 1891–1916.

LCL Loeb Classical Library

Primary Sources

Cicero. *Pro Quinctio. Pro Roscio Amerino. Pro Roscio Comoedo. Pro Tullio. De Lege Agraria.* Edited and translated by Andrew R. Dyck. Loeb Classical Library, 240. Harvard University Press, 2025.

Dionysius of Halicarnassus. *Roman Antiquities, Volume II: Books 3-4*. Translated by Earnest Cary. Loeb Classical Library, 347. Harvard University Press, 1939.

Frontinus. *The Stratagems and The Aqueducts of Rome*. Translated by Charles E. Bennett, Mary B. McElwain. Loeb Classical Library, 174. Harvard University Press, 1925.

Pliny. *Natural History, Volume V: Books 17-19*. Translated by H. Rackham. Loeb Classical Library, 371. Harvard University Press, 1950.

Pliny. *Natural History, Volume X: Books 36-37*. Translated by D. E. Eichholz. Loeb Classical Library, 419. Harvard University Press, 1962.

[82] Not to be confused with large-scale projects in arid regions that were commissioned by rulers to supply cities (notably baths), such as the Caesarea aqueduct in today's Israel (built under king Herod the Great in the late 1st century BCE and expanded under Hadrian in the early 2nd century CE), or the Zaghouan aqueduct (2nd century CE) in modern Tunisia that supplied Carthage, the Antonine Baths in particular. These aqueducts did not necessarily serve the needs of nearby local communities (and potentially even harmed them) and were built and managed by imperial, provincial or urban authorities. Moreover, impressive structures such as aqueduct bridges dominating the landscape were visible landmarks of the power of the ruler and/or ruling classes.

Pliny the Younger. *Letters, Volume II: Books 8-10. Panegyricus*. Translated by Betty Radice. Loeb Classical Library, 59. Harvard University Press, 1969.

Suetonius. *Lives of the Caesars, Volume I: Julius. Augustus. Tiberius. Gaius Caligula.* Translated by J. C. Rolfe. Introduction by K. R. Bradley. Loeb Classical Library, 31. Harvard University Press, 1914.

Vitruvius. *On Architecture, Volume II: Books 6-10*. Translated by Frank Granger. Loeb Classical Library, 280. Harvard University Press, 1934.

Secondary Sources

Bannon, Cynthia J. "A Short Introduction to Roman Water Law." *Memoirs of the American Academy in Rome* 66 (2021): 1–18.

Beltrán Lloris, Francisco. "An Irrigation Decree from Roman Spain: 'The Lex Rivi Hiberiensis'." *Journal of Roman Studies* 96 (2006): 147–97.

Berger, Adolf. *Encyclopedic Dictionary of Roman Law*. The American Philosophical Society, 1953.

Birks, Peter, Alan Rodger, and John S. Richardson. "Further Aspects of the Tabula Contrebiensis." *Journal of Roman Studies* 74 (1984): 45–73.

Breeze, David J. *Frontiers of the Roman Empire: The Roman Frontier in Egypt / Frontières de l'empire romain: la frontière romaine en Égypte*. Archeopress, 2021.

Brélaz, Cédric and Thomas Lau. "Introduction." In *Patterns in the History of Polycentric Governance in European Cities. From Antiquity to the 21st Century*, edited by Cédric Brélaz, Thomas Lau, Hans-Joachim Schmidt, and Siegfried Weichlein, 21–37. De Gruyter, 2024. https://doi.org/10.1515/9783111029054.

Bruun, Christer. *The Water Supply of Ancient Rome: a Study of Roman Imperial Administration*. Societas Scientiarum Fennica, 1991.

Bruun, Christer. "Water Legislation in the Ancient World (c. 2000 B.C.–c. A.D. 500)" in *Handbook of Ancient Water Technology*, edited by Örjan Wikander, 539–604. Brill, 2000.

Bruun, Christer. "Water Supply, Drainage and Watermills." In *The Cambridge Companion to Ancient Rome*, edited by Paul Erdkamp, 297–313. Cambridge University Press, 2013.

Crawford, Michael and Francisco Beltrán Lloris. "The *Lex Rivi Hiberiensis*." *Journal of Roman Studies* 103 (2013): 233.

Cuvigny, Hélène. "L'organigramme du personnel d'une carrière impériale d'après un ostracon du Mons Claudianus." *Chiron* 35 (2005): 309–53.

De Kleijn, Gerda. *The Water Supply of Ancient Rome. City Area, Water, and Population*. Gieben, 2001.

Eck, Werner. "Die Wasserversorgung im römischen Reich: Sozio-politische Bedingungen, Recht und Administration." In *Die Wasserversorgung antiker Städte* (Geschichte der Wasserversorgung 2), 49–101. Frontinus-Gesellschaft e.V., 1987.

Eck, Werner. "Die Verteilung des Mangels: Landwirtschaftliche Bewässerung in römischer Zeit in ariden Gebieten nach schriftlichen Quellen." In *Cura Aquarum in Jordanien*, edited by Christoph Ohlig, 227–38. Deutsche Wasserhistorische Gesellschaft e.V., 2008.

Einheuser, Vanessa. *Studien zur lex rivi Hiberiensis: Zur Rechtsdurchsetzung innerhalb einer Bewässerungsgemeinschaft im 2. Jh. n. Chr.* Harrassowitz Verlag, 2017.

Evans, Harry B. "*In Tiburtinum usum*: Special Arrangements in the Roman Water System (Frontinus, Aq. 6.5)." *American Journal of Archaeology* 97, no. 3 (July 1993): 447–55.

Fagan, Garrett G. *Bathing in Public in the Roman World*. University of Michigan Press, 1999.

Finley, Moses I. "The Ancient City: From Fustel de Coulanges to Max Weber and Beyond." *Comparative Studies in Society and History* 19, no. 3 (July 1977): 305–27.

Gates-Foster, Jennifer, Isabelle Goncalves, Bérangère Redon, Hélène Cuvigny, Mariola Hepa, and Thomas Faucher. "The Early Imperial Fortress of Berkou, Eastern Desert, Egypt." *Journal of Roman Archaeology* 34, no. 1 (2021): 30–74.

Gautier di Confiengo, Edoardo and Elettra Santucci. "Water Control in Phrygia (Asia Minor). The Examples of Castella Aquae of Hierapolis and Laodicea." In *Bathing Culture in Budapest. Proceedings of the International Frontinus Conference, Budapest, September 8-15, 2022*, edited by Gilbert Wiplinger, 305–26. Peeters Publishers, 2025.

Gerrard, James. "The Social Lives of Wells in Roman Britain and Beyond." In *Water in the Roman World. Engineering, Trade, Religion and Daily Life*, edited by Martin Henig and Jason Lundock, 77–104. Archaeopress, 2022.

Guizzi, Francesco. "An Edict of a Proconsul of Asia on the Aqueduct of Laodikeia (114/115 CE?)." In *15. Yılında Laodikeia (2003-2018)*, edited by Celal Şimşek, 143–64. Zero Books / Ege Yayınları, 2019.

Guizzi, Francesco. "Proteggere l'acquedotto. Un editto del proconsole d'Asia a Laodicea sul Lico (114/115 d.C.)." In *Terre Acque Diritto. Atti del Convegno del Centro Studi sui Fondamenti del Diritto Antico 30 novembre – 1 dicembre*, edited by Mariavaleria del Tufo, Francesco Fasolino, and Francesco Lucrezi, 151–72. Editoriale Scientifica, 2021.

Guizzi, Francesco. "Public Texts (Provincial administration, decrees, letters)." In *Laodikeia (Laodiceia on the Lycus). Greek and Latin inscriptions found in the excavation 2003-2021*, edited by Celal Şimşek, 7–15. Zero Books / Ege Yayınları, 2022.

Hayes, Edmund. "Land, Law and Water Politics: Ibn ʿAsākir's Chapter on Umayyad Canal Building in the Hinterland of Damascus." *Al-Masāq* 38, no. 1 (2025): 1–17. https://doi.org/10.1080/09503110.2025.2485572.

Hirt, Alfred Michael. *Imperial Mines and Quarries in the Roman World. Organizational Aspects 27 BC–AD 235*. Oxford University Press, 2010.

Hodge, Alfred Trevor. *Roman Aqueducts and Water Supply*. Duckworth, 1992.

Hodge, Alfred Trevor. "Qanats." In *Handbook of Ancient Water Technology*, edited by Örjan Wikander, 35–38. Brill, 2000.

Horden, Peregrine, and Nicholas Purcell. *The Corrupting Sea. A Study of Mediterranean History*. Blackwell Publishers, 2000.

Jansen, Gemma C.M. *Water in de Romeinse stad. Pompeji – Herculaneum – Ostia*. Peeters Publishers, 2002.

Murray, George William. "The Roman Roads and Stations in the Eastern Desert of Egypt." *Journal of Egyptian Archaeology* 11, no. 1 (1925): 138–50.

Oleson, John Peter. "Irrigation." In *Handbook of Ancient Water Technology*, edited by Örjan Wikander, 183–215. Brill, 2000.

Olsson, Richard. "Water quantities for public and private use in Pompeii." *Water History* 15 (2023): 363–92. https://doi.org/10.1007/s12685-023-00333-2.

Ostrom, Elinor. *Governing the Commons. The Evolution of Institutions for Collective Action*. Cambridge University Press, 1990.

Ostrom, Vincent, Charles M. Tiebout, and Robert Warren, "The Organization of Government in Metropolitan Areas: A Theoretical Inquiry," *American Political Science Review* 55, no. 4 (December 1961): 831–42.

Peachin, Michael, *Frontinus and the* curae *of the* curator aquarum. Franz Steiner Verlag, 2004.

Piras, Susanna A.G. "Cura Aquarum in Campania: eine Einführung," in *Cura Aquarum in Campania. Proceedings of the Ninth International Congress on the History of Water Management and Hydraulic Engineering in the Mediterranean Region*, edited by Nathalie de Haan and Gemma C.M. Jansen, 87–91. Peeters Publishers, 1996.

Pont, Anne-Valérie, "The Dynamics of Dual-level Governance in the Roman Empire, First-Third Centuries CE: Incremental Permeation and Occasional Intrusions of Roman Normativity in Local Life." In *Patterns in the History of Polycentric Governance in European Cities from Antiquity to the 21st Century*, edited by Cédric Brélaz, Thomas Lau, Hans-Joachim Schmidt, and Siegfried Weichlein, 21–37. De Gruyter, 2024. https://doi.org/10.1515/9783111029054.

Richardson, John S. "The *Tabula Contrebiensis*: Roman Law in Spain in the Early First Century B.C.: I." *Journal of Roman Studies* 73 (1983): 33–41.

Shaw, Brent. "Lamasba: An Ancient Irrigation Community." *Antiquités Africaines* 18 (1982): 61–103. https://doi.org/10.3406/antaf.1982.1085.

Tamburrino, Eugenio, Italo Riera, and Paola Zanovello. "Looking at Water Shortage from a Roman Point of View: A Survey of Literary Sources and Epigraphical Attestations from Italian *regiones*." In *Wasserwesen zur Zeit des Frontinus. Bauwerke – Technik – Kultur*, edited by Gilbert Wiplinger and Wolfram Letzner, 381–94. Peeters Publishers, 2016.

Wilson, Andrew. "Villas, Horticulture and Irrigation Infrastructure in the Tiber Valley." In Mercator Placidissimus. *The Tiber Valley in Antiquity. New Research in the Upper and Middle River Valley*, edited by Filippo Coarelli & Helen Patterson, 731–68. Edizioni Quasar, 2009.

CHAPTER 3

Water Management in Early Islamic Medina: Infrastructure, Labour and Politics

Harry Munt

Abstract

Water management has always been of particular importance to both the settled and (semi-) nomadic populations of the Hijaz. This was certainly true in the early Islamic period as the region came to relatively greater prominence as one of the centres of a newly established empire and as the principal holy land of the early Muslim community. This chapter explores what we know about water management in and around Medina, the Hijaz's main economic and administrative centre in the Umayyad and early Abbasid periods. It draws mostly on the information about this topic found in narrative sources, especially local histories of the region and geographical works. These sources are used to discuss some of the water management strategies that were deployed in early Islamic Medina along with what we know of the labour practices that were exploited to carry them out. The chapter also highlights how efforts to control water around early Islamic Medina were politically highly charged and put some of the most important families competing for power in the early Islamic empire in direct competition with each other.

Keywords: Medina; Hijaz; Arabian Peninsula; Umayyads; Abbasids; Zubayrids; water management

In the year 88/707, the hajj was led by the then governor of Medina (and future caliph), ʿUmar ibn ʿAbd al-ʿAzīz. When he and his party arrived at al-Tanʿīm,[1] a group of Qurashīs came to inform him that Mecca was experiencing a serious water shortage, caused by exceptionally low rainfall, and that they feared there would not be enough for all the pilgrims to drink. To ʿUmar, the solution was obvious: pray for rain. This they did and then proceeded on to Mecca. By the time they reached the Kaʿba, it was already raining. It ended up raining so much, in fact, that floods appeared, threatening to cut off important pilgrimage sites such as ʿArafa and Minā, causing general alarm. As a further result, we are told, there were even verdant plants growing in Mecca that year.[2] This anecdote highlights the obvious

[1] A location two *farsakh*s (roughly twelve kilometres) from Mecca; see al-Bakrī, *Muʿjam mā istaʿjam*, 1:321.

[2] al-Ṭabarī, *Taʾrīkh*, 2:1196–97.

environmental and climatic difficulties faced by inhabitants of and visitors to the Hijaz in the early Islamic period, most notably the potential precarity of the water supply. Yet it also highlights the importance of water and ideas about water to the operation of rule and government in the early Islamic empire.[3]

The history of the early Islamic empire has not often been considered holistically through the lens of environmental history. This is not to say that environmental, climatic, ecological and epidemiological matters have not been key to much scholarship on early Islamic history. There is no space for an exhaustive survey of such scholarship, but it is worth highlighting that for far more than a century now, studies on the emergence of the Islamic empire have often offered detailed discussions of the climate and environment of the Middle East, Arabia in particular, an understanding of which is seen as crucial to appreciating the events of the 1st/7th century in the region.[4] Surveys and studies of early Islamic history regularly offer some discussion of such matters, both of Arabia and of the other regions that became provinces of the Umayyad and Abbasid empires.[5] Climatic, environmental and ecological changes over Late Antiquity have increasingly come to be seen, often in impressively interdisciplinary studies, as important longer-term contexts behind developments in the practice of Roman imperial authority and subsequently of the early Muslim community's establishment of a new empire over an unprecedented expanse of territory.[6] And there are dedicated studies of environmental change and management over the *longue durée* in particular regions of the early Islamic world.[7] What I mean to emphasise here is that studies of the early Islamic empire to date have rarely prioritised analysing its operation as, to borrow Alan Mikhail's characterisation of the early modern Ottoman empire, an "ecosystem", "a complex nexus of relationships among resources, people, ideas, and places in which each element of the community is dependent on every other. A change or perturbation in any one part of the environment affects each component of the whole."[8]

[3] In this chapter, the "early Islamic empire" refers to the polity established from the early 1st/mid-7th century down to roughly the mid-3rd/9th century.

[4] A good, relatively early example is Lammens, *Berceau de l'Islam*.

[5] See, for example, Kennedy, *Early Abbasid Caliphate*, 18–34; and recently Marsham, *Umayyad Empire*, 31–40.

[6] A reasonably early example of such scholarship is Korotaev, Klimenko and Proussakov, "Origins of Islam". Among a number of more recent studies, see, for example, Izdebski et al., "Environmental, Archaeological and Historical Evidence"; Harper, *Fate of Rome*, esp. 283–87 in particular for the early Muslim community; and Haldon and Fleitmann, "Sixth-Century CE Drought".

[7] Examples include Bulliet, *Cotton, Climate, and Camels*; Verkinderen, *Waterways of Iraq and Iran*.

[8] Mikhail, *Animal in Ottoman Egypt*, 9–10 (quotation at 9).

That said, an ever-growing number of studies are taking the environmental history of the early Islamic empire and its operation seriously.[9] This chapter is a small contribution to these efforts. My intention here is to think about the political importance of water and the management of water and access to water in the early Islamic Hijaz, particularly the region around Medina. The Hijaz—at least beyond the period of the Prophet's lifetime and the early caliphs down to the first *fitna* (loosely, "civil war") in the mid-to-late 30s/650s—needs to be brought more fully into studies of the early Islamic empire, including its environmental history, for a number of reasons. (The same could be said of the Arabian Peninsula more broadly, but that is beyond the scope of this chapter.) These are to do, on the one hand, with our understanding of the mechanics and practice of empire in the early Islamic world more broadly and, on the other, with how we think about the Hijaz's history more specifically. Regarding the former, it is clear, despite this being questioned in some studies, that the Hijaz was a politically important province of the early Islamic empire, both because of the continuing political significance of Mecca and Medina even after the seat of caliphal government moved to Syria and then Iraq, and because many of the region's inhabitants worked hard to establish themselves among the elites who were building the new empire.[10] What this political importance meant for the particular ways in which the Hijaz and its inhabitants were tied into the projects of empire-building during the Umayyad and early Abbasid periods, however, is still far from clear.

To tie this into how we think about the Hijaz's history more specifically, it is worth remembering that the roughly two centuries or so from the mid-1st/7th century were the first time that large parts of the Arabian Peninsula, especially the Hijaz and Najd, had been incorporated meaningfully within a universalising empire.[11] For areas other than Mecca and Medina, as well as a few sites on the routes to those towns from Syria and Egypt, this was also really the last time that they had been a meaningful part of such an empire.[12] It was an unusual period in the Hijaz's and central Arabia's history. A second reason why it is important to think about the Hijaz's environmental history as a province of the early Islamic empire

[9] For example, Azad, "Ecology, Economy"; Marsham, *Umayyad Empire*, esp. 213–56.

[10] One stark example of what is being questioned would be Khalid Blankinship's labelling of the region as "the Western Arabian backwater" in his *End of the Jihad State*, at 73–75. My reasons for disagreeing with this assessment are outlined, with further references, in my *Holy City of Medina*, 149–61; and further in my "Caliphal Imperialism".

[11] By a region's "meaningful" incorporation into an empire, I suppose I mean that many of its inhabitants would have known of the regular presence of imperial officials, who were there both to extract resources but also to redistribute them and occasionally to bring in additional resources from other regions.

[12] This point has also been made recently in Cook, *History of the Muslim World*, 157.

is to help scholars to move away from old-fashioned environmentally determinist understandings of Arabia's history. Environmental determinism, as a means of studying the history of many places in many times, has of course been criticised heavily.[13] It has perhaps been particular pernicious in older studies of Arabia, in which an image of the peninsula as something of a timeless desert "has done much to marginalize" its study.[14] Much of Arabia is certainly very dry—apparently it has not always been quite so dry in the era of human settlement[15]—although this is not true of the whole peninsula; and even where it is, there are relatively few areas that humans over the last few millennia have not worked hard to settle, live in and control.[16] In the 1st/7th to 3rd/9th centuries, it was not only local inhabitants who sought to understand and adapt to Arabia's varied environments;[17] the rulers and administrators of a world empire sought to do so too.

Writing the environmental history of the early Islamic empire is a monumental task. This chapter's small contribution to the task is to look at practices of water management and their political importance in one particular area of the Hijaz, Medina and its hinterland. We will start with a brief survey of the methods of water management that were deployed in early Islamic Medina as well as of the labour employed to manage them. We will then turn to the question of the political significance of water management in and around Medina. As we saw in the perhaps unusual circumstances described in the opening anecdote of this chapter, the provision of adequate water could be seen as a thoroughly political act, as one of the duties owed by those who claimed office to their subjects, the successful fulfilment of which was a sign of the legitimacy of their office. After some initial discussion of the evidence for the political significance of water provision in Medina more broadly, the chapter will end by turning to one particular and seemingly important feature in early Islamic Medina's water-management landscape, the Wādī al-ʿAqīq. In this discussion, we will look at the actions of one of Medina's local families, the Zubayrids, to consolidate their position in the ʿAqīq.[18]

13 For example, Davis, "Power, Knowledge, and Environmental History"; Mikhail, *Under Osman's Tree*, 5–8.

14 Magee, *Archaeology of Prehistoric Arabia*, 14.

15 Dinies et al., "When the Desert Was Green".

16 A really good overview can be found in Magee, *Archaeology of Prehistoric Arabia*, 14–45.

17 For an interesting discussion of how some late pre- and early Islamic-era Arabian groups understood their environment, see Miller, "Seasonal Poetics".

18 Some of this expands upon material included in my "Caliphal Estates". The brief description of available sources in that chapter (at 433) and the methodological approach taken there still apply more or less to my approach in this chapter. A particularly important local historian for this study is ʿAlī ibn ʿAbd Allāh al-Samhūdī (d. 911/1506), who, although writing much later than the period covered in this chapter, is a great source of material from now-lost earlier local histories; for more on this question, see my "Writing the History"; and "Mamluk historiography". There has been some notable

Water Management in Medina and Its Labour

The Hijaz north of today's ʿAsīr province in Saudi Arabia does not normally see enough precipitation for solely rain-fed cultivation. Thanks to more plentiful rainfall in the Pleistocene and early Holocene, however, there are aquifers and a water table accessible in many places, although it is of varying depth, for example at four metres at Taymāʾ and seventeen metres at Madāʾin Ṣāliḥ; at Medina until later 20th-century developments, water was apparently accessible in places between four and ten metres deep.[19] It is primarily then these aquifers and the water table that had to be managed to allow the development of cultivation at oases, some of which could be relatively fertile.[20] It would be incorrect, however, to ignore the importance of rainfall for life and economic activity in the Hijaz. The anecdote set around the hajj of 88/707 at the start of the chapter makes clear how important rainfall could be and that even when average precipitation was meagre, a relative scarcity could still be problematic. The prayer for rain (*istisqāʾ*) in times of drought, led by ʿUmar ibn ʿAbd al-ʿAzīz in that anecdote, was a pre-Islamic practice often assumed to have been given prophetic sanction.[21] Complaints about the lack of rain feature fairly commonly in poetry about Medina's landscape and water resources preserved by successive local historians of the town.[22] Such anecdotes and the importance of prayer for rain as a practice also indicates how the problem of relatively low average rainfall was at times compounded by its sometimes erratic variability. In Medina, for example, between 1957 and 1978, annual rainfall varied from 0mm to 103.8mm.[23] Unusually high levels of rain, as at Mecca in 88/707, could cause flooding. Disparities in rainfall between different regions in Arabia led to different customs and migratory patterns among the peninsula's semi-nomadic inhabitants.[24] So water management practices required ensuring effective access

archaeological work in the northern Hijaz in recent years, though not really in the area immediately surrounding Medina itself. One volume which offers an overview of some of this work is Foote et al., eds., *Revealing Cultural Landscapes*. For a discussion of some of the archaeological evidence for water management across the Arabian Peninsula over several millennia, see Wilkinson, *Archaeological Landscapes*, 151–83.

19 Makki, *Medina*, 15.

20 Apart from the information about the depth of water at Medina, this is drawn from the survey of western Arabia in Magee, *Archaeology of Prehistoric Arabia*, 29–32. For the "Holocene Moist Phase", from *ca.* 9000–8000 BCE to *ca.* 4000 BCE, see *ibid.*, 42–44.

21 For some examples of the Prophet performing the prayer for rain, see Ibn Shabba, *Taʾrīkh al-Madīna*, 1:92–94.

22 A fairly standard example is offered by the anonymous poet cited in al-Samhūdī, *Wafāʾ al-wafā*, 4:27: "I passed by the ʿAqīq and its inhabitants who were complaining about the meagre spring rains."

23 Makki, *Medina*, 57.

24 Miller, "Seasonal Poetics".

to aquifers and the water table as well as careful preservation of as much rainfall as possible and its channelling in useful ways.

Sources that discuss Medina's land usage and environment—principally the town's local histories, but also some other historical and geographical works—refer to a number of different mechanisms for accessing, storing and channelling water. A passage by a non-local traveller and geographer, al-Yaʿqūbī (d. after 295/908), summarises the basics of water management practices in Medina nicely:

> [Medina] has four wadis, which run with water when it rains along with flood water that runs off from the mountains at the place known as Ḥarrat Banī Sulaym, a distance of about ten *farsakhs* from Medina.[25] The wadis are Buṭḥān, the Greater ʿAqīq, the Lesser ʿAqīq and Qanāt.[26] The water in these wadis comes during the time of floods and they all come together at a place called al-Ghāba,[27] which they leave in a [single] wadi called Wādī Iḍam.[28] The Greater ʿAqīq and the Lesser ʿAqīq also provide the water for wells (*ābār*), including Biʾr Rūma, which was dug by the Banū Māzin, and Biʾr ʿUrwa.[29] For the remainder of the year

[25] A *ḥarra* is a tract of volcanic rock. Ḥarrat Banī Sulaym is east of Medina on the road to al-Rabadha; see *Kitāb al-manāsik*, 334–35. A *farsakh* is roughly three (early Islamic-era) miles or six kilometres; see Hinz, *Islamische Masse und Gewichte*, 62.

[26] For a more expansive description of these wadis and others, from which much of the material in this chapter will be drawn, see especially al-Samhūdī, *Wafāʾ al-wafā*, 4:7–70.

[27] al-Ghāba was the name of a location somewhere between eight miles and one *barīd* (so roughly between sixteen and twenty-four kilometres) north of Medina; see al-Fīrūzābādī, *al-Maghānim al-muṭāba*, ed. Aḥmad, 3:979 [= ed. al-Jāsir, 299]. On calculating the metric equivalent of a *barīd* as a unit of distance in the Hijaz, see Munt, *Holy City of Medina*, 76–77. Several Medinan local historians, including Ibn Zabāla (wr. 199/814), al-Zubayr ibn Bakkār (d. 256/870) and ʿUmar ibn Shabba (d. 262/876), give the name of the location where all the wadis meet as Zaghāba, so the text of al-Yaʿqūbī has perhaps been corrupted here; see Ibn Shabba, *Taʾrīkh al-Madīna*, 1:110; al-Samhūdī, *Wafāʾ al-wafā*, 4:50–52, 68. Al-Samhūdī (*Wafāʾ al-wafā*, 4:308–9) also noted the tendency of some earlier scholars to substitute the toponym al-Ghāba elsewhere for the lesser-known Zaghāba, the latter's location being closer to Medina, by Uḥud.

[28] Buṭḥān runs through Medina from the southeast, Qanāt along the northern side of the oasis from the east and the ʿAqīq (sometimes considered in two parts distinguished as "Lesser" and "Greater", as by al-Yaʿqūbī) runs along the western side of the area from the south. Other wadis discussed frequently by local sources include Mahzūr and Mudhaynib (alternatively Mudhaynīb). Medina's wadis are located on most published maps of the town and its immediate surroundings, of which there are many. One that has been particularly important for my research is that produced and updated by Michael Lecker, a recent example of which can be found in his *Studies on the Life of Muhammad*, 38. For some further discussion of Medina's topography in general, as well as many studies by Michael Lecker, see al-ʿAlī, "Studies".

[29] The ʿUrwa in question is ʿUrwa ibn al-Zubayr (d. 94/712), to whose estate and its water infrastructure we will return later in this chapter. Biʾr Rūma was apparently established to provide water for the Muslim community, an act commonly credited elsewhere to the caliph ʿUthmān ibn ʿAffān (r. 23–35/644–56); see, for example, Ibn Shabba, *Taʾrīkh al-Madīna*, 1:98–99; al-Samhūdī, *Wafāʾ al-wafā*, 3:374–80.

> the Medinans drink from these two wells as well as from others which are not as famous as these two. There are also wells from which date palms and cultivated fields are watered by water-drawing animals, usually camels working along the irrigation infrastructure (*zarānīq*). In Medina there are also springs (*ʿuyūn*) from which water pours forth in specific locations, among them ʿAyn al-Ṣawrayn, ʿAyn Thaniyyat Marwān, ʿAyn al-Khāniqayn, ʿAyn Abī Ziyād, [ʿAyn] Khayf al-Qāḍī, ʿAyn Bard and ʿAyn Azwāj al-Nabī (ṣ). Most of its inhabitants' properties are date palm groves from which comes their sustenance and nourishment; the taxes they pay also come from the *ʿushr* on date palms alongside *ṣadaqa*.[30]

So, water came to Medinans through ground water, aquifers and water courses formed at times of rain. These supplies were primarily accessed through wells and "springs".[31] The Arabic word translated by the latter is *ʿuyūn*, singular *ʿayn*, which can refer to somewhere water naturally flows forth from the ground but also to the place where water from subterranean, manmade canals meets the surface;[32] Ṣāliḥ Aḥmad al-ʿAlī assumed that in early Islamic Medina's case the term *ʿayn* referred to an artesian well, but while this might have been true of some *ʿuyūn* we need not assume that this was the case for all instances where the term is used.[33] Wells are most commonly known, as to al-Yaʿqūbī, as *ābār*, singular *biʾr*, a somewhat generic term from which we cannot really draw any more specific meaning. Very occasionally, however, another term is used. Al-Yaʿqūbī's passage just cited mentions camels working along irrigation infrastructure known as *zarānīq*. This term denotes a system of canals that draw water from a type of well situated between two pillars known as *al-zurnūqān*, between which runs a beam from which a pulley operates.[34] A well in the Wādī al-ʿAqīq that had been excavated by the governor of Medina Hishām ibn Ismāʿīl al-Makhzūmī (in office 82–87/701–6) and then later restored by his grandson, the caliph Hishām ibn ʿAbd al-Malik (r. 105–25/724–43), in an account preserved by al-Samhūdī (d. 911/1506), was labelled a *jarr*, a term presumably related to the slightly better attested *jarūr*, a particularly deep well.[35]

Alongside wells and springs to access groundwater, we also hear about the construction of dams (both the terms *sadd* and *ḍafīra* are commonly used) to

30 al-Yaʿqūbī, *Buldān*, 312–13.

31 For a survey of wells, springs and other aspects of Medina's water infrastructure before 20th-century CE developments, see Makki, *Medina*, 56–73. For a pre-modern overview, see, alongside the other specific examples discussed in this chapter, al-Samhūdī, *Wafāʾ al-wafā*, esp. 3:339–402.

32 See in brief Lane, *Arabic-English Lexicon*, 5:2215.

33 al-ʿAlī, "Milkiyyāt al-arāḍī," 969. That Medina had some perennial and seasonal springs before the 1950s, see Makki, *Medina*, 16.

34 Lane, *Arabic-English Lexicon*, 3:1229; and, for example, al-Fīrūzābādī, *al-Qāmūs al-muḥīṭ*, 890.

35 al-Samhūdī, *Wafāʾ al-wafā*, 4:25. For *jarūr*, see Lane, *Arabic-English Lexicon*, 2:401; and al-Fīrūzābādī, *al-Qāmūs al-muḥīṭ*, 364.

manage the water that arrived when it did rain. As mentioned above, while generally low, rainfall in Medina could come in intense bursts that not only filled the wadis and turned them into fast-flowing rivers, however briefly, but could also lead to damaging flooding. Wadis that ran through or very close by what became the Islamic-era town of Medina, centred around the Prophet's Mosque, could pose a particular threat to the built environment. The Wādī Mahzūr, for example, caused flooding which threatened buildings in Medina, including the Prophet's Mosque, during ʿUthmān ibn ʿAffān's caliphate and then again in 156/772–73 or 158/774; only emergency action building dams and channels to divert the waters prevented total disaster.[36] The Wādī Buṭḥān apparently posed a flooding threat as late as 1966.[37] Dams, then, were in part necessary to control such flooding as much as possible, but they were clearly also used to preserve water from rainfall so it could be used by the local inhabitants. In the volcanic tracts that ran north and east of Medina, the Umayyad caliph Muʿāwiya ibn Abī Sufyān (r. 40–60/661–80) had a dam built to block a wadi and so trap its water to which people could then bring their flocks to drink.[38] Muʿāwiya seems to have been a relatively prolific constructor of dams around Medina; he was responsible for at least one more that still bears an inscription commemorating his patronage in the Wādī al-Khanaq, a little over ten kilometres southeast of Medina.[39] The owners of several estates around Medina discussed in local histories are also reported to have built dams to help manage the water supply.[40]

Water was frequently conveyed away from aquifers, springs, wells, wadis and pools formed by dams by excavated canals. Some of these were known as *qanāt*s and were possibly partially subterranean, but many seem to have been

[36] Ibn Shabba, *Taʾrīkh al-Madīna*, 1:108; al-Balādhurī, *Futūḥ al-buldān*, 11; al-Samhūdī, *Wafāʾ al-wafā*, 4:63–64.

[37] Makki, *Medina*, 22.

[38] Lughda al-Iṣfahānī, *Bilād al-ʿarab*, 401. A dam built by, or at least named after Muʿāwiya, east of Medina (so perhaps, but not necessarily, the same one) was also known to the author of the *Kitāb al-manāsik*, 330. Ibn Saʿd (d. 230/845) may have been aware of this dam/one of these dams as well. The text in the most recent edition of his *Kitāb al-ṭabaqāt al-kabīr* (2:28) locates a place against which the Prophet led a raid called Qarqarat al-Kudr near Maʿdin Banī Sulaym, "beyond the dam of Maʿūna". In the quotation of the passage about this raid from Ibn Saʿd in al-Samhūdī, *Wafāʾ al-wafā*, 4:443, however, it reads "beyond the dam of Muʿāwiya" instead, which was possibly the intended reading.

[39] See especially al-Rāshid, *Dirāsāt*, 32–60.

[40] See, for example, al-Samhūdī, *Wafāʾ al-wafā*, 4:16, 17, 26. For some discussion of the archaeological remains of an early Islamic-era dam in northwest Arabia, near Khaybar, see recently Depreux et al., "New Insights". The use of dams and other infrastructure to channel rain- and floodwaters in helpful ways was an ancient practice throughout the Arabian Peninsula, often referred to in modern scholarship as *sayl* irrigation; see, for example, Wilkinson, *Archaeological Landscapes*, 162–64.

uncovered.[41] We hear examples of canals known as a *khalīj* on several estates around Medina;[42] there is also al-Yaʿqūbī's previously discussed use of the term *zarānīq* to describe the system of canals bringing water from wells to Medina's date palm groves and cultivated fields. Finally, regarding channels used to convey water, we hear occasionally of the construction of a *sharj*, which seems to have indicated specifically a canal used to bring water from tracts of volcanic rock, such as famously surround Medina.[43] There are also occasional references in sources to pools, cisterns and reservoirs used to store water for future use, for example one (the term used is *ḥawḍ*) on an estate in the ʿAqīq.[44] The presence of visitors and pilgrims in Medina also required the construction of dedicated water sources for public drinking water. The caliph al-Walīd ibn ʿAbd al-Malik (r. 86–96/705–15) ordered his governor of Medina, ʿUmar ibn ʿAbd al-ʿAzīz, in 88/706–7 to build a fountain (*fawwāra*) which was to have attendants so it could provide water for people coming to the Prophet's Mosque, which he had just had expanded and lavishly decorated.[45] His brother and successor Sulaymān (r. 96–99/715–17) also built a public fountain (*siqāya*) in al-ʿArṣa, in the Wādī al-ʿAqīq northwest of the town, to provide for the pilgrims leaving Medina to head back towards Syria.[46] It is worth observing that while there is very little archaeological evidence for any early Islamic practices of water management in and immediately surrounding Medina, this evidence from literary sources does at least chime broadly with remains discovered at other sites in western and central Arabia from the early Islamic period. At the site of Fayd, for example, a little over 400 kilometres northeast of Medina, excavations revealed wells, dams, reservoirs and canals. At al-Rabadha, roughly 170 kilometres east of Medina, there are two reservoirs as well as subterranean cisterns.[47]

Despite Medina's inhabitants' best-laid plans, however, sometimes the collection of water was seemingly left to chance occurrences. A very good indication of this comes from perhaps an unlikely source, the chronicle of prophetic and caliphal history (mostly focused on the Umayyad era) composed by the Andalusī Ibn Raʾs

41 On *qanāts* in general, see recently, with further references, Schriwer, "Water and Technology," 525–28. For a recent archaeological study of the largely undated *qanāt* system at Dūmat al-Jandal in northern Arabia, see Charloux et al., "Mapping an Ancient Qanat System".

42 For example, Ibn Shabba, *Taʾrīkh al-Madīna*, 1:107; al-Samhūdī, *Wafāʾ al-wafā*, 4:17, 18.

43 For discussion of one *sharj* built by an estate manager of Muʿāwiya's, see Munt, "Caliphal Estates," 437; for another named after ʿUthmān ibn ʿAffān, see al-Samhūdī, *Wafāʾ al-wafā*, 4:63; for another dredged by al-Zubayr ibn al-ʿAwwām (d. 36/656), see al-Balādhurī, *Futūḥ al-buldān*, 12.

44 al-Samhūdī, *Wafāʾ al-wafā*, 4:24–25.

45 al-Ṭabarī, *Taʾrīkh*, 2:1196.

46 al-Samhūdī, *Wafāʾ al-wafā*, 4:29. Al-Samhūdī is clear (*ibid.*, 4:382) this toponym is vocalised as al-ʿArṣa.

47 al-Hawas et al., "Preliminary Report," 44–45, 49–50, 53–54; al-Rāshid, *al-Rabadhah*, 41–48.

Ghanama around 621/1224–25.[48] In his account of the causes of the Battle of the Ḥarra near Medina in 63/683, he explained:

> The reason the Medinans expelled the Umayyads and besieged them in the Dār Marwān,[49] why they expelled their governor and rejected Yazīd ibn Muʿāwiya [r. 60–64/680–83], was that a torrent of water came to Medina, so people went out to channel it towards their properties. Muʿāwiya's clients also went out so the Medinans fought with them. This coincided with the population's general agitation against Yazīd, with Ibn al-Zubayr having already rejected him. The people in the marketplace hoisted a banner and fought with Muʿāwiya's clients, killing them. The Medinans thought little of this and so also forced out their governor, ʿUthmān ibn Muḥammad ibn Abī Sufyān.[50]

This anecdote does refer to a relatively early period in Medina's post-prophetic history, a time in which its size as a town increased as did the area surrounding it under cultivation as estates, and so it is possible that this is just the result of adequate water infrastructure not having been built yet; but it does reveal an understanding of how water supply in Medina still remained tied to the ad hoc exploitation of rainfall. To end this brief survey, it is important to note that Medinans in the early Islamic period do seem to have been aware that even beyond the vagaries of rainfall, water was a resource vulnerable to overexploitation. In an anecdote originally from a work by al-Zubayr ibn Bakkār (d. 256/870), on which more is to come in this chapter, the Prophet's famous companion Abū Hurayra (d. *ca.* late 50s/670s) discusses a plot of land that had been granted to him by the then governor of Medina Marwān ibn al-Ḥakam (d. 65/685), to which he proceeded to draw water with dams. This anecdote eventually led to the observation that, "The ʿAqīq remained verdant until all the water infrastructure (*ʿuyūn*) was built."[51]

What kind of labour practices made Medina's water management systems work in the early Islamic centuries is even more obscure than the nature of those systems themselves. We hear a little bit about labour on estates around Medina

48 This is the date explicitly mentioned by Ibn Ra's Ghanama as the time of writing in *Manāqil al-durar*, 536. The editors also note, however, that some material was added in 626/1228–29 as well; see their introduction to this edition, at 102.

49 The residence of the Umayyad Marwān ibn al-Ḥakam that became an administrative centre for Umayyad government in Medina; see Ibn Shabba, *Ta'rīkh al-Madīna*, 1:157; al-Samhūdī, *Wafāʾ al-wafā*, 3:48–49; al-ʿAlī, "Studies," 79.

50 Ibn Ra's Ghanama, *Manāqil al-durar*, 372–73. This passage is also discussed in Kister, "Battle of the Ḥarra," 45–46. Ibn Ra's Ghanama, who seems quite scrupulous in his citation of his sources, appears to suggest that this comes from al-Ṭabarī's (d. 310/923) famous *Ta'rīkh*, but as both Kister and the text's editors note, there is no parallel for this passage in the extant version of that work.

51 al-Samhūdī, *Wafāʾ al-wafā*, 4:47.

in the Umayyad and early Abbasid periods across various different sources, with those involved seemingly ranging from small property owners working their own land to larger estates, with either present or absentee owners, run by managers, the owners' clients (*mawālī*) and slaves.[52] In terms of the operation of the water systems themselves, we should no doubt assume that most Medinans would have had to take themselves to the wells and springs and to dig canals to draw water to their properties, be it for cultivation or washing or drinking. The anecdote just given from Ibn Raʾs Ghanama's history suggests this, and also that absentee owners of large estates had local clients to do this work for them.

The question of slavery in early Islamic Medina is a slightly tricky one to address: there certainly were slaves there, but we do not often hear a great deal about what tasks they performed.[53] In terms of water management specifically, we do hear of slaves at work in this area from some sources. To give a few examples: the hostages taken from Transoxiana by Saʿīd ibn ʿUthmān (d. after 56/675–76) during the conquest of that region and then taken to Medina were apparently put to work there "with the water infrastructure and the animals involved in that (*al-saqy wa-l-sawānī*) among other tasks".[54] Another Medinan, ʿAbd Allāh ibn ʿAnbasa ibn Saʿīd, also apparently had a Sogdian slave (*ghulām ṣughdī*) whose job was to bring water to his household.[55] Two 4th/10th-century travellers and geographers also specifically noted that Medina "has many date palms, the water for which, and for their cultivated estates, is drawn from wells by slaves (*ʿabīd*)".[56] One further possible example comes from al-Ṭabarī's (d. 310/923) account of a revolt in 200/815–16, which resulted in its leader, Muḥammad ibn Jaʿfar al-Ṣādiq, setting up his "caliphate" in Mecca (according to al-Ṭabarī's version; other sources have Medina). This account notes that Muḥammad ibn Jaʿfar found support among several groups including "the black Africans who worked the water infrastructure" (*sūdān ahl al-miyāh*).[57]

Finally on labour, as some examples already discussed demonstrate, it is clear that animals were vital to the functioning of Medina's water infrastructure, as they

[52] See the discussion in Munt, "Caliphal Estates," 442–45; and also al-ʿAlī, "Milkiyyāt al-arāḍī," 964, 968–69.

[53] Alongside the discussion in the two references cited in n. 52, see also, for example, Kister, "Battle of the Ḥarra," 44–47; and Trabelsi, "L'esclavage domanial," 314–18.

[54] al-Balādhurī, *Futūḥ al-buldān*, 412.

[55] al-Samhūdī, *Wafāʾ al-wafā*, 4:26. The precise date of ʿAbd Allāh ibn ʿAnbasa's death is unknown, but from an anecdote discussed below it is clear he was alive at the time of Hishām ibn ʿAbd al-Malik. His father apparently studied with the Prophet's companion Abū Hurayra; see Ibn Saʿd, *Ṭabaqāt*, 7:235.

[56] al-Iṣṭakhrī, *Masālik*, 18; Ibn Ḥawqal, *Ṣūrat al-arḍ*, 1:30.

[57] al-Ṭabarī, *Taʾrīkh*, 3:992. Hugh Kennedy interpreted this as two groups, "water carriers" and "black slaves"; see his *Early Abbasid Caliphate*, 211. Abū al-Faraj al-Iṣfahānī (d. 356/967) has Muḥammad ibn Jaʿfar receive the pledge of allegiance initially in Medina rather than Mecca, although he still places many of his subsequent activities in Mecca; see his *Maqātil al-ṭālibiyyīn*, 438–41.

were elsewhere in the premodern Middle East.[58] The passage from al-Yaʿqūbī cited earlier mentioned the key role played by animals, specifically camels (*nawāḍiḥ*, sing. *nāḍiḥa*), in moving water along Medina's irrigation networks; and we have seen in the previous paragraph that the Transoxanian prisoners of war taken to Saʿīd ibn ʿUthmān's estates oversaw the labour there of the *sawānī* (sing. *sāniya*), sometimes used as a synonym for *nawāḍiḥ* and sometimes more specifically to denote camels used to draw water from deep wells.[59] When the caliph Hishām ibn ʿAbd al-Malik granted ʿAnbasa ibn Saʿīd and his aforementioned son ʿAbd Allāh an estate, the supposed beneficiaries complained about the cost of getting the land into a state for cultivation until the caliph also granted them 20,000 *dīnār*s and forty Bactrian camels to assist with bringing water to the fields and cistern (*ṣahrīj*).[60] Camel labour used to draw water from wells for it to be used for cultivation can be seen across much of late pre- and early Islamic Arabia, as for example by a couple of Najdī poets from Tamīm who compared their weeping to a leather water sack (*gharb*) drawn from a well by an ash-coloured camel (*adham, dahmāʾ*).[61] Sometimes the roles played by camels in such quotidian agricultural tasks in Arabia are overlooked, partly due to lack of evidence, in favour of their use for milk and in interregional trade;[62] their role in the water infrastructure of early Islamic Medina, however, comes through clearly despite drawing relatively little attention in extant sources.[63]

The Politics of Water Management

We saw right at the very start of this chapter how important water management could be to ideas about effective and legitimate rule in the early Islamic period. This was no doubt in part due to the need for careful supervision of its collection, preservation and distribution, all efforts that could be costly and require considerable investment of funds and labour resources. Early Islamic Arabia was, in many respects, little different to late pre-Islamic Arabia in this regard. The Abbasid family claimed that their ancestors had overseen the office of providing water for the

58 See, for example, Mikhail, *Animal in Ottoman Egypt*, 30–31, on the use of camels for digging and dredging work on water infrastructure in Ottoman Egypt.

59 See Lane, *Arabic-English Lexicon*, 4:1450.

60 al-Samhūdī, *Wafāʾ al-wafā*, 4:26.

61 Miller, "Seasonal Poetics," 20–21, 27.

62 A point made, for example, in Magee, *Archaeology of Prehistoric Arabia*, 197–213.

63 Richard Bulliet noted some time ago now that, "It is particularly unfortunate that more data is not available relating to camels being used for turning mills and irrigation devices;" see his *Camel and the Wheel*, 213.

public and visitors (an office known as *siqāya*) in pre-Islamic Mecca.[64] After they became the caliphal family from the mid-2nd/8th century, they claimed that their provision of water along pilgrim routes to Arabia, especially the famous Darb Zubayda from Baghdad/Kufa to Mecca, was a continuation of this ancestral office.[65] We have already seen that the Umayyad caliph Muʿāwiya was a keen builder of dams around Medina, as well as elsewhere in the Hijaz.[66] Through such building work he was following in the footsteps of earlier Arabian rulers, notably the South Arabian Abraha (r. *ca.* 525–65 CE), famous for his construction of the dam at Marib, east of Sanaa, on which he had inscribed a lengthy inscription announcing his deeds dated to March 548 CE.[67]

Other anecdotes also reveal that Umayyad caliphs' activities in providing water were compared, often favourably, to their pre-Islamic predecessors'. One fraught question in the Umayyad period concerned the relative status of prophets or messengers and caliphs, and one Umayyad governor apparently sought to address this question using the relative merits of one caliph's and one prophet's efforts at water provision.[68] In the anecdote in question, al-Ṭabarī cites via the Medinan al-Wāqidī (d. 207/822) an episode in which Khālid al-Qasrī, while governor of Mecca between 89/707–8 or 91/709–10 and 96/715, gave the following sermon:

> O people, who is more venerable? Is it a man's deputy (*khalīfa*, 'caliph') over his people or his messenger (*rasūl*) to them? By God, were you not to know the merits of the deputy, [then know] that Abraham, the friend of the Merciful, prayed for water and He gave him brackish, salty water, while the caliph asked Him for water and He gave him sweet water.

Al-Wāqidī's source continues to explain that Khālid was referring here to a well excavated on the orders of the Umayyad caliph al-Walīd ibn ʿAbd al-Malik, from which water was transported to a leather container next to the famous well of Zamzam in Mecca (often closely associated with Abraham's son, Ishmael), so that its superiority over Zamzam's water would become well known.[69]

64 For a discussion of the "offices" in pre-Islamic Mecca, see Hawting, "Sacred Offices".

65 Zadeh, "Early Hajj," 44–45, 49–51, 56–57, 59–60, 62–64. On the Darb Zubayda, see especially al-Rāshid, *Darb Zubaydah*; Webb, "Bedouin, Bandits, and Caliphal Disappearance"; and see the contribution of Brown in this volume.

66 For his famous dam near al-Ṭāʾif, see Miles, "Early Arabic Inscriptions".

67 See the English translation of the inscription in Robin, 'Ḥimyar, Aksūm, and Arabia Deserta," 164–67. For some observation of the connection between Muʿāwiya's and Abraha's dam constructions, see al-Rāshid, *Dirāsāt*, 56.

68 On this question, see most famously Crone and Hinds, *God's Caliph*; and, more recently, Marsham, "'God's Caliph' Revisited".

69 al-Ṭabarī, *Taʾrīkh*, 2:1199–1200.

In Medina, we hear of caliphs and their representatives making similar efforts to provide water to the public. We have already seen ʿUthmān ibn ʿAffān's supposed patronage of Biʾr Rūma to provide water for the Muslim community in Medina, al-Walīd's building of a public fountain (*fawwāra*) for visitors to the Prophet's Mosque, and Sulaymān ibn ʿAbd al-Malik's funding of another public fountain (*siqāya*) for pilgrims heading back to Syria from Medina.[70] The Umayyads also paid for the upkeep and maintenance of other water resources in Medina. Al-Zubayr ibn Bakkār, for example, quoted the Medinan scholar ʿAbd Allāh ibn Dhakwān (d. 130/748) that, "The Umayyads used to assign a stipend from the *dīwān* for whoever looked after the cistern (*ḥawḍ*) of Marwān ibn al-Ḥakam in the ʿAqīq, for its upkeep, and for whenever the rope and bucket of al-Mughīra's well needed repair." He also noted that Hishām ibn ʿAbd al-Malik set aside some money in the treasury for the deep well (*jarr*) he was told had been dug by his grandfather Hishām ibn Ismāʿīl.[71]

Some Umayyad caliphs and their family members were known to have excavated other wells and springs around Medina. This was perhaps not always with the intention to improve their public image, but rather to seize new land and properties for themselves. Muʿāwiya was particularly famous for this, although he was not alone among the Umayyads.[72] Control over water resources, natural ones as well as built ones, could lead to physical conflicts between Umayyad authorities and locals. We saw this, for example, in Ibn Raʾs Ghanama's account of Medinans competing with Muʿāwiya's *mawālī* for access to flood waters that came to Medina one day.[73] Outside of Medina, we hear of Medinans collaborating with Umayyad officials to take control over water resources from (semi-) nomadic groups. At Ḍariyya, for example, about seven days journey from Medina on the road between Mecca and Basra, more than two hundred Medinans apparently joined forces with armed guards (*ḥuwwāṭ*) employed by Hishām ibn ʿAbd al-Malik's uncle and governor of Medina, Ibrāhīm ibn Hishām al-Makhzūmī (in office 106/724–25 to 114/732–33), to seize a watering spot from the local tribe, Ghanī.[74]

A particularly good example of conflicts over the control of land and water infrastructure in Medina comes from the range of reports, many of them conveniently collected by al-Samhūdī, about one of Medina's most important water sources, the Wādī al-ʿAqīq. This was an important area of settlement and exploitation of water

[70] See above, nn. 29, 45–46.

[71] al-Samhūdī, *Wafāʾ al-wafā*, 4:24–25.

[72] Umayyad efforts at controlling land and estates in Medina is the focus of my "Caliphal Estates"; and see further discussions in al-ʿAlī, "Milkiyyāt al-arāḍī"; Kister, "Battle of the Ḥarra"; and al-Rāshid, *Dirāsāt*, 57–59.

[73] See above, n. 50.

[74] al-Samhūdī, *Wafāʾ al-wafā*, 4:87–88 (and 86 for the location of Ḍariyya). For the dates of Ibrāhīm ibn Hishām's term as governor, see al-Ṭabarī, *Taʾrīkh*, 2:1471, 1487, 1561–62.

resources for cultivation for Medinans throughout the early Islamic centuries and a number of Medina's important families, Umayyads and others, had properties along the ʿAqīq;[75] sometimes, their interests came into conflict.[76] The Umayyads apparently went to considerable efforts at times to control access to at least some parts of the ʿAqīq. The Medinan local historian Ibn Zabāla (wr. 199/814), for example, noted that they prevented people undertaking construction works in al-ʿArṣa and that no estates could be granted in that area without the permission of the caliph until the time of al-Walīd ibn ʿAbd al-Malik.[77] They came into particularly noteworthy conflict in the ʿAqīq, however, with one of their main rival families, the Zubayrids.

The Zubayrids were the descendants of the Prophet's Meccan companion al-Zubayr ibn al-ʿAwwām (d. 36/656), who had settled with the Prophet after the *hijra* in Medina, where his family went on to establish themselves as an important political, social and economic presence.[78] A number of well-known scholars came from this family, including from the Umayyad period ʿUrwa ibn al-Zubayr and his son, Hishām (d. 146/763).[79] ʿUrwa's brother, ʿAbd Allāh ibn al-Zubayr (d. 73/692), had also famously challenged the Umayyads for control of the caliphate—and was indeed accepted by many Muslims across the empire as the legitimate caliph—during the second *fitna*.[80] A number of Zubayrids settled in the ʿAqīq and were known for their properties there. Among them were ʿUrwa ibn al-Zubayr and his descendants, to whose properties we will return shortly. Other Zubayrids were also known for managing properties and estates along the wadi, including ʿAbd Allāh ibn al-Zubayr's grandson, Muṣʿab ibn Thābit (d. 157/773–74), whose estate was known for a villa (*qaṣr*) and pool for collecting rainwater (*qarāra*);[81] his son ʿAbd Allāh (d. 184/800)

[75] A quick, albeit partial overview of some of those with properties there can be gleaned from Abū ʿAlī al-Hajarī's (probably died first quarter of 4th/10th century) list of those who settled there; see al-Jāsir, *al-Taʿlīqāt wa-l-nawādir*, 3:1535; also in al-Samhūdī, *Wafāʾ al-wafā*, 4:29. A later local historian of Medina, Ibn al-Najjār (d. 643/1245), noted that the ʿAqīq was unpopulated and in ruins by his day, but part of it at least had apparently still been a flourishing area with a good water supply when al-Muqaddasī (d. after 381/991–92) visited the region; see Ibn al-Najjār, *al-Durra al-thamīna*, 98; and al-Muqaddasī, *Aḥsan al-taqāsīm*, 82.

[76] The situation would seem to be similar to that in Khurasan observed by Azad, "Ecology, economy," 338: "Water was the major source of wealth, prosperity, and power in the region and so also of rivalry."

[77] al-Samhūdī, *Wafāʾ al-wafā*, 4:30. Apparently the granting of estates and undertaking of construction work in the area was much more liberal under the Abbasids; see *ibid.*, 4:35.

[78] For a still useful survey of the family in the early Islamic centuries, see Wüstenfeld, "Familie el-Zubeir".

[79] On ʿUrwa's scholarly activities in particular, see Görke and Schoeler, *Ältesten Berichte*.

[80] See, for example, Robinson, *ʿAbd al-Malik*, 31–48.

[81] al-Zubayr ibn Bakkār, *Jamharat nasab Quraysh*, 1:169–70. For this meaning of *qarāra*, see Dozy, *Supplément aux dictionnaires arabes*, 2:320.

was also known for a villa in the ʿAqīq, which seems to have been a property he purchased rather than inherited and so it was not the same as his father's.[82]

Some claimed that the entire ʿAqīq had been granted as an estate to al-Zubayr ibn al-ʿAwwām by the second caliph ʿUmar ibn al-Khaṭṭāb (r. 13–23/634–44).[83] Others provided slightly more complicated stories about prophetic and early caliphal land grants that did not confer such exclusive rights on the Zubayrids.[84] These varying narratives, some of which were transmitted by members of the Zubayrid family, reveal how important the ʿAqīq and conflicts over access to its resources were to those involved in them. Its importance is also demonstrated by some of the traditions transmitted, often by Zubayrids, in which the prophet recognised the wadi's merits and distinctions (*faḍāʾil*), particularly frequently labelling it "a blessed wadi".[85]

The ʿAqīq was so important to the Zubayrids that one member of the family dedicated a monograph to it, the well-known scholar al-Zubayr ibn Bakkār. A descendant of ʿAbd Allāh ibn al-Zubayr, al-Zubayr ibn Bakkār was born in Medina in 172/788–89.[86] His immediate family was politically quite important: his father, for example, was one of Hārūn al-Rashīd's (r. 170–93/786–809) governors of Medina and was close enough to the caliph to have arranged an introduction at the court at al-Raqqa for a young, ambitious Medinan historian, al-Wāqidī.[87] Al-Zubayr spent some time in Iraq—we know he was in Baghdad and Samarra at times[88]—but was also appointed *qāḍī* of Mecca sometime after 241/855–56, the town in which he died in 256/870 after falling off a roof.[89]

Al-Zubayr ibn Bakkār was a fairly prolific scholar and author and at least two of his works survive, a collection of anecdotes known as *al-Muwaffaqiyyāt* and his major genealogical history of Quraysh, *Jamharat nasab Quraysh*. He is also credited with a few works on Medina's history, including an *Akhbār al-Madīna*, which

[82] al-Samhūdī, *Wafāʾ al-wafā*, 4:27–28.

[83] al-Balādhurī, *Futūḥ al-buldān*, 12, 21.

[84] Ibn Shabba, *Taʾrīkh al-Madīna*, 1:96–97; al-Samhūdī, *Wafāʾ al-wafā*, 4:14–16.

[85] See, for example, within Ibn Shabba, *Taʾrīkh al-Madīna*, 1:94–97; al-Samhūdī, *Wafāʾ al-wafā*, 4:7–13.

[86] Important overviews of al-Zubayr ibn Bakkār's life can be found, for example, in Ibn al-Nadīm, *Fihrist*, 1/ii:340–43; al-Khaṭīb al-Baghdādī, *Taʾrīkh Baghdād*, 9:486–92; Ibn Khallikān, *Wafayāt al-aʿyān*, 2:311–12; al-Mizzī, *Tahdhīb al-Kamāl*, 9:293–99; al-Dhahabī, *Siyar aʿlām al-nubalāʾ*, 12:311–15.

[87] On Bakkār as a governor of Medina, see al-Ṭabarī, *Taʾrīkh*, 3:739; on the introduction at al-Raqqa for al-Wāqidī, see Ibn Saʿd, *Ṭabaqāt*, 7:605–6.

[88] According to Ibn al-Athīr (d. 630/1233), al-Zubayr had to leave Medina for Baghdad after a quarrel with the Alid family in the former; see his *al-Kāmil fī al-taʾrīkh*, 6:526. Ibn al-Nadīm (d. before 388/998) mentions that his last trip to Baghdad was in 253/867; see his *Fihrist*, 1/ii:341. We know he was in Samarra in 246/860–61 from the introductory *riwāya-isnād* to the extant selections from his recension of Ibn Zabāla's *Azwāj al-nabī*; see the editor's introduction to Ibn Zabāla, *Muntakhab*, 17.

[89] For al-Zubayr's appointment as *qāḍī* of Mecca, see Wakīʿ, *Akhbār al-quḍāt*, 1:269.

given it is only very rarely cited by pre-modern authorities may simply refer to his recension of Ibn Zabāla's work generally known by the same title.[90] He is more securely credited with a *Kitāb al-ʿAqīq*, which was an important source for several later historians of Medina, especially al-Fīrūzābādī (d. 817/1415) and al-Samhūdī.[91] That al-Zubayr ibn Bakkār composed this work dedicated to the ʿAqīq suggests its importance to his family and their efforts to write its history and describe its geography in a way that emphasised the Zubayrids' prominence there. To a large extent, these efforts worked; since a great deal of what later historians like al-Fīrūzābādī and al-Samhūdī have to say about the ʿAqīq is drawn from al-Zubayr ibn Bakkār, their often quite extensive descriptions of that area emphasise Zubayrid-owned properties and water infrastructure.

We largely, therefore, get the Zubayrid perspective on perhaps the most important indication of the political importance of water management in Medina and its immediate hinterland, which comes from the story of the estate owned by ʿUrwa ibn al-Zubayr and his descendants.[92] ʿUrwa first bought the land, apparently running roughly between the third and fourth milestones from Medina, from one ʿAbd Allāh ibn ʿAyyāsh ibn ʿAlqama. ʿUrwa' son, Hishām, reported that:

> ʿUrwa bought the place [that would become] his villa, his land and his wells from ʿAbd Allāh ibn ʿAyyāsh. He built there, excavated irrigation works, set up boundaries and built dams. He was told that it was not a good location for sowing grain, but he replied that, "God will provide for it from al-Naqīʿ."[93] A flood came and entered the cultivated fields, which

[90] It is mostly modern scholars who credit al-Zubayr with an *Akhbār al-Madīna* based on quite rare references from sources no earlier than the 9th/15th century; see, for example, al-ʿAlī, "al-Muʾallafāt al-ʿarabiyya," 134; al-Jāsir, "Muʾallafāt," 386; ʿUsaylān, *al-Madīna al-munawwara*, 31. For his recension of Ibn Zabāla's work, see Munt, "Writing the History," 14–15, 18, 24–25.

[91] Various premodern sources mention this work (either by the title *Kitāb al-ʿAqīq* or something very similar to that); see, for example, Ibn al-Nadīm, *Fihrist*, 1/ii:342; Yāqūt, *Irshād al-arīb*, 4:219. Al-Fīrūzābādī cites it explicitly as a source in, for example, *al-Maghānim al-muṭāba*, 3:1083, 1121 [= ed. al-Jāsir, 377, 409]; as does al-Samhūdī in his *Wafāʾ al-wafā*, 1:147; 3:375; 4:44. Yāqūt (d. 626/1229) also used the work as an important source in his geographical encyclopaedia; for discussion, see Heer, *Die historische und geographische Quellen*, 29–30. For some modern discussions, see al-ʿAlī, "al-Muʾallafāt al-ʿarabiyya," 134; al-Jāsir, "Muʾallafāt," 386; ʿUsaylān, *al-Madīna al-munawwara*, 115.

[92] There is some discussion of this estate and these conflicts in Munt, "Caliphal estates," 448–50. The most important source for this estate is al-Samhūdī, *Wafāʾ al-wafā*, 4:16–22, from which the following is drawn unless otherwise stated; but see also al-Fīrūzābādī, *al-Maghānim al-muṭāba*, esp. 3:1038–43 [= ed. al-Jāsir, 342–45]. Between the citations given in these two sources, it is clear that a great deal of this comes from al-Zubayr ibn Bakkār, presumably from his *Kitāb al-ʿAqīq*.

[93] al-Naqīʿ was an area, about twenty *farsakh*s or 120 kilometres from Medina, which the Prophet was reported to have declared protected pastureland (*ḥimā*), and which was known to be the starting point of the Wādī al-ʿAqīq; see al-Samhūdī, *Wafāʾ al-wafā*, 4:72.

> he watered with a canal that he had dredged. The buildings there were *janābidh*—i.e. the plural of *junbadh*, with a *ḍamma* on the *jīm*—which are elevated and circular [structures] like domes.[94]

ʿUrwa, assisted by his son Yaḥyā and grandson ʿUmar ibn ʿAbd Allāh, built a number of water resources for the estate, including several wells, one of which was for public use (*biʾr al-siqāya*).[95] After ʿUrwa died, the estate was inherited by a few of his sons in turn, including Yaḥyā, ʿAbd Allāh, Hishām and ʿUbayd Allāh.

ʿUrwa's estate was the source of conflict with at least two Umayyad governors of Medina as well as with several of his neighbours. The first major conflict came when one of those neighbours, an Umayyad, ʿAbd Allāh ibn ʿAmr ibn ʿUthmān ibn ʿAffān (d. 96/714–15), started dredging a canal for his estate, called al-Marājil. When it looked like that canal would infringe upon ʿUrwa's property, he opposed it, so ʿAbd Allāh ibn ʿAmr appealed to his Umayyad relative, the governor of Medina ʿUmar ibn ʿAbd al-ʿAzīz.[96] The latter then started tearing down the buildings on the estate and destroyed its dams and wells. What is particularly interesting for our purposes is that ʿAbd Allāh ibn ʿAmr claimed that ʿUrwa, through the building works on his estate, "had infringed upon the rights of the state" (*ḥamala ʿalā ḥaqq al-sulṭān*). More specifically, he told ʿUmar ibn ʿAbd al-ʿAzīz that:

> He has undertaken building and dredging work where he has no right. He has seven *janābidh*. Riders used to alight by Marwān's Well, but then when ʿUrwa dug his well, which has sweeter water, they chose ease and sweetness and stopped alighting at Marwān's Well.

So ʿAbd Allāh ibn ʿAmr's initial dispute with ʿUrwa was over the dredging of a canal and he tried to get the assistance of the Umayyad authorities (who were also his relatives) in his conflict with his Zubayrid neighbour by claiming that the water infrastructure on ʿUrwa's estate was a threat to the Umayyad family's own claim to provide water through their well (Marwān's Well) for visitors to Medina. In a poem attributed to ʿUrwa about his estate, he also drew attention to the perceived threat his well posed to the Umayyads and their claim to the duty of providing water for pilgrims to the Hijaz, in this case those on their way from Medina to Mecca:

[94] al-Samhūdī, *Wafāʾ al-wafā*, 4:16–17.

[95] al-Samhūdī, *Wafāʾ al-wafā*, 4:18.

[96] The Zubayrids' involvement in disputes surrounding canal-dredging in Medina apparently went back to al-Zubayr ibn al-ʿAwwām himself, who was ordered by the Prophet to allow water to pass through his land (in an unspecified part of Medina) via a canal and then onto his neighbour's; see al-Balādhurī, *Futūḥ al-buldān*, 12.

We built it, and we built it well
with God's blessing in the best of the 'Aqīq.
You can see them looking at it askance
as it gleams before them on the road.
It grieves the haters and is a source of ire
for my enemies, while my friend is cheered by it.
It is seen by all who might profit, who travel overnight,
who are performing the 'umra to the ancient house.

Eventually, the caliph al-Walīd ibn 'Abd al-Malik intervened on 'Urwa's behalf in this dispute and he rebuilt his estate and its water infrastructure.[97]

It would seem that the value of the well on 'Urwa's estate in providing water as a public service carried on, since on inheriting the estate his son 'Ubayd Allāh recited:

Were the shaykh to know my journey first thing in the morning
towards the public fountain (*siqāya*) that he had excavated,
With youths blazing like *dīnār*s,
may God preserve them from hypocrisy and discontent,
Among Abū Bakr and Zayd and 'Umar
and then the apostle (*al-ḥawārī*),[98] whose grandfather was most esteemed,
Who are there in the evenings and mornings,
offering water to those who come and not a soul is harmed,
He would increase his thanks though he has already given thanks.[99]

The water infrastructure on 'Urwa's estate also seems to have been behind the second major conflict that emerged, while it was under the ownership of his sons 'Abd Allāh and Yaḥyā, during the caliphate of Hishām ibn 'Abd al-Malik. The governor of Medina, Ibrāhīm ibn Hishām, was trying to seize some land belonging to 'Urwa's family elsewhere and because the Zubayrids opposed him, he took his revenge by destroying again the buildings on 'Urwa's estate and violating the now famous well by having a camel coated in tar thrown into it. Again, however, the caliph

97 The narrative of this dispute here is drawn from two separate narratives provided in al-Samhūdī, *Wafā' al-wafā*, 4:17–18.

98 The term 'apostle' (*ḥawārī*) was often used to refer to the Zubayrids' eponymous ancestor, al-Zubayr ibn 'Awwām; see, for example, Ibn Sa'd, *Ṭabaqāt*, 3:98–99.

99 Ibn Sa'd, *Ṭabaqāt*, 4:19–20. For a slightly different text of this poem, see al-Dhahabī, *Siyar a'lām al-nubalā'*, 4:428–29.

intervened on the Zubayrids' side and so the estate and its water infrastructure were returned to use.[100]

The fame of ʿUrwa's well on this estate in the ʿAqīq continued into the Abbasid period. His descendant (ʿAbd Allāh ibn Urwa's grandson) ʿĀmir ibn Ṣāliḥ, a poet who died during the caliphate of Hārūn al-Rashīd, recited thus about the estate:

How perfect is the villa, with its loftiness and its well, at the bottom of the well-watered ʿAqīq.
Water from rain clouds from which ʿUrwa desires nothing
    except God's protection for the lands granted as fiefs,
In a delightful place in the ʿAqīq,
    with cool shade and beautiful early mornings.

Al-Zubayr ibn Bakkār, our source for much of the preceding information about this estate, said that he personally witnessed pilgrims who had to be sparing with their water as they travelled through the Hijaz top up plentifully from ʿUrwa's well. They would also take water from that well to carry to their families back wherever they came from. His father Bakkār had sent water from the well as a gift to the caliph Hārūn in al-Raqqa in the Jazīra.[101]

ʿUrwa's grandson, ʿUmar ibn ʿAbd Allāh, also seems to have been in dispute with another Umayyad neighbour, ʿĀṣim ibn ʿAmr, the great-grandson of the caliph ʿUthmān ibn ʿAffān. This dispute, so far as we can tell, did not result in physical conflict and destruction, but did lead to a poetic war of words, with ʿUmar ibn ʿAbd Allāh and an associate, the famous late Umayyad-era rebel ʿAbd Allāh ibn Muʿāwiya (d. 130/747–48), disparaging ʿĀṣim's estate and its water resources:[102]

O ʿĀṣim's villa, if you could speak out perspicaciously
    we would appeal for protection from the commander of the faithful.
You mention the tribulations you have faced,
    and indeed you have met with sadness at times.
You were built along the road that all the people use
    and yet all those of nobility and upright faith vilify you.
You were not placed in some lowland where you could be hidden
    nor on some flat, smooth land.

100 For this dispute, see al-Samhūdī, *Wafāʾ al-wafā*, 4:20; and the broader discussion of ʿAbd Allāh ibn ʿUrwa's problems with Ibrāhīm ibn Hishām in al-Zubayr ibn Bakkār, *Jamharat nasab Quraysh*, 1:300–2.

101 al-Samhūdī, *Wafāʾ al-wafā*, 4:21.

102 On this rebellion, see Bernheimer, "Revolt of ʿAbdallāh b. Muʿāwiya".

> Drought is seen in you like nothing else,
> so you have been called the imposter among springs.[103]

ʿĀṣim responded in kind, but that is less important than the fact that the Zubayrids continued to work hard to cultivate estates in the ʿAqīq and promote the virtues of their water infrastructure there, including as a public service for Medinans and for visitors to the town, at the expense of those belonging to their rivals.

Conclusion

Pre-modern Muslims were no strangers to the idea that environmental and climatic changes could have significant ramifications for social and political life, nor that water management in particular was one key area of concern. Among several reasons given for the legendary migration of South Arabia's Azd tribe throughout the Arabian Peninsula, the disaster caused by the breaching of the famous Marib dam was a particularly popular one. As al-Yaʿqūbī told the story:

> The reason the Yemenis dispersed throughout the lands and left their homelands was the Flood of al-ʿArim.[104] According to what the transmitters have passed down, this all went back to ʿAmr ibn ʿĀmir ibn Ḥāritha ibn Imruʾ al-Qays ibn Thaʿlaba ibn Māzin ibn al-Azd. He was the leader of his people and a soothsayer. He foresaw that the lands of Yemen would be inundated. He made a show of being angry towards one of his children and then sold his spring pastureland and left with his family. They travelled to the lands of ʿAkk and then on to Najrān where Madhḥij fought them. So they moved on from Najrān and passed by Mecca, settled at that time by Jurhum, who fought them until they expelled them from the town. So they travelled to al-Juḥfa and then on to Yathrib…[105]

That control of water in the early Islamic Hijaz, then, was not only seen as a practical resource issue, but also a matter of politics and rulership, should not trouble us. In this chapter, we have seen some of the ways that Medinans in particular managed the collection and distribution of water over the first couple of centuries. While certainly scarce at times and unpredictable, Medina's water supply in this

103 al-Samhūdī, *Wafāʾ al-wafā*, 4:23.

104 The fame of this event, identified as a flood that came with the breaching of the Marib dam, was due to its appearance in the Qurʾan, in Q34.16. *ʿrm* in Sabaic can mean a dam; see Beeston et al., *Sabaic Dictionary*, 19.

105 al-Yaʿqūbī, *Taʾrīkh*, 1:231–32.

period was plentiful enough to sustain cultivation, but it required constant and, no doubt, relatively expensive labour and management to do so.

As a system that required considerable expense and complex organisation, and so offered opportunities for control and the exercise of authority, it is little surprise that the early Islamic empire's governing powers, particularly relatives and local representatives of the Umayyad caliphs (two groups that frequently overlapped in Medina), became heavily involved. Less of a surprise still is the fact that local landowners opposed their involvement at times, especially of course when they were seen to be infringing on locals' rights. In this chapter, we have focused primarily on one particular region of Medina's hinterland, the Wādī al-ʿAqīq, and on the rivalry in that region between the Umayyads and the Zubayrids. Such attempts to manage relatively substantial estates with wells, springs, dams and irrigation canals, as well as the resulting conflicts between neighbours and rival families, can also be seen in other regions in and around Medina. In the Wādī al-ʿAqīq, however, we can see particularly clearly how one family, that in the late 1st/7th century had famously claimed caliphal authority for themselves, could also use, or be perceived to be using, water management to challenge the ruling caliphs' authority and legitimacy. The Zubayrids had to defend their interests in the Wādī al-ʿAqīq, which they did not only through petitioning, legal disputes and occasional acts of resistance, but also through attempting to control the historical narrative of land ownership and water management in the region. Al-Zubayr ibn Bakkār's *Kitāb al-ʿAqīq* has not survived, but through other sources' citations from it, the work stands as a monument to the political importance of water management in early Islamic Medina.

Bibliography

Primary Sources

Abū al-Faraj al-Iṣfahānī. *Maqātil al-ṭālibiyyīn*. Edited by Aḥmad Ṣaqr. 4th ed. Muʾassasat al-Aʿlamī li-l-Maṭbūʿāt, 1427/2006.

al-Bakrī. *Muʿjam mā istaʿjam*. Edited by Muṣṭafā al-Saqqā. 4 vols. Maṭbaʿat Lajnat al-Taʾlīf wa-l-Tarjama wa-l-Nashr, 1364–71/1945–51.

al-Balādhurī. *Kitāb futūḥ al-buldān*. Edited by M. J. de Goeje. Brill, 1866.

al-Dhahabī. *Siyar aʿlām al-nubalāʾ*. Edited by Shuʿayb al-Arnāʾūṭ et al. 25 vols. Muʾassasat al-Risāla, 1401–9/1981–88.

al-Fīrūzābādī. *al-Maghānim al-muṭāba fī maʿālim Ṭāba*. Edited by Ḥabīb Maḥmūd Aḥmad. 4 vols. Markaz Buḥūth wa-Dirāsāt al-Madīna al-Munawwara, 1423/2002. Also partially edited by Ḥamad al-Jāsir. Dār al-Yamāma, 1389/1969.

al-Fīrūzābādī. *al-Qāmūs al-muḥīṭ*. 6th ed. Muʾassasat al-Risāla, 1426/2005.

Ibn al-Athīr. *al-Kāmil fī al-taʾrīkh*. Edited by C. J. Tornberg. Brill, 1851–76; reprinted with different pagination in 13 vols. Dār Ṣādir, 1965–67.

Ibn Ḥawqal. *Kitāb ṣūrat al-arḍ*. Edited by J. H. Kraemers. 2nd ed. 2 vols. Brill, 1938–39.
Ibn Khallikān. *Wafayāt al-aʿyān wa-anbāʾ abnāʾ al-zamān*. Edited by Iḥsān ʿAbbās. 8 vols. Dār Ṣādir, 1397–98/1977–78.
Ibn al-Nadīm. *Kitāb al-fihrist*. Edited by Ayman Fuʾād Sayyid. 2 parts in 4 vols. Muʾassasat al-Furqān li-l-Turāth al-Islāmī, 1430/2009.
Ibn al-Najjār. *al-Durra al-thamīna fī taʾrīkh al-Madīna*. Edited by Muḥammad Zaynahum Muḥammad ʿAzab. Maktabat al-Thaqāfa al-Dīniyya, 1416/1995.
Ibn Raʾs Ghanama. *Manāqil al-durar wa-manābit al-zahar*. Edited by Rāʾid Amīr ʿAbd Allāh al-Rāshid and Khālid ʿAbd al-Jabbār Shayt al-Rāshid. Dīwān al-Waqf al-Sunnī, Markaz al-Buḥūth wa-l-Dirāsāt al-Islāmiyya, 1429/2008.
Ibn Saʿd. *Kitāb al-Ṭabaqāt al-kabīr*. Edited by ʿAlī Muḥammad ʿUmar. 11 vols. Maktabat al-Khānjī, 1421/2001.
Ibn Shabba. *Taʾrīkh al-Madīna al-munawwara*. Edited by ʿAlī Muḥammad Dandal and Yāsīn Saʿd al-Dīn Bayān. 2 vols. Dār al-Kutub al-ʿIlmiyya, 1417/1996.
Ibn Zabāla. *Muntakhab min Kitāb azwāj al-nabī: riwāyat al-Zubayr ibn Bakkār*. Edited by Akram Ḍiyāʾ al-ʿUmarī. Maṭbaʿat al-Jāmiʿa al-Islāmiyya, 1401/1981.
al-Iṣṭakhrī. *Kitāb al-masālik wa-l-mamālik*. Edited by M. J. de Goeje. Brill, 1870.
al-Khaṭīb al-Baghdādī. *Taʾrīkh Baghdād*. Edited by Bashshār ʿAwwād Maʿrūf. 17 vols. Dār al-Gharb al-Islāmī, 1422/2001.
Kitāb al-manāsik wa-amākin ṭuruq al-ḥajj wa-maʿālim al-jazīra. Edited by Ḥamad al-Jāsir. Dār al-Yamāma, 1389/1969.
Lughda al-Iṣfahānī. *Bilād al-ʿarab*. Edited by Ṣāliḥ Aḥmad al-ʿAlī and Ḥamad al-Jāsir. Dār al-Yamāma, 1388/1968.
al-Mizzī. *Tahdhīb al-Kamāl fī asmāʾ al-rijāl*. Edited by Bashshār ʿAwwād Maʿrūf. 35 vols. Muʾassasat al-Risāla, 1402–13/1982–92.
al-Muqaddasī. *Aḥsan al-taqāsīm fī maʿrifat al-aqālīm*. Edited by M. J. de Goeje. 2nd ed. Brill, 1906.
al-Samhūdī. *Wafāʾ al-wafā bi-akhbār dār al-muṣṭafā*. Edited by Qāsim al-Sāmarrāʾī. 5 vols. Muʾassasat al-Furqān li-l-Turāth al-Islāmī, 1422/2001.
al-Ṭabarī. *Taʾrīkh al-rusul wa-l-mulūk*. Edited by M. J. de Goeje *et al.* 3 parts in 13 vols. Brill, 1879–1901.
Wakīʿ. *Akhbār al-quḍāt*. Edited by ʿAbd al-ʿAzīz Muṣṭafā al-Marāghī. 3 vols. Maṭbaʿat al-Saʿāda, 1366–69/1947–50.
al-Yaʿqūbī. *Kitāb al-buldān*. Edited by M. J. de Goeje. Brill, 1892.
al-Yaʿqūbī. *Taʾrīkh*. Edited by M. T. Houtsma. 2 vols. Brill, 1883.
Yāqūt. *Irshād al-arīb ilā maʿrifat al-adīb*. Edited by D. S. Margoliouth. 7 vols. Luzac & Co., 1907–27.
al-Zubayr ibn Bakkār. *Jamharat nasab Quraysh wa-akhbārihā*. Edited by Maḥmūd Muḥammad Shākir. 2nd ed. 2 vols. Dār al-Yamāma, 1419/1999.

Secondary Sources

al-ʿAlī, Ṣāliḥ Aḥmad. "Milkiyyāt al-arāḍī fī al-Ḥijāz fī al-qarn al-awwal al-hijrī." *Majallat al-ʿArab* 3 (1389/1969): 961–1005.
al-ʿAlī, Ṣāliḥ Aḥmad. "al-Muʾallafāt al-ʿarabiyya ʿan al-Madīna wa-l-Ḥijāz." *Majallat al-majmaʿ al-ʿilmī al-ʿirāqī* 11 (1384/1964): 118–57.
al-ʿAlī, Ṣāliḥ Aḥmad. "Studies in the Topography of Medina (During the 1st Century A.H.)." *Islamic Culture* 35, no. 2 (1961): 65–92.

Azad, Arezou. "Ecology, Economy, and the Conquest of Khurasan." In *The Umayyad World*, edited by Andrew Marsham, 332–54. Routledge, 2021.

Beeston, A. F. L., M. A. Ghul, W. W. Müller, and J. Ryckmans. *Sabaic Dictionary (English-French-Arabic)*. Peeters and Librairie du Liban, 1982.

Bernheimer, Teresa. "The Revolt of ʿAbdallāh b. Muʿāwiya, AH 127–130: A Reconsideration through the Coinage." *Bulletin of the School of Oriental and African Studies* 69, no. 3 (2006): 381–93.

Blankinship, Khalid Yahya. *The End of the Jihad State: The Reign of Hisham Ibn ʿAbd al-Malik and the Collapse of the Umayyads*. State University of New York Press, 1994.

Bulliet, Richard W. *The Camel and the Wheel*. Harvard University Press, 1975.

Bulliet, Richard W. *Cotton, Climate, and Camels in Early Islamic Iran: A Moment in World History*. Colombia University Press, 2009.

Charloux, Guillaume, Paul Courbon, Olivier Testa, and Matthieu Thomas. "Mapping an Ancient Qanat System in a Northern Arabian Urbanized Oasis." *Water History* 10, no. 1 (2018): 31–51.

Cook, Michael. *A History of the Muslim World: From Its Origins to the Dawn of Modernity*. Princeton University Press, 2024.

Crone, Patrica, and Martin Hinds. *God's Caliph: Religious Authority in the First Centuries of Islam*. Cambridge University Press, 1986.

Davis, Diana K. "Power, Knowledge, and Environmental History in the Middle East and North Africa." *International Journal of Middle East Studies* 42, no. 4 (2010): 657–59.

Depreux, Bruno, Munirah AlMushawh, Stephen McPhillips, Guillaume Chung-To, Sylvain Colin, Elora Chambraud, Kévin Guadagnini, Saifi AlShilali, Mashael AlDughailbi, Fatimah AlRashidi, Khalid AlKhaldi, Charlotte Faiers, Rémy Crassard, and Guillaume Charloux. "New Insights into Early Islamic Hydro-Agricultural Strategies in Northwest Arabia: A Geoarchaeological Study of al-Bint Dam (Sadd al-Bint)." *Geoarchaeology* 40, no. 5 (2025): 1–22.

Dinies, Michèle, Birgit Plessen, Reinder Neef, and Harald Kürschner. "When the Desert Was Green: Grassland Expansion During the Early Holocene in Northwestern Arabia." *Quaternary International* 382 (2015): 293–302.

Dozy, Reinhart. *Supplément aux dictionnaires arabes*. 2nd ed. 2 vols. Brill, 1927.

Foote, Rebecca, Maria Guagnin, Ingrid Périssé, and Steve Karacic, eds. *Revealing Cultural Landscapes in North-West Arabia*. Archaeopress, 2023.

Görke, Andreas, and Gregor Schoeler. *Die ältesten Berichte über das Leben Muḥammads: das Korpus ʿUrwa ibn az-Zubair*. Darwin Press, 2008.

Haldon, John, and Dominik Fleitmann. "A Sixth-Century CE Drought in Arabia: New Palaeoclimate Data and Some Historical Implications." *Journal of Late Antique, Islamic and Byzantine Studies* 3, nos 1–2 (2024): 1–45.

Harper, Kyle. *The Fate of Rome: Climate, Disease, and the End of an Empire*. Princeton University Press, 2017.

al-Hawas, Fahd bin Saleh, Al-Syed Anis Hasham, Jahaz Birjis Abdullah al-Shamri, Aʿjab Mohammad Al-Utaibi, Maher Khalifah Al-Musa, Saad Abdulrahman Al-Rawaisan, and Abdullah Abdulmohsin Al-Khalil. "Preliminary Report on the Excavations of the Islamic Site of Faid, Hail Region, First Season, 1427 H/2006 AD." *al-Aṭlāl* 20 (1431/2010): 39–74.

Hawting, Gerald R. "The 'Sacred Offices' of Mecca from Jāhiliyya to Islam." *Jerusalem Studies in Arabic and Islam* 13 (1990): 62–84.

Heer, F. Justus. *Die historischen und geographischen Quellen in Jāqūt's geographischem Wörterbuch*. Karl J. Trübner, 1898.

Hinz, Walther. *Islamische Masse und Gewichte umgerechnet ins metrische System*. Brill, 1955.

Izdebski, Adam, Jordan Pickett, Neil Roberts, and Tomasz Waliszewski. "The Environmental, Archaeological and Historical Evidence for Regional Climate Changes and Their Societal Impacts in the Eastern Mediterranean in Late Antiquity." *Quaternary Science Reviews* 136 (2016): 189–208.

al-Jāsir, Ḥamad. "Mu'allafāt fī ta'rīkh al-Madīna." *Majallat al-'arab* 4 (1389–90/1969–70): 97–100; 262–66; 327–34; 385–88; 465–67.

al-Jāsir, Ḥamad. *al-Ta'līqāt wa-l-nawādir 'an Abī 'Alī Hārūn ibn Zakariyyā al-Hajarī: dirāsa wa-mukhtārāt.* 4 vols. No publisher, 1413/1992–93.

Kennedy, Hugh. *The Early Abbasid Caliphate: A Political History*. Croom Helm, 1981.

Kister, M. J. "The Battle of the Ḥarra: Some Socio-Economic Aspects." In *Studies in Memory of Gaston Wiet*, edited by Myriam Rosen-Ayalon, 33–49. Hebrew University of Jerusalem, 1977.

Korotaev, Andrey, Vladimir Klimenko, and Dmitry Proussakov. "Origins of Islam: Political-Anthropological and Environmental Context." *Acta orientalia academiae scientiarum hungaricae* 52, nos 3–4 (1999): 243–76.

Lammens, Henri. *Le berceau de l'Islam: l'Arabie occidentale à la veille de l'hégire*. Volume 1. Sumptibus Pontificii Instituti Biblici, 1914.

Lane, Edward W. *Arabic-English Lexicon*. 8 vols. Williams and Norgate, 1863–93.

Lecker, Michael. *Studies on the Life of Muhammad and the Dawn of Islam: Idol Worshippers, Christians and Jews in Pre- and Early Islam*. Routledge, 2024.

Magee, Peter. *The Archaeology of Prehistoric Arabia: Adaptation and Social Formation from the Neolithic to the Iron Age*. Cambridge University Press, 2014.

Makki, M. S. *Medina, Saudi Arabia: A Geographic Analysis of the City and Region*. Avebury, 1982.

Marsham, Andrew. "'God's Caliph' Revisited: Umayyad Political Thought in Its Late Antique Context." In *Power, Patronage, and Memory in Early Islam: Perspectives on Umayyad Elites*, edited by Alain George and Andrew Marsham, 3–37. Oxford University Press, 2018.

Marsham, Andrew. *The Umayyad Empire*. Edinburgh University Press, 2024.

Mikhail, Alan. *The Animal in Ottoman Egypt*. Oxford University Press, 2014.

Mikhail, Alan. *Under Osman's Tree: The Ottoman Empire, Egypt and Environmental History*. University of Chicago Press, 2017.

Miles, George C. "Early Arabic Inscriptions near Ṭā'if in the Ḥijāz." *Journal of Near Eastern Studies* 7, no. 4 (1948): 236–42.

Miller, Nathaniel A. "Seasonal Poetics: The Dry Season and Autumn Rains among Pre-Islamic Naǧdī and Ḥiǧāzī Tribes." *Arabica* 64, no. 1 (2017): 1–27.

Munt, Harry. "Caliphal Estates and Properties around Medina in the Umayyad Period." In *Authority and Control in the Countryside: From Antiquity to Islam in the Mediterranean and Near East (Sixth–Tenth Century)*, edited by Alain Delattre, Marie Legendre and Petra M. Sijpesteijn, 432–63. Brill, 2018.

Munt, Harry. "Caliphal Imperialism and Ḥijāzī Elites in the Second/Eighth Century." *al-Masāq* 28, no. 1 (2016): 6–21.

Munt, Harry. *The Holy City of Medina: Sacred Space in Early Islamic Arabia*. Cambridge University Press, 2014.

Munt, Harry. "Mamluk Historiography Outside of Egypt and Syria: 'Alī b. 'Abd Allāh al-Samhūdī (d. 911/1506) and His Histories of Medina." *Der Islam* 92, no. 2 (2015): 413–41.

Munt, Harry. "Writing the History of an Arabian Holy City: Ibn Zabāla and the First Local History of Medina." *Arabica* 59, nos 1–2 (2012): 1-34.

al-Rāshid, Sa'd. *Darb Zubaydah: The Pilgrim Road from Kufa to Mecca*. Riyadh University Libraries, 1980.

al-Rāshid, Sa'd. *Dirāsāt fī al-āthār al-islāmiyya al-mubakkira bi-l-Madīna al-munawwara*. Mu'assasat al-Ḥuzaymī li-l-Tijāra wa-l-Tawkīlāt, 1421/2000.

al-Rāshid, Saʿd. *al-Rabadhah: A Portrait of Early Islamic Civilisation in Saudi Arabia*. King Saud University, and Harlow: Longman, 1986.

Robin, Christian Julien. "Ḥimyar, Aksūm, and Arabia Deserta in Late Antiquity." In *Arabs and Empires before Islam*, edited by Greg Fisher, 127–71. Oxford University Press, 2015.

Robinson, Chase F. *ʿAbd al-Malik*. Oneworld, 2005.

Schriwer, Charlotte. "Water and Technology in the Islamicate World." In *Routledge Handbook on the Sciences in the Islamicate World*, edited by Peter Barker, Sonja Brentjes and Rana Brentjes, 523–37. Routledge, 2022.

Trabelsi, Salah. "L'esclavage domanial dans le paysage agraire musulman au Moyen Age." In *Esclavage et dépendances serviles: historie comparée*, edited by Myriam Cottias, Alessandro Stella, and Bernard Vincent, 305–24. L'Harmattan, 2006.

ʿUsaylān, ʿAbd Allāh. *al-Madīna al-munawwara fī āthār al-muʾallifīn wa-l-bāḥithīn qadīman wa-ḥadīthan*. ʿAbd Allāh ʿUsaylān, 1418/1997.

Verkinderen, Peter. *Waterways of Iraq and Iran in the Early Islamic Period: Changing Rivers and Landscapes of the Mesopotamian Plain*. I.B. Tauris, 2015.

Webb, Peter. "Bedouin, Bandits, and Caliphal Disappearance: A Reappraisal of the Qarāmiṭa and Their Success in Arabia." In *The Historian of Islam at Work: Essays in Honor of Hugh N. Kennedy*, edited by Maaike van Berkel and Letizia Osti, 254–82. Brill, 2022.

Wilkinson, T. J. *Archaeological Landscapes of the Near East*. The University of Arizona Press, 2003.

Wüstenfeld, Ferdinand. "Die Familie el-Zubeir." *Abhandlungen der königlichen Gesellschaft der Wissenschaften zu Göttingen*, Hist.-Phil. Classe, 23 (1878): 3–112.

Zadeh, Travis. "The Early Hajj: Seventh–Eighth Centuries CE." In *The Hajj: Pilgrimage in Islam*, edited by Eric Tagliacozzo and Shawkat M. Toorawa, 42–64. Cambridge University Press, 2016.

CHAPTER 4

An Environmental History of the Darb Zubayda: Towards a First Approximation*

Peter J. Brown

Abstract

This chapter considers the Darb Zubayda pilgrimage route, along which pilgrims made the journey from Kufa in Iraq to the holy cities of Mecca and Medina. Particularly under the early Abbasid caliphs, substantial investment was devoted to the facilitation of hajj traffic through, for example, the establishment of waystations, and the provision of water facilities at intervals along the route. While in most existing scholarship, these phases of investment have been interpreted as responses to political and socio-economic factors, this chapter assesses the extent to which an environmental component, primarily linked to the seasonality of hajj pilgrimage, might have played a role in influencing the timing and intensity of Abbasid investment and patronage associated with the Darb Zubayda.

Keywords: Darb Zubayda; water scarcity; seasonality; Abbasid Caliphate

Introduction

Since the beginnings of Islam, the practice of hajj pilgrimage has formed one of the most important aspects of religious practice for Muslims around the world. Today, more than a million pilgrims undertake the journey to Mecca and Medina each year and this influx has a significant impact on the economy of the Kingdom of Saudi Arabia. While most pilgrims now travel using modern transport methods, including air travel and motorized transport, the journey was considerably more difficult and time-consuming in the past. While many pilgrims travelled by sea, this option predominantly brought non-Arabic speaking pilgrims from other lands.[1] Pilgrims from the Arabian Peninsula itself and the immediately adjacent regions,

* This publication is part of the project "Source of Life: Water Management in the Premodern Middle East (c. 7th–15th centuries)" with file number VI.C.181.016 of the research programme Vici SGW which is financed by the Dutch Research Council (NWO).

[1] Tagliacozzo, "The Hajj by Sea," 129–30.

however, predominantly made long and hazardous overland journeys, either by foot or borne by animals—most commonly camels.

Several long-distance routeways facilitating pilgrims' travel to Mecca and Medina from neighbouring regions of the Islamic world, including Syria, Egypt, Iraq, and Yemen, were developed from an early date.[2] These routeways sometimes incorporated pre-existing tracks, paths, and infrastructure. The Syrian route, for example, grew out of earlier trade routes dating back, at least, to the Nabatean era and was the most important route to the Hijaz during the Umayyad era—when Damascus was the centre of the Islamic world.[3] Though the evidence is limited, there are hints that it was during this period that many of the rituals surrounding the hajj pilgrimage were refined and standardized.[4] When the Abbasids came to power, as a new ruling dynasty, demonstrating their legitimacy as caliphal rulers became extremely important.[5] Pilgrimage offered a particularly public avenue to showcase both the piety of the Abbasid caliphs as well as their role as religious leaders and stewards of the Islamic religion.[6] With the shift in political geography that accompanied the Umayyad/Abbasid transition, Damascus and Syria lost their pre-eminent position within the caliphate in favour of Mesopotamia and Iraq. From the onset of the Abbasid period, therefore, the pilgrimage route from Iraq gained particular significance, as it connected the new caliphal capital, Baghdad (founded in 145/762), with the holy cities of the Hijaz.[7] This route, accordingly, attracted significant patronage from members of the Abbasid elite, including the caliphs themselves, alongside members of the caliphal family and high officials.

The route from Iraq, which formally began in Kufa, stretched across over 1200 km, traversing an inhospitable path across southern Iraq before passing through the Saudi Arabian regions of Najd and Hijaz (Fig. 4.1), which now fall within the modern provinces of Ḥā'il, Medina and Mecca. Though a route from Kufa existed to some extent in earlier times, Abbasid period investment transformed it into a sophisticated network, provisioned with waystations at regular intervals where pilgrim caravans could rest and resupply. At these waystations, as well as at other points along the route, hydraulic infrastructure, such as wells and reservoirs, were established to provide regular access to water. The passage of pilgrims was further facilitated by the clearance of paths through difficult terrain and the erection of milestones and way markers to guide the pilgrim traffic through an unfamiliar

[2] Petersen, "The Archaeology of the Syrian and Iraqi Hajj Routes," 48–49; Blair and Ulrich, "From Iraq to the Hijaz in the Early Islamic Period," 44–45.

[3] Petersen, "The Archaeology of the Syrian and Iraqi Hajj Routes," 49.

[4] Donner, "Umayyad Efforts at Legitimization," 199–201.

[5] El-Hibri, *The Abbasid Caliphate*, 67; Brown, "The "Cup of Pharaoh" from Samarra," 71–74.

[6] Webb, Bedouin, Bandits, and Caliphal Disappearance, 263.

[7] Webb, Bedouin, Bandits, and Caliphal Disappearance, 263–69.

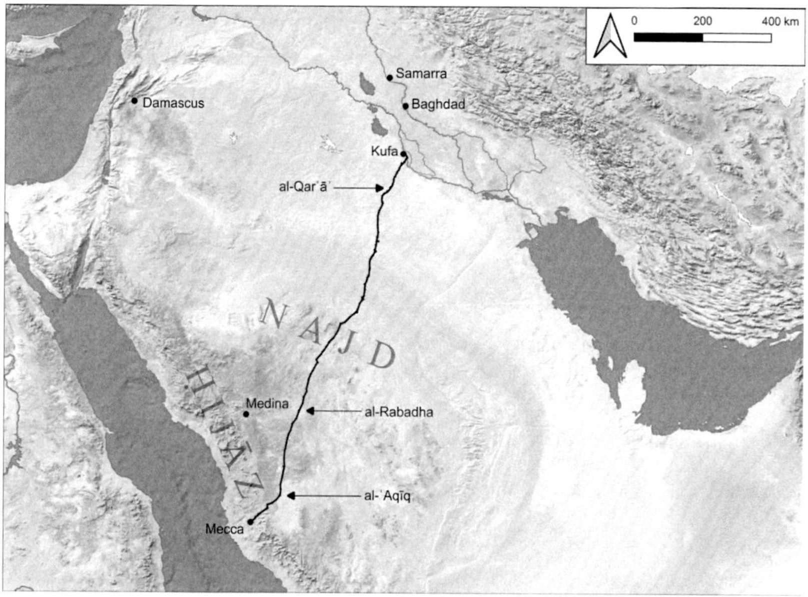

Figure 4.1: The Darb Zubayda route between Kufa and Mecca. Key locations discussed in the text are marked. Created by the author. Spatial data for the northern section of the route courtesy of W.M. Deadman. Basemap from Natural Earth.

landscape. The need for such infrastructure is underscored by the difficulties Dr Saad al-Rashid encountered in accessing sites along the route from Iraq when he began his archaeological survey of the route in the 1970s, many of which were far removed from occupied settlements where provisions could be obtained, set against challenging topography.[8] Based on the textual evidence, major phases of work on the route from Iraq occurred during the late 8th and early 9th centuries, especially during the caliphates of al-Mahdī (r. 158–69/775–85) and Hārūn al-Rashīd (r. 170–93/786–809). One of the latter's wives, the princess Zubayda (d. 216/831) is renowned in later sources for her patronage and donations which supported the establishment of facilities, including wells and other hydraulic infrastructure, along the route to assist pilgrims making the journey. As a result, Zubayda was commonly associated with the route,[9] so much so that it has come to be known as the "Trail (or Road) of Zubayda" or the Darb Zubayda.

The Abbasid infrastructure of the Darb Zubayda was not erected upon a blank canvas. While some form of pilgrimage route existed through the same area during Umayyad times, the lands through which the route passed were also inhabited

[8] al-Rashid, *Darb Zubaydah*, 73.

[9] See for example Ibn Baṭṭūṭa, *The Travels of Ibn Baṭṭūṭa Volume I*, 254.

by resident populations. During the Umayyad era, these groups were somewhat marginalised, and the opportunity to participate in the construction and running of a major state project brought by the Abbasid rejuvenation of the Darb Zubayda must have sent a significant shockwave through these populations. The project brought many benefits such as employment opportunities, enhanced food security, and commercial contacts from far and wide.[10] While Hārūn al-Rashīd invested heavily in infrastructure along the route, and traversed it himself as a pilgrim on several occasions, his successors showed less interest in either making the pilgrimage or maintaining the facilities along the route. Al-Muʿtaṣim's (r. 218–27/833–42) foundation of (yet) another new caliphal capital upriver from Baghdad at Samarra, for example, diverted attention and funds that might otherwise have been devoted to projects such as the Darb Zubayda. Limited maintenance was carried out along the pilgrim route under al-Muʿtaṣim's sons, al-Wāthiq (r. 227–32/842–47) and al-Mutawakkil (r. 232–47/847–61), but with the death of al-Mutawakkil in 247/861 and the anarchic period that followed, investment and maintenance along the route appears to have become even more infrequent. Throughout the 3rd/9th century, the irregular nature of interest and investment from the Abbasid elite in the Darb Zubayda introduced profound challenges for both the pilgrims travelling along the route as well as the local populations occupying the lands through which it passed, who had benefited from the initial phase of investment. The textual sources, for example, record food and water shortages, corrupt officials, and Bedouin raids.[11] Though it never again saw the same degree of interest as during the early Abbasid era, pilgrims continued to traverse the Darb Zubayda for many centuries—Ibn Baṭṭūṭa, for example, followed the route in the mid-8th/14th century.[12]

The trajectory of Abbasid investment in the pilgrimage infrastructure that facilitated the flow of pilgrims along the Darb Zubayda has received attention from both archaeologists and historians.[13] For the early Abbasid period, a variety of textual sources exist[14] which help to understand the route as it was at that time, while several archaeological surveys and excavations have been carried out along the Darb Zubayda since the 1970s.[15] Many other aspects of the route, however, and

[10] Webb, "Bedouin, Bandits, and Caliphal Disappearance," 263–69.

[11] Webb, "Bedouin, Bandits, and Caliphal Disappearance," 273–75.

[12] Ibn Baṭṭūṭa, *The Travels of Ibn Baṭṭūṭa Volume I*, 249–56.

[13] Webb, "Bedouin, Bandits, and Caliphal Disappearance," 262–75; al-Rashid, *Darb Zubaydah*, 17–58.

[14] See for example the descriptions of Ibn Khordadbeh and al-Yaʿqūbī. Ibn Khurdādbih, *Le Livre des routes et des provinces*, 500–502; al-Yaʿqūbī, *The Works of Ibn Wāḍiḥ al-Yaʿqūbī*, 149.

[15] The most extensive being the Darb Zubayda Survey Program carried out in the late 1970s and early 1980s. Knustad, "The Darb Zubayda Project: 1396 / 1976," 41–68; al-Dayel and al-Helwa, "Preliminary Report on the Second Phase of the Darb Zubayda Reconnaissance," 51–64; al-Dayel et

its operation are not known in detail. The reality of pilgrimage through the Arabian environment, for example, and how pilgrims were impacted by environmental conditions has received relatively limited attention. In particular, the shifting seasonality of hajj pilgrimage combined with both regular meteorological cycles and fluctuating climatic conditions must have introduced significant challenges for the pilgrim caravans traversing Arabia throughout the early Islamic period. While only scratching the surface of this important topic, this chapter considers the various ways in which environmental realities impacted the flow of people along the Darb Zubayda pilgrim road, focusing on the early Abbasid period, as well as the various ways through which these risks and stressors were minimized and mediated.

The Seasonality of Hajj Pilgrimage

Each year, hajj occurs during the first ten days of the final month of the Islamic, or hijri, calendar: Dhū al-Ḥijja. Since the hijri calendar follows the lunar cycle, the dates of hajj shift by ten or eleven days each year relative to the solar year. Over the years, therefore, the timing of hajj occurs at different times of the year and shifts through different seasons relative to the Gregorian calendar. While there are important local variations, the stark seasonality of the climate of modern-day Saudi Arabia is immediately obvious from average temperature and precipitation records between 1901 and 2020 (Fig. 4.2). On average, less than 5mm of precipitation occurs between the months of June and October while between June and September the average mean surface temperature soars to above 30°C—with the average maximum during that time 10°C warmer still. Though outside June–September the climate remains relatively warm and dry, the summer months would have imposed specific challenges for large groups crossing the desert with negligible rainfall, high temperatures and rapid evaporation. Today, even with fast, air-conditioned transport, such seasonal variation between years poses a challenge to be overcome for those making the journey to Mecca and Medina.[16] In the past, however, when it was necessary for pilgrims to cover long distances, either on foot or borne by camels, the timing and seasonality of the hajj pilgrimage significantly impacted the difficulty of the journey, as is widely attested in both textual and oral accounts

al., "Darb Zubayda Survey Project. Preliminary Report on the Survey of Darb Zubaydah, third season," 43–54; Mackenzie and al-Helwa, "Darb Zubaida Architectural Documentation Program. Darb Zubaida – 1979," 37–50; Morgan and al-Helwa, "Preliminary Report on the Fifth Phase of Darb Zubayda Reconnaissance," 85–107; al-Helwa et al., "Preliminary Report on the Sixth Phase of Darb Zubaydah Reconnaissance," 39–62.

16 Ahmed et al., "Health Risks at the Hajj," 1013.

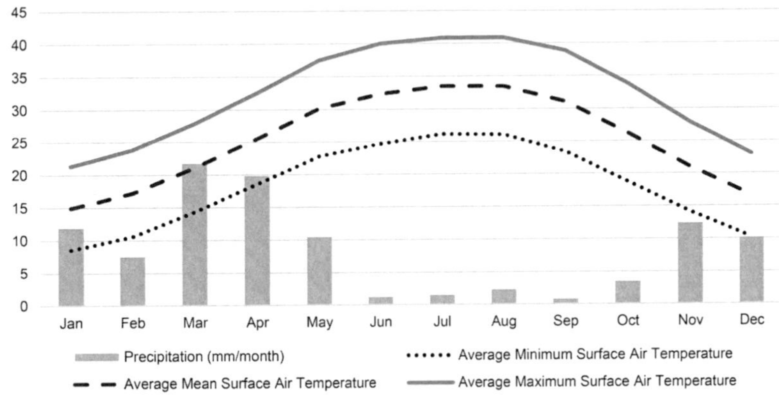

Figure 4.2: Monthly average temperature and precipitation in Saudi Arabia, 1318–1442/1901–2020. Temperature is measured in degrees Celsius. Compiled by the author based on data from Harris, Osborn, Jones and Lister, "Version 4."

relating to pilgrim travel. For example, an oral tradition related to the Finnish traveller Georg August Wallin[17] in 1848, concerning how an entire caravan of pilgrims died of thirst when travelling on the Darb al-Bakrah, an ancient route from Jordan to northwestern Saudi Arabia, emphasises the potentially fatal consequences of hajj pilgrimage.

Over the course of the early Abbasid period, between 132–c.288/750–c.900, the timing of the hajj pilgrimage went through almost five full annual cycles (Fig. 4.3). Periods when the hajj occurred during the colder, wetter and more pleasant winter months, therefore, were approximately between 140–59/758–75, 173–92/790–807, 206–26/822–40, 240–59/855–72, and 274–93/888–905. On the other hand, the pilgrimage season coincided with the hot summer months between 159–73/776–89, 192–206/808–21, 226–240/841–54, and 259–74/873–87. Of course, there is room for debate in relation to the margins of these ranges; natural variability may have meant that, for example, conditions during one pilgrimage in March in a particular year were more challenging than those that affected a pilgrimage in April a few years later. In general, however, the summer pilgrimages experienced significantly more extreme conditions than those that occurred during the winter months. It is also important to note that the dates of the occurrence of hajj itself—depicted by the bars in Figure 4.3—relate to much longer pilgrimage journeys to and from Mecca either side of these dates. In a given year, therefore, pilgrim travel began and ended weeks and months before and after the dates of hajj itself, depending on each pilgrim's specific start and end points and how far they had to travel.

17 Wallin, *Notes Taken during a Journey Through Part of Northern Arabia*, 321.

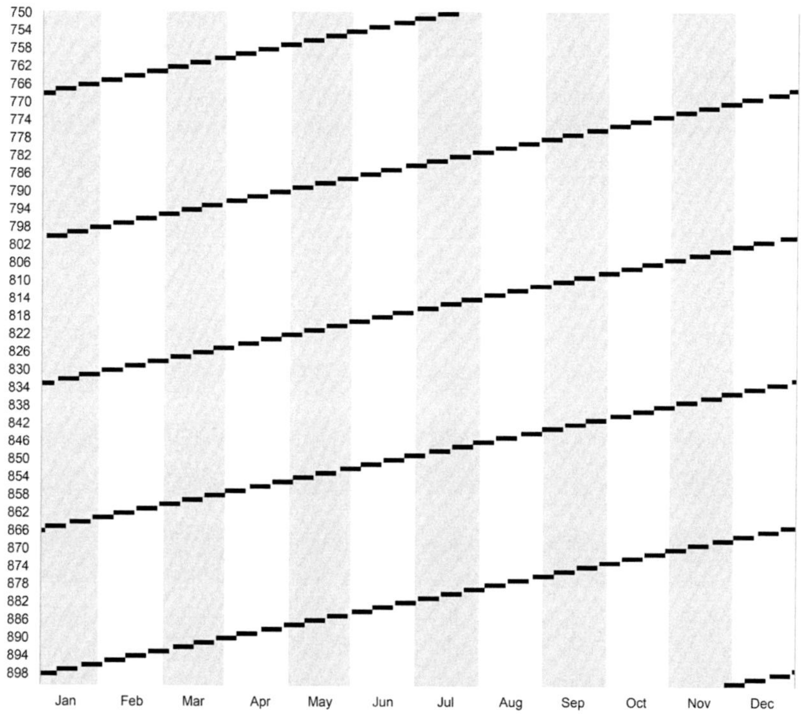

Figure 4.3: The timing of hajj between 132–288/750–900. Due to the Islamic lunar calendar, the timing of hajj shifts each year relative to the solar year meaning that, in different years, the pilgrimage season takes place in different seasons.

The phases in which passage along the Darb Zubayda was likely made more or less challenging, through the seasonality of its occurrence, are of interest when contrasted against the picture gleaned from the textual sources. The first phase of favourable years during the Abbasid era, for example, 140–59/758–75, closely matches the reign of al-Manṣūr (r. 136–58/754–75) who himself made the pilgrimage to Mecca six times.[18] The reign of his successor al-Mahdī (r. 158–69/775–85), however, coincided with a phase of pilgrimage seasons between 159–73/776–89 when the hajj fell during the summer months. In comparison to the relatively favourable hajj conditions experienced under his father, the difficulties experienced along the hajj road may have compelled al-Mahdī to improve facilities—including wells, reservoirs, and other hydraulic infrastructure—to try and mitigate the more difficult conditions pilgrims faced during his reign. Indeed, reportedly during a pilgrimage in 164/781, al-Mahdī himself was forced to turn back due to a lack

[18] al-Rashid, *Darb Zubayda*, 18.

of water—an experience which perhaps spurred greater investment in water facilities.[19] It is notable that one of the periods of greatest interest and investment in the Darb Zubayda, the caliphate of Hārūn al-Rashīd (r. 170–93/786–809), closely coincides with a period when the hajj occurred at the most favourable times of the year—between 173–92/790–807. As well as the caliph himself, his consort Zubayda sponsored the establishment of many facilities along the route, especially hydraulic infrastructure.[20] Hārūn al-Rashīd made the pilgrimage himself on multiple occasions and invested heavily in facilities including cisterns, reservoirs, and wells. After Hārūn al-Rashīd's reign, investment and interest in the road largely abated until limited initiatives were carried out under al-Wāthiq, whose reign (r. 227–32/842–47) fell during a period of, more challenging, summer pilgrimage seasons between 226–240/841–54.

The period between Hārūn al-Rashīd's reign and that of al-Wāthiq is conspicuous for the apparent lack of caliphal interest in the Darb Zubayda.[21] Certainly, the construction of the new capital at Samarra, which was founded in 221/836,[22] significantly absorbed caliphal capacity to enact large-scale infrastructure projects. Another factor which may go some way to explain the relative lack of interest shown by the caliphs during this interval, however, is that the investments in facilities along the route made under al-Mahdī and Hārūn al-Rashīd may have been sufficient to keep the route at close to an acceptable level over the next few decades. The unfavourable cycle between 192–206/808–21 may have been largely ameliorated through the presence of relatively newly constructed facilities that made it easier to make the journey. Then, when more favourable hajj seasons began to occur between 206–26/822–40 the now slightly older facilities were still in a good enough state of repair to meet the needs of most pilgrims. When the next phase of challenging hajj years kicked in, however, coinciding with the ascension of al-Wāthiq, the need to repair and improve the route must have been much more apparent. Though primarily from economic and political perspectives, Peter Webb sees a comparable process in the interval between the reigns of Hārūn al-Rashīd and al-Wāthiq. By the mid-3rd/9th century, the declining state of the route's infrastructure, combined with the lack of new investment in the region, led to a growing sense of desperation within the local populations that had grown accustomed to caliphal charity, investment and the employment opportunities relating to the route in earlier times—leading to raiding activity and attacks on pilgrim caravans.[23]

19 al-Rashid, *Darb Zubayda*, 18–19; Webb, "Power and Money on the Hajj," 128, fn. 108.

20 Webb, "Power and Money on the Hajj," 135.

21 al-Rashid, *Darb Zubayda*, 21.

22 Northedge, *The Historical Topography of Samarra*, 81.

23 Webb, "Bedouin, Bandits, and Caliphal Disappearance," 278–79.

Al-Wāthiq himself was aware of the growing problems on the Darb Zubayda. In 231/845–46, intending to make the pilgrimage himself, he sent one of his officials, ʿUmar ibn Faraj, to scout out the route. ʿUmar returned to tell the caliph that the route suffered from a lack of water causing al-Wāthiq to cancel his plans for that year.[24] The following year, 232/846–847, the situation seems to have been even worse as pilgrims travelling near the station of al-Rabadha were unable to find enough water, and what water was available skyrocketed in price, resulting in many deaths.[25] Hajj during these two years fell at almost the worst conceivable time: in 231/845–46 this was between 29th July–7th August while in 232/846–847, hajj occurred between 19th–28th July. In both years, therefore, the pilgrimage caravans were most likely crossing Arabia in June and July and returning in late August and September when temperatures were at their hottest, and precipitation was virtually non-existent.

In the following period it becomes more difficult to understand the degree of investment made by the caliphs in the Darb Zubayda. Al-Mutawakkil (r. 232–47/847–61) appears to have taken a limited interest in restoring conditions along the route, including repairing wells, but, as with his father, his grand plans for Samarra seem to have diverted the greater part of his attention and funds.[26] After al-Mutawakkil's reign, the many internal and external challenges facing the caliphate—including the Samarran anarchy, the Zanj Rebellion,[27] and in the territory through which the Darb Zubayda itself passed, the rise of the Qarāmiṭa[28]—significantly impacted the flow of pilgrims from Iraq to the Hijaz. Another incident relating to water on the Darb Zubayda is reported from 258/872 when al-Ṭabarī relates that most of the pilgrims turned back after reaching the waystation of al-Qarʿāʾ due to concern for the lack of water. Those pilgrims that did continue, however, were able to make the journey safely.[29] While this episode occurred during one of the more favourable intervals, between 240–59/855–72, it fell at the extreme end of that range. The hajj that year occurred between 8th–17th October meaning the pilgrims were likely travelling during late August and September. Moreover, natural variation may have made this a drier and/or hotter year than average.

Though there is a significant amount of historical data, it is difficult to understand the trajectory of Abbasid investment between c. 132–288/750–900 in detail. While his account may not be exhaustive, al-Qāḍī Wakīʿ (d. 306/918) in his *Kitāb*

24 al-Ṭabarī, *Incipient Decline*, 35–36.

25 al-Ṭabarī, *Incipient Decline*, 51.

26 Webb, "Bedouin, Bandits, and Caliphal Disappearance," 273; al-Rashid, *Darb Zubayda*, 23.

27 An uprising against the Abbasid state of African slaves and local peasants in southern Iraq between 255–70/869–83.

28 An Ismaʿīlī group who attracted a considerable following in north-eastern Arabia and opposed the Abbasids.

29 al-Ṭabarī, *The Revolt of the Zanj*, 149.

al-Ṭarīq lists a multitude of sites along the Darb Zubayda where different facilities were patronised by the caliphs and other members of the Abbasid elite.[30] Taking the number of sites for each caliphate offers a crude visualisation of the amount of investment and interest in the pilgrimage route over the course of the early Abbasid period—which may be easily contrasted against the phases of favourable and less-favourable environmental conditions for hajj pilgrimage (Fig. 4.4). While the timing of the hajj alone does not explain either trends in the pattern of Abbasid patronage of facilities along the route, the flow of pilgrim traffic, and/or the socio-economic situation of the lands through which the route passed, it did impose important constraints which, in tandem with the multi-year seasonal hajj cycle, impacted these matters to some degree.

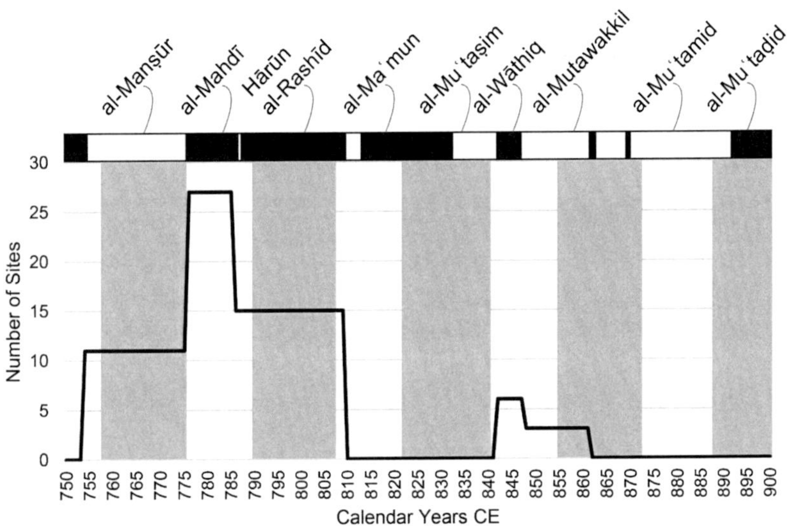

Figure 4.4: An index of early Abbasid investment in the Darb Zubayda between 132–288/750–900. The grey bars indicate periods when the hajj fell in the winter months, meaning the pilgrimage should have been less challenging. The black line shows the number of sites along the Darb Zubayda patronised by the Abbasid elite based on the list given by al-Qāḍī Wakīʿ.[31] The durations of major caliphates, and those discussed in the text, are indicated.

[30] Webb, "Power and Money on the Hajj," 134–35.

[31] See note 30.

Traversing the Environment of the Darb Zubayda

Despite the challenges of traversing the Darb Zubayda, many pilgrims still made the journey even when the hajj season fell during the hot summer months. Though some pilgrims travelled by foot, large numbers of camels were vital to the pilgrim caravans. To prepare them for a long journey, camels ideally require access to good pasture in the preceding weeks and months. Pasturage along the Darb Zubayda was strictly controlled through the creation of reserves, or *aḥmāʾ* (sg. *ḥimā*), which ensured the availability of pasture for horses and camels required for official purposes.[32] The *ḥimā* of al-Rabadha, for example, which was established from the time of ʿUmar ibn al-Khaṭṭāb (r. 13–23/634–644), was particularly renowned for its lush vegetation including the *ḥamḍ* plant (*Haloxylon salicornicum*) which was believed to provide excellent nourishment for camels.[33] Though such areas were especially verdant, the availability of suitable pasture waxed and waned in response to changes in precipitation in the dry summer and wetter winter and spring periods (Fig. 4.5). The nutrition and preparedness of the camels and other livestock which supported and enabled the pilgrim caravans, therefore, was another area impacted by the seasonality of hajj. When limited pasture was available, the camels of the pilgrim caravans would have been weaker and less able to cope with limited water availability. Certainly, in later times, the provision of camels for hajj gave rise to business speculation through which investors were incentivised to acquire their camels for the lowest possible price so they could make a considerable profit hiring them out to pilgrims.[34] Conversely, the pilgrims who hired the camels were primarily motivated to reach their destination as quickly as possible rather than looking out for the longer-term interests of their steeds. The long-distance daily treks required to traverse the Darb Zubayda extorted a considerable toll on the camels, which were rarely given sufficient time to graze and recover their strength.[35] Therefore, because of their respective priorities, neither camel owners nor pilgrims necessarily prioritized the welfare and fortitude of the camels involved in the hajj caravans.

To avoid the intense heat, especially during the summer months, it would have been preferable to travel in the early mornings, the evenings, and at night. Beacons or fire signals, which guided travellers during the hours of darkness, are attested in the textual sources.[36] Their existence raises questions about how fuel was sourced

32 Webb, "Bedouin, Bandits, and Caliphal Disappearance," 258–59.

33 Al-Rashid, *Al-Rabadhah*, 3-4.

34 Grant, *The Syrian Desert*, 229.

35 Blunt, *A Pilgrimage to Nejd*, 92.

36 al-Rashid, *Darb Zubayda*, 22, 149–50; Webb, "Bedouin, Bandits, and Caliphal Disappearance," 268. Prior to its collapse, the tower at Umm al-Qurūn in southern Iraq, which is located on a topographic

Figure 4.5: View across the landscape of the *ḥimā* of al-Rabadha, Medina Province, in August 2025. Photograph by the author.

to keep them burning over a lengthy period, as well as how the labour to light and tend the fires was organised. During the day, shielding from the sun must have been essential to prevent severe heatstroke. Though she did not believe it, Lady Anne Blunt, a 13th/19th-century English traveller, relates a tradition that the remains of a wall alongside one section of the route had been erected through the patronage of Zubayda herself with the intention that awnings could be hung from it to provide shade for passing pilgrims.[37] In the early 14th/20th century, photographs of pilgrim caravans show that covered howdahs, carriages mounted on the backs of camels providing two berths per camel, were used to protect travellers from the sun (Fig. 4.6). The English traveller Richard Burton, who made the journey from Cairo, via Suez and Yanbu, to Medina and Mecca in 1853 describes the advantages and drawbacks of travelling in this manner. The howdah he encountered (which he knew as a *shugduf*) had a wicker framework, covered by matting and carpets, providing shade for the traveller within which cooled water and provisions were available from the camel's pack. He also describes, however, how the travellers in each berth, either side of the camel, had to be of similar weights to balance the howdah which was liable to being blown about by the wind, damaged by thorns as the camel brushed past trees, or broken if the camel fell under the awkward weight it carried.[38] Ibn Baṭṭūṭa describes a similar double berthed camel litter in which he travelled when he set out to Mecca with a pilgrim caravan from Baghdad

highpoint, may have hosted a beacon to guide travellers. See Al-Maiyali and Wilding, "The Rest-stop and Minara of Umm Al-Qurun."

[37] Blunt, *A Pilgrimage to Nejd*, 68.

[38] Burton, *Personal Narrative of a Pilgrimage*, 233–34, note 4.

Figure 4.6: A caravan transporting pilgrims to Mecca in 1328/1910. Matson Photograph Collection. (Library of Congress, Prints and Photographs Division, [LC-DIG-matpc-04654].)

in 727/1327.[39] Notably, Ibn Baṭṭūṭa's trip, which probably occurred in September and October, was marred by health problems, likely exacerbated by the extreme climate. Although howdahs were certainly available during the early Islamic period[40] and, by shielding travellers from the harsh sun during the day, must have made the long journey significantly more bearable, they were likely only available to higher status pilgrims.

The caravans would also have had to transport water with them throughout the journey. Not only would some water have been necessary to sustain travellers during the intervals between water facilities, when other sources were unavailable, but also to provide a safety net in case something went wrong along the way. If the caravan became lost, for example, or expected water sources along the route were either depleted or could not be located, or if the caravan could not proceed for some other reason, having a supply of water on hand could mean the difference between life and death. When a caravan arrived at a source of water, such as a reservoir or well, therefore, not only would the pilgrims, and their camels, see to their immediate needs, they must have also filled bags and/or vessels, with water

[39] Ibn Baṭṭūṭa, *Travels in Asia and Africa Volume II*, 355.

[40] Indeed, they are commonly mentioned in early Islamic poetry e.g. Hussein, "The Fat Bride and the Foolish Messengers,", 256–59.

to take with them on the onward journey, further exacerbating the stress such a caravan placed on the water facilities along the route. Such water supplies were most likely carried in waterskins, often goatskins, which were commonly used to transport water throughout the Middle East in pre-modern times.[41] Ancient Egyptian travellers crossing the Libyan desert carried water in ceramic jars, often using donkeys,[42] and sometimes relying on caches of these jars, containing water, which had been pre-emptively placed along the route for later retrieval.[43] While evidence for similar practices has not yet been uncovered on the hajj roads, it is certainly possible that the larger, better organised, caravans may have been provisioned in a similar way.

Water Resources along the Darb Zubayda

The waystations along the Darb Zubayda were what made the journey a viable prospect for massed groups of pilgrims. By providing safe havens, at which passing caravans could stop to rest and resupply, the arduous journey became significantly more achievable, especially for large groups of people. Perhaps the most important facilities provided by the waystations were sources of water, principally cisterns, reservoirs, and wells (Fig. 4.7). Evidently, however, even the many different tanks and well-shafts that were created to provide for the hajj traffic were sometimes insufficient. In the textual sources, one of the most frequently mentioned problems experienced by pilgrims on the Darb Zubayda is access to water and, as recounted above, this was particularly a problem during the summer months. As explored already, to try to ensure sufficient water was available for the pilgrim caravans, the caliphs repeatedly took measures to either build new hydraulic infrastructure, expand water capacity along the route, or repair existing facilities, to increase the efficiency of what already existed.

As with other pre-modern water supply solutions,[44] the water facilities of the Darb Zubayda were informed by, and tailored to, the natural environments in which they were constructed. Reservoirs are common at the waystations along the route, a key feature of which is the collection of surface runoff water during seasonal rains. Typically, the reservoir is situated close to a wadi channel through

41 A near contemporary waterskin of this type comes from the mid-7th/mid-8th century CE Ma-agan Mikhael B shipwreck. See: Cohen and Cvikel, "Ma-agan Mikhael B, Israel," 196–97.

42 Förster, "With Donkeys, Jars and Water Bags," 2–3.

43 Förster, "With donkeys, Jars and Water Bags," 3–4.

44 See for example the different water supply solutions in classical Mediterranean cities within the volume: Bonnie and Klingborg, *Water in Ancient Mediterranean Households*.

Figure 4.7: The larger of the two *birkas* (reservoirs) at al-Kharāba. Mecca Province, Ṭā'if Governorate, Saudi Arabia. Photograph by the author.

which water flows when the rains arrive. The rise in water level in the wadi allows this discharge to be captured and directed into the reservoir, often through an inlet. In a number of reservoirs, the water first enters a settling tank, allowing sediment to sink to the bottom so that the water entering the reservoir itself does not lead to significant sedimentation. If the flow of water is sufficient, this allows the reservoir to be filled. For a period after the rains have passed, this stored water within the structure may then be accessed by local communities, and passing traffic. The precise siting of these pieces of infrastructure in relation to both the expected runoff from wadi flows, as well as the local topography, therefore, was essential in ensuring their ability to successfully capture sufficient water to allow the route to remain passable.

Considerable ecological knowledge and technical skill is discernible in the siting of the reservoirs[45] along the Darb Zubayda. Though the underlying structures have, in some cases, been restored in modern times, many continue to hold water to this day, signifying the sustainability of the sites chosen for their construction. At least in the few he studied, Tony Wilkinson highlighted that the reservoirs along the Darb Zubayda were usually located on smaller wadi tributaries or distributaries. Compared to the main branch channels of wadis, which carried much greater

[45] Wilkinson, "Darb Zubaida," 51–67.

volumes of water and would have required larger control structures such as dams, the flows in these channels would have been significantly easier to manage.[46] Furthermore, reservoirs were often situated on the margins of wadi flood zones rather than within wadi beds themselves—perhaps to reduce the exposure of reservoirs to both the full force of water flowing down the wadi and the associated sediments they transported.[47] In addition, the design of the water inlets through which the reservoirs were filled likely also reduced sedimentation meaning that, since sediment likely accumulated at only a very gradual rate, relatively little ongoing maintenance was required during the Abbasid period.[48] That these complex problems may have engendered a degree of experimentation and adaptation to local conditions in the design of these pieces of infrastructure may be suggested by the presence of settling tanks at some cisterns/reservoirs while these are absent at others.[49] Though there are no details in the textual accounts relating to the survey, design, or construction of these facilities, it is clear that both local knowledge about the best places to construct such facilities, as well as expertise relating to the design of hydraulic infrastructure, were available during these phases.[50]

Two problems associated with the reservoirs on the Darb Zubayda are, firstly, that they were unable to store captured water for indefinite periods and, secondly, that they were at the mercy of unpredictable rainfall patterns. At the waystation of al-ʿAqīq, (Fig. 4.8) for example, the maximum capacity of the reservoir can be estimated at just under 1,000 m^3, or one million litres.[51] As the reservoirs on the Darb Zubayda are open to the elements, as soon as the supply of water in the wadi subsided, and the reservoir ceased filling, the water it held would have begun to evaporate. Estimating the rate of evaporation is not simple as it depends on variables, such as wind speed, humidity, and air temperature, which are highly dynamic. A conservative estimate, with relatively high humidity (50%) and low wind speed (1 km/hr), using the mean temperatures from the hottest period in the summer (Fig. 4.2) points toward a daily loss of water of around 19,500 litres. If the reservoir had been filled to its maximum, this would have meant that all the

46 Wilkinson, "Darb Zubaida," 55.

47 Wilkinson, "Darb Zubaida," 55.

48 Wilkinson, "Darb Zubaida," 55–56. This suggestion challenges interpretations based on the textual evidence. See for example: Webb, "Bedouin, Bandits, and Caliphal Disappearance," 270.

49 Wilkinson, "Darb Zubaida," 56.

50 It is likely that, in common with other caliphal projects such as the construction of Samarra, experts were brought in specifically for this project. See for example: al-Yaʿqūbī, *The Works of Ibn Wāḍiḥ al-Yaʿqūbī*, 98.

51 This is based on the square reservoir measuring approximately 49m x 49m narrowing to a smaller base of 35.8m x 35.8m at a depth of 5.5m. In simple terms, this suggests a volume of 9748 m^3. These measurements are based on figures given by al-Rashid, *Darb Zubayda*, Plate of Site No. 37.

Figure 4.8: The *birka* (reservoir) at the waystation of al-ʿAqīq, Ṭāʾif Governorate, Mecca Province, Saudi Arabia. Al-ʿAqīq Archaeological Project.

water would have evaporated after only 52 days. In reality, humidity was probably significantly lower along much of the route, with 20% more typical for the summer months, and windspeed much more variable, both of which would have increased the rate of evaporation. On the other hand, the stepped nature of the reservoir tank would have meant that, as its contents evaporated, a smaller surface area would have been exposed which, to a limited extent, would have slowed the rate of further evaporation. These figures, however, fail to account for actual use of the reservoir by humans and animals. Camels can drink large quantities of water, often more than 100 litres in a single session. A large caravan of pilgrims, including both humans and camels, therefore, must have been able to significantly deplete a reservoir such as the one at al-ʿAqīq. Though the rains that fed the wadis, along which the reservoirs were sited, were seasonal, their occurrence was still highly variable and there were undoubtedly some years when no rains occurred at all in some catchments. As a result, beyond the seasonality of the occurrence of hajj itself, a further layer of temporal variability affected the pilgrimage in regard to the availability of water. The fact that the occasions on which pilgrims suffered from thirst, or were forced to turn back, were noteworthy enough to be documented by contemporaries suggests that, by and large, most pilgrimages were successful. However, especially during the summer months, there must always have been a high degree of uncertainty that water would be available within a particular catchment at the specific moment a pilgrim caravan arrived.

In providing water for passing pilgrims, wells faced a different problem. Rather than relying on water from wadi runoff, wells tapped groundwater aquifers. While these are also ultimately recharged through rainfall, because the water is stored underground it is much less vulnerable to evaporation. If used sustainably, therefore, it is possible for wells to supply water continuously throughout periods of low rainfall. In comparison to reservoirs, however, accessing water from wells was significantly more labour intensive. Where reservoirs could usually be accessed directly by people and animals, a well required water to be lifted in relatively small quantities from depth, either using human or animal power. Both the operation of wells to raise water for the pilgrim caravans, as well as their initial construction, are perhaps the types of jobs that the inhabitants of the lands through which the Darb Zubayda passed could have participated in, bringing a new level of prosperity to these regions compared to preceding periods.[52] Even though it was more labour intensive to supply large volumes of water using wells, over-exploitation of groundwater aquifers, causing the well to run dry, remained a real risk. While expanding the overall availability of water on the Darb Zubayda obviously remained the general aim behind digging new wells, the exhaustion of already exploited aquifers perhaps provides another reason for the continued need to dig new wells during the early Abbasid period.

Conclusions

An important aspect of travel along the Darb Zubayda pilgrimage route, often overlooked in existing studies, was the timing of hajj in any given year. Because the Islamic calendar was a lunar one, the shift in timing of the departure of pilgrim caravans year by year meant that hajj pilgrimages went through a multi-year cycle. As a result, although one half of the cycle, during the winter months, was likely to be relatively unproblematic for the passage of pilgrims, in the other half, the timing of hajj clashed with the hottest months of the year when water scarcity, in an already arid region, was amplified. At least at the beginning of the Abbasid era,

52 These employment opportunities likely became unstable when local unrest threatened the safety of pilgrim caravans, creating a vicious cycle that fuelled more economic hardship, desperation, caravan raiding and banditry. See Webb, "Bedouin, Bandits, and Caliphal Disappearance," 278.

there seems to have been some relationship between interest in the route and these cycles. Al-Manṣūr benefited from favourable hajj seasons, but al-Mahdī faced much more challenging conditions for the hajj during his reign, possibly spurring greater investment in facilities along the route. Hārūn al-Rashīd continued investment at a similar scale and was able to make the pilgrimage himself numerous times, in part, due to the favourable seasonality of hajj during his reign. The cycle of unfavourable and favourable hajj seasons that followed, until the reign of al-Wāthiq, coincided with other projects which distracted caliphal attention, while the state of decay of the Darb Zubayda infrastructure had likely not yet reached a tipping point that would have triggered renewed interest in the route. This point was reached during the caliphate of al-Wāthiq, when a new phase of unfavourable hajj seasonality set in, stimulating limited attempts to restore conditions along the route. His successor al-Mutawakkil adopted a similar strategy, though never reaching the level of interest or investment of his earlier caliphal predecessors. After the caliphate of al-Mutawakkil, the plethora of other challenges facing the caliphate brought an end to significant investment in the facilities along the Darb Zubayda and, as a result, hajj pilgrimage along this route, even during the winter months, likely became much more challenging. Notably, however, pilgrims did continue to make the journey in the centuries that followed.

To mitigate the challenges of traversing the Darb Zubayda, especially during periods of extreme summer heat with scarce rainfall, there were some practical measures that may have helped to make transit viable. The provision of shade of some description was paramount, though all pilgrims likely covered themselves with appropriate clothing and headwear to shield their bodies from the sun, those with sufficient funds likely opted to travel in a shaded howdah on the back of a camel. Water was carried by the caravans for use along the way but had to be regularly refilled either at wells or reservoirs which were largely provided, at waystations and in isolation, through the patronage of the Abbasid elite. The reservoirs and wells along the route faced their own challenges in supplying large caravans of people and animals—either through the rapid evaporation of water from reservoirs or, in the case of wells, the depletion of groundwater aquifers. Especially in the case of reservoirs, the unpredictability of where and when rains would fall in any given year meant that there must have always been great uncertainty about whether water would be available at waystations along the route. On top of the timing of hajj itself, the seasonality of rainfall patterns exerted a further constraint that determined where and when water for passing caravans would be available. Future work should consider in greater detail the seasonality of individual catchments and the variability of rainfall along the route at a higher spatial resolution to enhance our understanding of the challenges faced by pilgrim traffic traversing the road between Kufa and Mecca.

This chapter has primarily focused on the environmental challenges faced by pilgrim traffic, making little mention of commercial interaction along the Darb Zubayda. It is important to emphasise, however, that the Darb Zubayda was not solely a conduit for those undertaking hajj. In linking Iraq and the Hijaz, and during its heyday, providing infrastructure and security enabling travel over this route, the Darb Zubayda also facilitated trade and commerce.[53] Correspondingly, periods during which it was more difficult to traverse the Darb Zubayda, either for environmental or political reasons, must have resulted in a drop in trade and economic activity between Iraq and the Hijaz, as well as interconnected areas. Importantly, however, unlike pilgrims, commercial traffic was not restricted to travelling during the hajj season and was likely, therefore, less affected by seasonality along the route.[54] Even so, understanding the factors which made traversing the Darb Zubayda more challenging, of which the multi-year cycle discussed here was one, is of wider significance beyond the religious practice of Muslims during the early Abbasid era.

Bibliography

Primary Sources

Blunt, Anne. *A Pilgrimage to Nejd, The Cradle of the Arab Race, Volume II, (2nd Edition).* John Murray, 1881.

Burton, Richard F. *Personal Narrative of a Pilgrimage to Al-Madinah & Meccah, Vol. I.* G. Bell & Sons, Ltd, 1914.

Ibn Baṭṭūṭa. *The Travels of Ibn Baṭṭūṭa. A.D. 1325–1354. Volume I.* Translated by H. A. R. Gibb. Cambridge University Press, 1958.

Ibn Baṭṭūṭa. *Ibn Battúta: Travels in Asia and Africa 1325–1354, Volume II.* Translated by H. A. R. Gibb. Cambridge University Press, 1962.

Ibn Khurdādbih. "*Kitāb al Masālik w'al Mamālik.* Le Livre des routes et des provinces." *Journal Asiatique*, Mai-Juin (1865): 440–527.

al-Ṭabarī. *Ta'rīkh al-rusul wa-l-mulūk, The History of al-Ṭabarī. Volume XXXIV, Incipient Decline. The Caliphates of al-Wāthiq, al-Mutawakkil, and al-Muntaṣir A.D. 841–863/A.H. 227–248.* Translated and edited by Joel L. Kraemer. State University of New York Press, 1989.

al-Ṭabarī. *Ta'rīkh al-rusul wa-l-mulūk, The History of al-Ṭabarī. Volume XXXVI, The Revolt of the Zanj. A.D. 869–879/A.H. 255–265.* Translated and edited by David Waines. State University of New York Press, 1992.

53 Power, *The Red Sea*, 120–24.

54 Indeed, caravans from Mecca to Syria during the early Islamic period routinely travelled in the winter when the availability of water and pasturage was most certain. See Grant, *The Syrian Desert*, 127.

Wallin, George August. "Notes Taken during a Journey Through Part of Northern Arabia, in 1848." *The Journal of the Royal Geographical Society of London* 20 (1850): 293–344.

al-Yaʿqūbī. *The Works of Ibn Wāḍiḥ al-Yaʿqūbī. An English Translation. Volume 1.* Translated and edited by Matthew S. Gordon, Chase F. Robinson, Everett K. Rowson, and Michael Fishbein. Brill, 2018.

Secondary Sources

Ahmed, Qanta A., Yaseen M. Arabi, and Ziad A. Memish, "Health Risks at the Hajj." *The Lancet* 367, no. 9515 (2006): 1008–1015. https://doi.org/10.1016/S0140-6736(06)68429-8.

Blair, Andrew, and Brian Ulrich. "From Iraq to the Hijaz in the Early Islamic Period: History and Archaeology of the Basran Hajj Road and the Way(s) through Kuwait." In *The Hajj: Collected Essays*, edited by Venetia Porter and Liana Saif, 44–51. The British Museum, 2013.

Bonnie, Rick, and Patrik Klingborg, eds. *Water in Ancient Mediterranean Households: Global Perspectives on Ancient Mediterranean Archaeology*. Routledge, 2024.

Brown, Peter J. "The "Cup of Pharaoh" from Samarra and the Reuse of Ancient Spolia as Water Features in the Medieval Islamic World." *Journal of Islamic Archaeology* 9, no. 1 (2022): 59–82. https://doi.org/10.1558/jia.23645.

Cohen, Maayan, and Deborah Cvikel. "Ma-agan Mikhael B, Israel: A preliminary report of a Late Byzantine–Early Islamic period shipwreck." *International Journal of Nautical Archaeology* 48, no. 1 (2019): 189–207. https://doi.org/10.1111/1095-9270.12331.

al-Dayel, K., and S. al-Helwa. "Darb Zubayda Survey Project. Preliminary Report on the Second Phase of the Darb Zubayda Reconnaissance." *Atlal* 2 (1978): 51–64.

al-Dayel, K., S. al-Helwa and N. Mackenzie. "Darb Zubayda Survey Project. Preliminary Report on the Survey of Darb Zubaydah, third season." *Atlal* 3 (1979): 43–54.

Donner, Fred M. "Umayyad Efforts at Legitimization: The Umayyad's Silent Heritage." In *Umayyad Legacies. Medieval Memories from Syria to Spain*, edited by Antoine Borrut and Paul M. Cobb, 185–211. Brill, 2010.

El-Hibri, Tayeb. *The Abbasid Caliphate: A History*. Cambridge University Press, 2021.

Förster, Frank. "With Donkeys, Jars and Water bags into the Libyan Desert: The Abu Ballas Trail in the Late Old Kingdom/First Intermediate Period." *British Museum Studies in Ancient Egypt and Sudan* 7 (2007): 1–36.

Grant, Christina Phelps. *The Syrian Desert: Caravans, Travel and Explorations*. Routledge, 2010.

Harris, Ian, Timothy J. Osborn, Phil Jones, and David Lister. "Version 4 of the CRU TS monthly high-resolution gridded multivariate climate dataset." *Scientific Data* 7, no. 109 (2020): 1–18. https://doi.org/10.1038/s41597-020-0453-3.

al-Helwa, S., A. A. al-Shaikh and A. S. Murad. "Preliminary Report on the Sixth Phase of Darb Zubaydah Reconnaissance 1981 (1401)." *Atlal* 6 (1982): 39–62.

Hussein, Ali Ahmad. "The Fat Bride and the Foolish Messengers: Humorizing the Love Theme in an Early Islamic Poem." *HUMOR* 35, no. 2 (2022): 253–274. https://doi.org/10.1515/humor-2021-0060.

Knustad, J. "The Darb Zubayda Project: 1396 / 1976. Preliminary Report on the First Phase." *Atlal* 1 (1977): 41–68.

Mackenzie, N., and S. al-Helwa. "Darb Zubaida Architectural Documentation Program. Darb Zubaida – 1979: A Preliminary Report." *Atlal* 4 (1980): 37–50.

Al-Maiyali, Rajwan, and Richard Wilding. "2024 Shepard Urgent Action Grant: The Rest-stop and Minara of Umm Al-Qurun on the Darb Zubaydah." ASOR, January 06, 2025. https://www.asor.org/news/2025/01/shepard-grant-umm-al-qurun.

Morgan, C., and S. al-Helwa. "Preliminary Report on the Fifth Phase of Darb Zubayda Reconnaissance. 1400-1980." *Atlal* 5 (1981): 85–107.

Northedge, Alastair. *The Historical Topography of Samarra, [Samarra Studies 1]*. British School of Archaeology in Iraq, 2007.

Petersen, Andrew. "The Archaeology of the Syrian and Iraqi Hajj Routes." *World Archaeology* 26, no. 1 (1994): 47–56.

Power, Tim. *The Red Sea from Byzantium to the Caliphate: AD 500–1000*. The American University in Cairo Press, 2012.

Al-Rashid, Saad. *Darb Zubaydah. The Pilgrim Road from Kufa to Mecca*. Riyadh University Libraries, 1980.

Al-Rashid, Saad. *Al-Rabadhah: A Portrait of Early Islamic Civilization in Saudi Arabia*. King Saud University, 1986.

Tagliacozzo, Eric. "The Hajj by Sea." In *The Hajj. Pilgrimage in Islam*, edited by Eric Tagliacozzo and Shawkat M. Toorawa, 113–30. Cambridge University Press, 2016.

Webb, Peter. "Bedouin, Bandits, and Caliphal Disappearance: A Reappraisal of the Qaramita and Their Success in Arabia." In *The Historian of Islam at Work. Essays in Honor of Hugh N. Kennedy*, edited by Maaike van Berkel and Letizia Osti, 254–82. Brill, 2022.

Webb, Peter. "Power and Money on the Hajj: Connections between Caliphate and Pilgrimage in Early Islam." In *Land & Trade in Early Islam*, edited by Hugh Kennedy and Fanny Bessard, 100–39. Oxford University Press, 2025.

Wilkinson, Tony J. "Darb Zubaida: The Water Resources." *Atlal* 4 (1980): 51–67.

CHAPTER 5

The Infrastructure of Purity: Plumbing Advice from Hadith and Early *Fiqh**

Edmund Hayes

Abstract

This chapter focuses on hadith and early *fiqh* as evidence for how early Muslims were guided in producing infrastructures of ritual purity. It focuses on a small number of juristic statements, from the 2nd/8th and 3rd/9th centuries, attributed to the Sunni imam Aḥmad ibn Ḥanbal (d. 241/855), and the Shiʿi imams, Jaʿfar al-Ṣādiq (d. 148/765) and one of his successors (exact identity unclear). These reports focus on the spatial separation of fresh water supplies like wells from sources of pollution such as sewers, drainpipes and latrines. They are remarkable in the unusual specificity of their description of infrastructural solutions to domestic plumbing arrangements. Likewise, they diverge from most early *fiqh* and hadith in that they imply the planning of infrastructure to prevent the occurrence of instances of pollution, rather than being *ex post facto* responses to pollution that has already occurred, as is the case in the bulk of rulings on water purity. As such, it is argued that these reports provide evidence that early Islamic juristic attitudes were engaged with society, responding to questions drawn from the messy experience of domestic purity regulations, and that they contributed to the creation of infrastructures of purity that must be expected to have had impacts on the way urban and domestic spaces developed.

Keywords: wells; latrines; *fiqh*; hadith; imams; pollution; seepage; ground water; urban planning; domestic spaces

* I would like to acknowledge the inspiration provided by the rich academic environment and searching questions of Maaike van Berkel and the rest of the Source of Life team, which sparked many of the ideas (though none of the faults) present here. This publication is part of the project "Source of Life: Water Management in the Premodern Middle East (c. 7th–15th centuries)" with file number VI.C.181.016 of the research programme Vici SGW which is financed by the Dutch Research Council (NWO). This article was also written during the course of the ERC project supported by the ERC Horizon Starting Grant Project, "Embodied Imamate: Mapping the Development of the Early Shiʿi Community 700-900 CE (ImBod)", grant number 101077946. Thanks also are due to my colleague and teammate Dr Aila Santi for the diagrams.

> From Abū Hurayra: the Messenger of God (SAAS) said: "…Certainly, none of you should urinate in standing water then ablute or drink from it."[1]

This quotation appears as a proof text for practice in numerous early works of Islamic law. On first sight, it might seem to be stating the obvious: perhaps even common sense. However, the fact that it was deemed relevant to preserve this statement suggests precisely the opposite: it was necessary to enshrine this position in textual form to provide it with authority.[2] In the realm of human culture and history, one cannot assume that anything is mere common sense, or a universal human attitude to dirt in one's environment. Certainly, humans, like other animals, can detect the presence of pathogens in their environments through the senses of smell, sight and taste, but as historians of emotion have vigorously suggested in the last few decades, the visceral sensations of the body cannot be expressed and systematised but through language, and therefore through cultural systems. In the case of systems of purity and pollution, Mary Douglas famously noted that the very idea of dirt, "matter out of place," presupposed a system of categorisation of what constituted dirt or pollution and why, much of which depends on ideological and cosmological systems rooted in religious discourse and practice. She also cautioned against the attempt to draw overly distinct boundaries between modern conceptions of scientific "hygiene" and pre-modern conceptions of pollution and impurity.[3] Modern and pre-modern people alike deploy their knowledge through cultural systems. In thinking about how people in the pre-modern Middle East planned and operated their water systems, then, we need to consider the cultural and religious framework within which the meanings of water were created. In understanding the infrastructure and social systems of water management, a key consideration is not just the provision of water, but the safeguarding of water: implementing systems to keep water supplies separate from sources of potential pollution. In this chapter I interrogate the role of Islamic authorities in guiding these systems and ask whether we can trace historical processes through which Islamic attitudes led to the creation of infrastructures of purity.

[1] al-Bukhārī, *Ṣaḥīḥ* 1:94.

[2] Furthermore, ultimately, this position was not deemed acceptable at face value, inasmuch as it appeared to conflict with another Prophetic ruling that water was always pure especially if present in large quantity. Thus, al-Shāfiʿī "says that this hadith does not mean that still water is made impure by urine, but that people dislike it, just as it is prohibited to urinate on a public road, under a tree or any other place where people gather, because they are annoyed by such an act, rather than because it is prohibited to urinate on the ground. It is certain that he interprets this hadith in accordance with the principle that more than two *qulla*s of water does not become impure." Yanagihashi, "Water to Be Used for Ablution," 69.

[3] Douglas, *Purity and Danger*, 32.

Early Muslims were, of course, heirs to Graeco-Roman thinking about science and medicine, as well as to monotheistic attitudes to ritual purity, with attitudes especially comparable to Jewish purity law. So, did the Arab-Muslim conquests change the way in which people dealt with their water? On one hand, the new regime that spread across the Middle East and North Africa in the 7th and 8th centuries clearly allowed old arrangements to continue. With something as important as water provision, the old maxim "if it ain't broke don't fix it" has a particular urgency. But Muslim rulers introduced certain regimes of behaviour that became hegemonic in society. Over time, the attitudes and institutions that the Arab-Muslim elites brought with them were influential on the way things were done in their domains. This has been studied in particular in the realm of bureaucracy—a field of enterprise which is relatively easy to track as it involves education in a particular language, which was increasingly the sacred language of Arabic.[4] However, it is clear that various other technologies and cultural forms were diffused along with the Arabs.

What about water? A certain amount of ink has been spilled on the question of the role of the Arabs in the diffusion of irrigation technologies.[5] The sewer systems of al-Andalus underwent different stages of development from Roman to Visigothic to Muslim rule.[6] While such changes can be traced by archaeologists, it is hard to propose reasons for them, a rare exception being Blanke, who makes a connection between the disappearance of Roman-style multi-seater communal latrines and Jewish and Muslim (and to a lesser extent, Christian) attitudes to modesty.[7] Literary texts like historical chronicles are not usually interested in water infrastructure, and what exists must be carefully identified and analysed.[8] Geographical and medical literature contains a certain amount of discussion of water and waters.[9] In looking at distinctively Muslim contributions to infrastructure, relatively untapped sources can be found among the closely related corpora of hadith and early *fiqh*.[10]

[4] See, for example, Garosi, *Projecting a New Empire*.

[5] See the debate regarding the diffusion of agricultural techniques, crops and technology, Blyn, "The Green Revolution Revisited," 705–25.

[6] See Reklaityte, "Les latrines en al-Andalus," 59–75.

[7] Blanke, "A Farewell to *Foricae*", 177-192.

[8] For example, on the question of water infrastructure, power and investment see Kennedy, "Landholding and Law in the Early Islamic State," 159–81; Lucas, "Le pouvoir de la terre".

[9] al-Nābulusī, *The Villages of the Fayyum*.

[10] In a similar vein, but dealing with juristic and para-juristic issues surrounding latrines, defecation and excrement, see Benhima, "Usages de bienséance et règles de pureté," 13–23.

Hadith reports represent the normative statements and exemplary narratives of authoritative figures like the Prophet Muḥammad and his companions, and the Shiʿi imams. *Fiqh*, the tradition of Islamic positive law and legal thinking, includes within it and builds upon hadith; it consists of juristic thinking about what the laws of Islam are. In this chapter I will look at a small number of reports, both "*fiqh*" and "hadith", that I have encountered among the statements of 2nd/8th- and 3rd/9th-century experts on Muslim law and tradition: the Sunni imam Aḥmad ibn Ḥanbal (d. 241/855), and the Shiʿi imams, Jaʿfar al-Ṣādiq (d. 148/765) and one of his successors, either Mūsā al-Kāẓim or ʿAlī al-Hādī. While these reports purport to date from their lifetimes, they were passed down and may have undergone processes of literary transmission, so one should be cautious about dating them too precisely. However, it is clear that they belong to a phase in which juristic positions were still being hammered out and before school positions had crystallised.

The reports of the Shiʿi imams are considered as hadith in the Twelver Shiʿi tradition, while the statement of Aḥmad ibn Ḥanbal is not considered hadith by the Ḥanbalis. Nonetheless, the reports studied here do in fact effectively belong to the same type: the juristic opinion. It is deceptive to think of Ibn Ḥanbal's statement as a juristic responsum (*masʾala*) as being therefore generically distinct from the others, because in the Shiʿi tradition they are regarded as hadith. They are all comparable *fiqhī* responses, given by a religious leader (an imam) to their followers, who wrote them down and preserved them for posterity. Likewise, these reports do not represent especially sectarian understandings of Islam. Rather, they engage with broad Muslim debates arising from the contemplation of Prophetic hadith, local Muslim practice, and the text of the Quran, engaging with how these should be understood and harmonised, and my discussion is therefore couched in a broad reading of the hadith and Quranic material they respond to. It is true that they may well represent particular regional traditions and juristic logics.[11] Aḥmad ibn Ḥanbal lived in Iraq, although he travelled in pursuit of knowledge, including to Arabia, while the earlier Shiʿi imams lived in and around Medina in the Hijaz, before later imams were brought to Iraq, where many early legal specialists lived.

The regional embedding is important. In such reports we sometimes see questions arising about specific locations that had particular environmental peculiarities:

– The prison mentioned below (probably in Baghdad)
– The cisterns/basins (*hiyāḍ*) between Mecca and Medina[12]

[11] See Yanagihashi, "Water to Be Used for Ablution," for a more detailed account of the intersection between region and *madhhab* in the early legal debates on purity law.

[12] al-Kulaynī, *al-Kāfī*, 3:4.

- The well of Buḍāʿa[13]
- A basin in Majanna[14]

Although such sites may have become tropic in representing particular juristic problems, it appears that they preserve memories of particular environmental problems that intersected in interesting ways with Muslim purity law: interesting enough to ensure their preservation in a juristic context.

But can we learn really learn about infrastructural planning from such early juristic statements? Hadith and *fiqh* might be considered as somewhat abstracted from reality. Certainly, hadith and early *fiqh* do not furnish comprehensive instructions for managing water infrastructure. That is not their aim. However, in this chapter I aim to show that, in the small set of reports analysed here, at least, we do see the implication that jurists were actively sought out to provide advice on planning and executing the infrastructure of domestic purity.

There are numerous studies that deal with aspects of purity law in Islam.[15] Far fewer deal with purity infrastructure in such a way as to explicitly interrogate the relationship between built infrastructure and the rules and procedures for guaranteeing purity. A notable exception is the landmark special issue on latrines and ablution facilities edited by Patrice Cressier, Sophie Gilotte, and Marie-Odile Rousset in 2016,[16] which contains contributions dealing with both textual and archaeological evidence, as well as a few that make an attempt to bridge the gap between text and infrastructure.[17] The major problem in tracing this issue through the evidence is that our textual sources tend to be uninterested in the details of the infrastructural planning and implementation, while archaeological sources do not come with an instruction manual or a give explicit indications as to intended usage.

My contribution to this subject, then, is to focus on a small number of juristic reports that are unusually (for this genre) explicit in their treatment of the

[13] The well of Buḍāʿa was a well in Arabia, belonging to the Banū Sāʿida tribe, which became the focus of a question to the Prophet regarding abluting with its water, even though it "[was] a well into which menstrual cloth… dead dogs (*laḥm al-kilāb*) and… stinking things (*natn*) were thrown." The Prophet's response was that, "water is pure and nothing makes it impure." The ramifications of this statement and the way it was harmonised with other, apparently contradictory prophetic hadith are discussed in Yanagihashi, "Water to Be Used for Ablution," especially 68–69, 95.

[14] "[When] ʿUmar arrived at a basin in Majanna he was told that a dog had lapped from it. Majanna was located one *barīd* (about 4 km) distant from Mecca." Yanagihashi, "Water to Be Used for Ablution," 82.

[15] For an assessment of some aspects of the field see Gauvain, "Ritual Rewards," 333–93.

[16] Cressier, Gilotte, and Rousset, eds., "Lieux d'hygiène et lieux d'aisance en terre d'Islam." Special issue.

[17] See especially Reklaityte, "Les latrines en al-Andalus," 59–75; Benhima, "Usages de bienséance et règles de pureté," 13–23.

infrastructural dimensions of purity law. In particular these reports focus on the issue of the distance between clean water supplies and waste pipes and sumps. In this way, then, these reports allow us to join up the dots between theory and practice for one aspect of the development of the infrastructure of purity in early Muslim contexts.

It is something of a truism in the study of early Islamic purity regulations that ritual purity is not the same as physical hygiene. As Reinhart puts it:

> the issue here is not cleanliness in the hygienic sense but, rather, in the ritual sense. Something may be quite septic or merely dirty and at the same time from the perspective of the *ṭahārah* system be "cleansed," or even "cleansing"; if the person or thing is *ṭāhir* (i.e., having done *ṭahārah*), it is ritually valid. Similarly, persons may be squeaky clean, but if "their ablution has been negated," or they (as a result of some act) "are obligated to perform lustration," [that is *ghusl*, the major ablution] they may not ritually worship, handle a Qur'ān, or participate in the pilgrimage, among other things.[18]

Reinhart's logic here cannot be faulted, but there is something more to be said on the practical interaction between conceptions of cleanliness and ritual purity in daily life. As Cressier, Gilotte and Rousset note, ritual and hygienic function can become intertwined in practice, as in the case of latrines provided at mosques.[19] In the foundational source of Islamic law, the prophetic hadith, the divisions between ritual and mundane conceptions of cleanliness are not as clearly demarcated as the later jurists that Reinhart mainly focuses on would sometimes wish. Thus, in the quotation at the head of this chapter, the Prophet Muḥammad's statement connects ablution water with drinking water. For the purpose of mature jurisprudence, the ability to drink water and to use it for ablution are distinct, but in the pre-systematisation era represented by hadith and early jurisprudence, such categories can sometimes seem to bleed into each other. In the practice of daily life, even more so, it would be difficult to clearly separate drinking and ablution water infrastructures, and likewise to mentally conceive of them as distinct. One could well imagine that habituation to preserving water sources for the purposes of ablution would affect the way drinking supplies were thought of. As such, while Reinhart uses the more precise term "defilement" to refer to matter and acts which break ritual purity, I will throughout use the word "pollution" to refer to contexts that I see as allowing for a conflation of dangers to ritual purity and to drinking water as a human need.

18 Reinhart, "Impurity/No Danger," 6.

19 "Fonctions rituelles et hygiéniques se confondent dans les latrines publiques, dont le nombre des fidèles fréquentant les mosquées rend la construction indispensable." Cressier, Gilotte, and Rousset, Introduction to "Lieux d'hygiène et lieux d'aisance en terre d'Islam," 5–12.

Conceptions of hygiene and ritual purity, then, while not identical, must be expected to be related and to inform each other. In the context of this volume, this is relevant, because it means we can expect to find the logic of ritual purity coming to invest infrastructural forms and processes in urban planning. As such, it is legitimate to think of "Muslim cities" (if not of "*the* Islamic city" in the singular) as far as regimes of water and waste management are concerned. That is to say, paying attention to Abu Lughod's insight that, "cities are processes, not products,"[20] while there is no one template that makes a city Islamic, there are Islamic logics that might spur certain kinds of development, albeit reacting differently to local ecologies and socio-political regimes, as Benhima's work hints at in his conception of a differentiated *adab* of excrement responding to different ecological and economic landscapes.[21] While a large proportion of decisions made in planning water and waste infrastructure might have been identical between e.g. Roman, Muslim Middle Eastern, Chinese and Latin Western cities, culture and religious values must have been partially responsible for certain forms of urban development.

Before I turn to the case studies, I should outline the broad context of purity regulations that relate to the question of the preservation of the purity of a water supply—mainly wells—from pollution. Pollutants come in many forms, but the main worries expressed in hadith on purity concern the following:

- Human waste: faeces, urine, blood
- Products and waste of animals prohibited for eating:[22] dead bodies, hair etc.
- The blood of animals permitted for eating
- Products of immoral and infidel humans[23]

Meanwhile, water is itself the agent *par excellence* for removing a state of ritual impurity from a person, whether by the minor ablution (*wuḍū'*) required for certain cases, or the major ablution (*ghusl*) required by others. This means that the purity status of water must be safeguarded so as to allow Muslims to restore their own personal purity status.

The major aim in discussing these issues, within texts of *fiqh* and legal hadith, then, is to give Muslims a guide for determining personal behaviour regarding

20 See Abu-Lughod's famous critique of the orientalist rollout of this concept, "The Islamic City: Historic Myth, Islamic Essence, and Contemporary Relevance."

21 Benhima, "Usages de bienséance et règles de pureté," 13–23.

22 For example, pigs or wild predators.

23 There are other, more ambiguous cases which jurists devote much discussion to including *su'r* (water touched by mouth/saliva); water already used in ablution; and "innocent dirt" such as algae, sediment and organically generated slime. For a more detailed view of the logic behind such judgements, see Reinhart, "Impurity/No Danger."

ritual purity. It is not to provide instructions for building infrastructure. However, tangentially, in the course of discussing how to deal with these sources of pollution, these texts do cover issues that either directly mention infrastructures, or perhaps imply the use of infrastructures:

- What is the relative purity of types of water?: rainwater, well-water, rivers, seawater etc.
- What are the dangers posed to water-storage? (especially animals falling in)
- How to purify water once polluted? (e.g. by removing a certain number of bucketfuls, or all of the water)
- Water that is resistant to pollution, such as: running water; large volumes of water
- What volume of water is sufficiently large to resist pollution?
- How much movement or stasis guarantees purity?
- Spatial separation from source of pollution

I am now going to focus on reports related to the spatial separation of water supplies from sources of pollution. This topic has not, to my knowledge, been studied. The topic of separation is interesting for the question of the planning of infrastructure, because it requires forethought, rather than merely being an *ex post facto* response to a case of pollution that has already occurred, as is the case in the bulk of rulings on water purity. The few scattered reports I have gathered here are likewise rare in that they refer to quite specific infrastructural situations. However, they are embedded within the ubiquitous logic of the broader ritual purity laws of Islam. Their very scarcity and specificity make them particularly valuable: they are less likely to be catechistic rehearsals of standard juristic positions, and more likely to be specific cases that arose in fact.

Report 1: Ibn Ḥanbal on a Problem of Polluted Water from a Prison

The following report is about how to understand the dangers to one's ritual purity due to pollution of a newly-dug domestic well. The report derives from the *Masāʾil* (responsa) of the famous hadith transmitter, Aḥmad ibn Ḥanbal, in the recension of his son, ʿAbd Allāh (d. 290/903). In the report, ʿAbd Allāh asks his father Ibn Ḥanbal a question:

> I asked my father about a man who digs a well in a house. A distance of 15 armspans[24] separates the well and a ditch close to a prison, and the water comes out changed in colour. What do you think about this?
>
> [Ibn Ḥanbal] said: If there is a good smell [then that is not a problem], and if there is not a good smell, and then [proceed to check] the taste.[25]
>
> Then [Ibn Ḥanbal] said: if the change in the water is due to the dirt (*najāsa*) of the prison, then he should not touch [lit. approach] this water, he should block off the well. However, if this water, when removed, returns to a state in which there is no change in colour or smell or taste, then I hope that would be unproblematic.[26]

The specificity of this case is fascinating. Given that the question is asked directly by ʿAbd Allāh to his father, we can perhaps place this as a prison in Baghdad, where Ibn Ḥanbal and his son lived.

The infrastructural situation described in Ibn Ḥanbal's report is visible in Figure 5.1. Apparently, the prison was served by a ditch, a *khandaq*, which was close-by, probably running along the street. Water from the prison's sewage system would have flowed into this ditch in the street, presumably under the prison walls. The petitioner's well was 15 armspans (cubits) away—so around 10 metres. The petitioner detected something nasty in the water and assumed that it was due to underground seepage from the ditch. The solution was to check and see if, when removed, the water in the well replenished itself with unobjectionable water.

Given the major interests of the legal literature, the implied context for this report is whether this water can be used for ablution, but this likely extends to drinking water, as in the prophetic hadith at the top of this chapter. The typical *fiqhī* tests for water purity, as also here, are colour, taste and smell. However, the type of pollutant here is clearly important. Ibn Ḥanbal makes a point of specifically referring to a change in the water brought about by the dirt (*najāsa*) of the prison. This is important, because not all matter is created equal. Certain types of pollutant like wine, or the bodily products of certain animals like dogs, pigs and beasts of prey, are classified as more dangerous to one's state of ritual purity than others, like birds and mice, which, in their turn are still more polluting than creatures without blood like cockroaches. This is where the pollution of the prison is brought in. The mention of a prison is intriguing, giving a sense of a rather atypical environmental hazard that might be encountered in an urban environment. My assumption is that the mention of the prison is relevant because it is a source of human dirt,

[24] Or "cubit" = 65.6 cm. See Nābulusī, *Villages of the Fayyūm*, 24.

[25] Or just "smell or taste". Or does this mean that there is a not a good smell, but there *is* a good taste? In either case it does not really influence the answer.

[26] Ibn Ḥanbal, *Masāʾil*, 5.

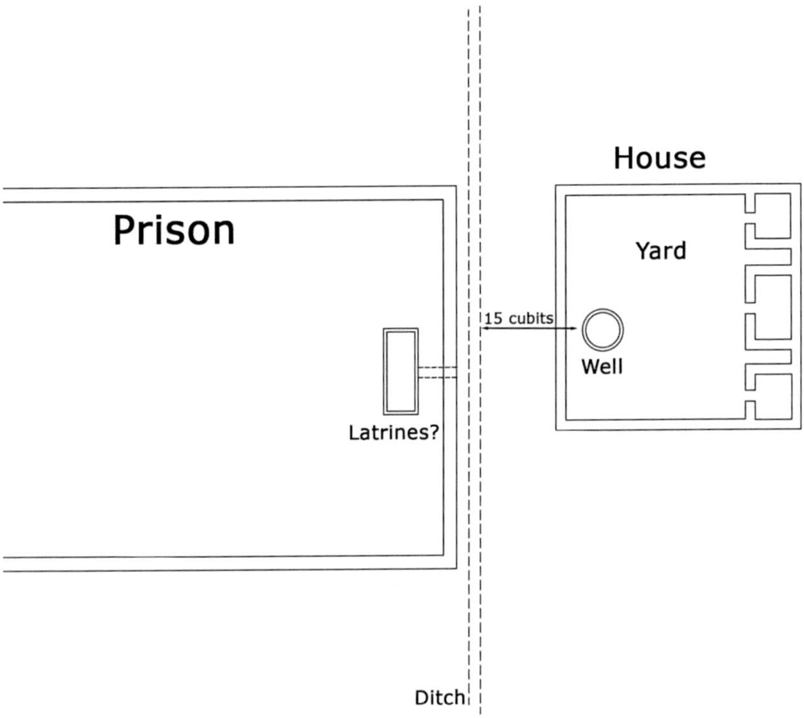

Figure 5.1: Prison ditch polluting a domestic well. Image created by Dr Aila Santi.

and furthermore, dirt from a particular kind of human: those with a moral deficiency. The moral standard of people was thought to have a direct implication for their purity status. As it says in the Quran 9:28: "Oh you who believe: the pagans (*mushrikūn*) are *najas* (dirty, unclean), therefore, do not let them come near the Masjid al-Haram after this year's pilgrimage."

For a comparative case, it is worth noting that there was a discussion about the possibility of reuse of runoff water from bathhouses. A Shiʿi hadith report indicates that such water cannot be used for ablution because it might have been used to wash non-Muslims, heretics and bastards, all of whose moral deficiencies might affect the purity of their water after using it.[27] Prison water might be expected to raise similar qualms: not only because of the transgressions of the inmates, but perhaps more importantly because it was a mixed population of non-Muslims and potentially heretical Muslims.

What kind of evidence does this report provide us? Well, we should think a little about the context of the *Masāʾil*. This work, and others like it were above all records

[27] See, for example, Hayes, "Encountering the Internal Other," 291.

of a process of studentship and scholarship. This work preserves questions posed to the scholar by his son, perhaps based on his own interest or experience, perhaps drawn from questions that were forwarded to him by others.[28] His son then took down his responses and preserved them. Some of these questions may be merely casuistry: juristic "what-ifs" that have only a tangential bearing on reality. Here, however, the case is written with very specific parameters. The general rules about impurity are not being questioned, but rather a specific case with very specific physical interactions between infrastructural entities. It seems likely that this emerged from a specific case, then, that Aḥmad's correspondent had encountered and upon which he wished to gain an authoritative opinion.

Report 2: Jaʿfar al-Ṣādiq on a Well That Is next to a Sewer Pipe

In the next case we turn from a Sunni collection to a Shiʿi collection, but a case that is nonetheless highly comparable. In it, a Kufan follower[29] of the Shiʿi imam, Jaʿfar al-Ṣādiq, who lived in Medina, asked the imam a question. Given these geographical details, it is likely that the setting of the meeting is Medina. However, it is very possible that the questioners bring with them a Kufan understanding of their built environment when they ask their imam about a plumbing issue:

> Al-Ḥasan ibn Ribāṭ from Abū ʿAbd Allaḥ [Jaʿfar al-Ṣādiq], he said: I asked about a drain (*bālūʿa*), can it be situated above a well?
>
> [Jaʿfar al-Ṣādiq] said: if it were above the well, then [there should be a separation of] 7 armspans, and if it were below the well, then 5 armspans on each side. And that is plenty.[30]

This report is illustrated in Figure 5.2. Purity law demands that dirty water from washing, urination and defecation, and other pollutants are kept apart from a water source. Here there is no specification of the usage of the water: ablution, drinking or something else, though in the context of *fiqhī* discussions, we can usually assume it is ablution unless specified. This hadith is again somewhat unusual in that it does not just give a categorical conceptual definition of what is pure or impure, nor gives an ex post facto solution, but rather suggests preventative measures for designing infrastructure so as to avoid impurity.

[28] In another of the *Masāʾil* written down by a student of Aḥmad ibn Ḥanbal, the aspect of correspondence at a distance is more pronounced. See, Al Sarhan, "The Responsa of Aḥmad Ibn Ḥanbal," 15.

[29] al-Najāshī, *Rijāl*, 46.

[30] al-Kulaynī, *al-Kāfī*, 3:7.

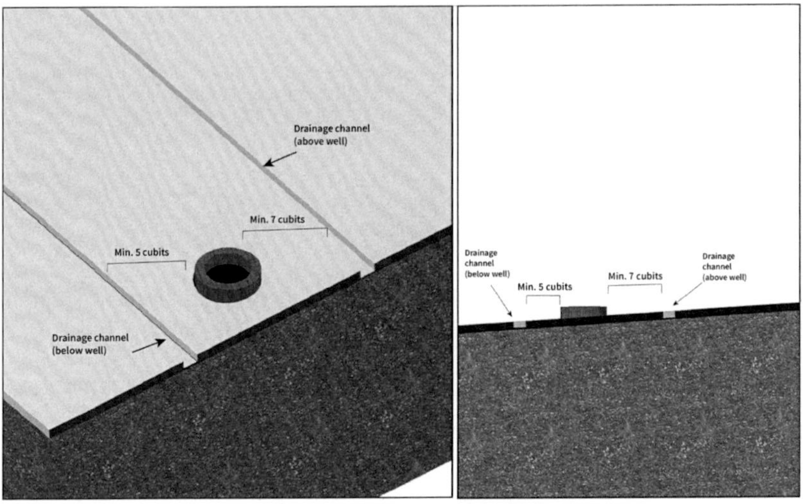

Figure 5.2: Recommended separation of a well from a sewer pipe. Image created by Dr Aila Santi.

The idea of separating supplies of water implies seepage through the ground. There is no specific logic given for the distances suggested. The mention of specific distances between well and gutter recalls the question to the report from Ibn Ḥanbal above. In the Ibn Ḥanbal case the distance between pollutant and water source was greater: 15 armspans (cubits) separation, and even then the water appeared to be polluted due to ground seepage. For Jaʿfar al-Ṣādiq, 7 armspans seems plenty.

Report 3: Jaʿfar al-Ṣādiq on the Dangers of Urine Flowing along Sloping Ground

The following report comes from the same chapter as the previous report. Again, the questioners, this time three of them, are based in Kufa. The imam is unnamed, but it is almost certainly again Jaʿfar al-Ṣādiq who is intended, for the men who are positioned as reporters of the story were followers of Jaʿfar,[31] and the compiler of these hadiths assumed that the readers would recognise that the second report was addressed to the same imam as the former:

> [A number of transmitters:] We said to [Abū ʿAbd Allāh Jaʿfar al-Ṣādiq]: a well which is abluted from, and the urine flows close to it, does it make it impure (*najis*)?

[31] Also his father, Muḥammad al-Bāqir.

[The transmitter] said: He said: If the well were at a high point in the *wādī*, and the urine flowed into the *wādī* from beneath it, and there were 3 or 4 armspans [of separation], then that would not impurify anything. But if it was less [separation] than that, it would impurify it. And if the well were at the bottom of the *wādī*, and the water passed over it, and there were 9 armspans between it and the well, then it would not impurify it, and whatever is less than that, then do not ablute from it.

Zurāra said: and I said to him: and what if the channel (*majrā*) of the urine is close to [the well], but it is not attached to the ground?

He replied: whatever does not have a base then there is no problem with that, and if a small part of [the conduit] rests on but does not pierce the ground, and it does not have depth (*lā qaʿr lahu*) until it reaches the well and there is no problem [i.e. of colour, smell or taste] with the well, then one can ablute with it. That [problem] is only if it all settles (*istanqaʿa*) [on the ground].[32]

This anecdote differs slightly from report 2, although the authority appears to be the same. In this version the variability is not from 5 to 7 armspans, but from 3 or 4 up to 9 armspans. As we see in Figure 5.3, the greater distance of 9 armspans is reserved for a well at the bottom of the valley. The logic is not made explicit, but reading in conjunction with the second half of the report, it seems that a well at the bottom of a valley is more liable to be surrounded by water that gathers and settles, allowing for seepage through the ground, rather than the urine being carried safely away past the well.

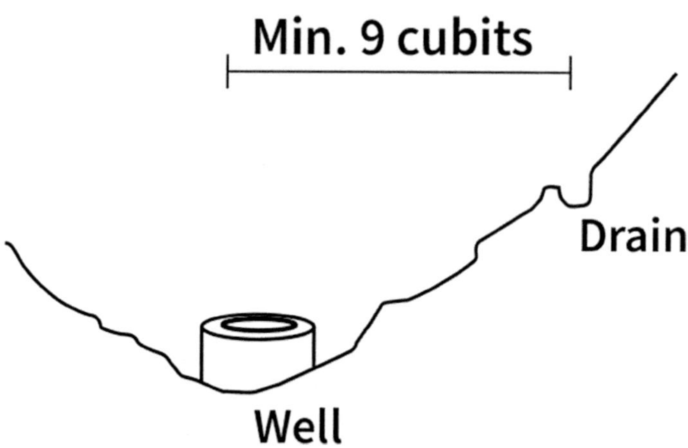

Figure 5.3: Urine flowing past a well at the bottom of a valley. Image created by Dr Aila Santi.

[32] al-Kulaynī, *al-Kāfī*, 3:8.

Yet more interesting is the second issue treated here: the penetration of the guttering into the ground, as illustrated in Figure 5.4. It appears that the spread of pollution is worse if the guttering has *qaʿr*: i.e. depth. The reason suggested is that the dirty water is likely to gather and settle (*istanqaʿa*), impregnating the ground with foul water that can seep to the well.

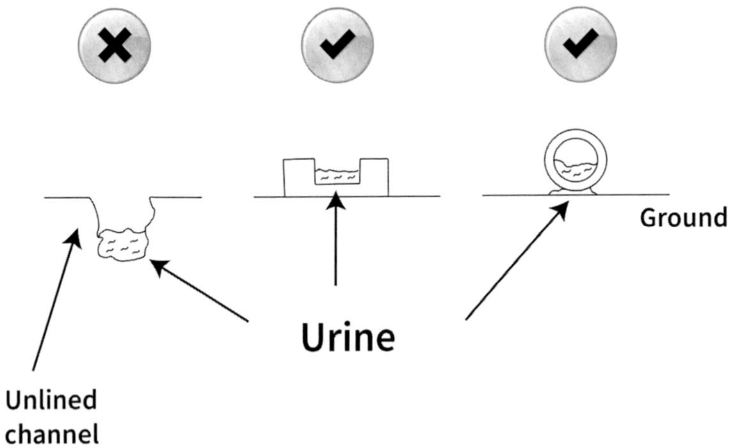

Figure 5.4: Suitable and unsuitable channels for carrying urine. Image created by Dr Aila Santi.

The wording of the report suggests that this is not based on a single concrete case brought to the imam, but rather is a more general discussion in which the imam's follower attempts to extract principles for a number of related situations, which do appear, however, to relate to lived experience, rather than from mere intellectual curiosity.

Report 4: Jaʿfar al-Ṣādiq on Water Pollution on Flat Ground

In the same chapter, we may read a further report from Jaʿfar al-Ṣādiq:

> From one of our brethren (*aṣḥāb*) from Abū ʿAbd Allāh [Jaʿfar al-Ṣādiq] (AS): he said:
>
> I asked him what is the smallest distance that can be between the well (a well of water) and a sewer pipe? And he said: if it is flat [ground] (*sahl*), then 7 armspans, and if it is a hill, then 5 armspans.
>
> Then he said: the water runs in the direction of the *qibla* to the right of the *qibla*, and flows from the right of the *qibla* to the left of the *qibla* and flows from the left of the *qibla* to

> the right of the *qibla*, but never flows in reverse, away from the *qibla* (*lā yajrī min al-qibla ilā dubr al-qibla*).[33]

Here, the imam's calculation is somewhat different. Rather than a hill being a danger to a water supply, because the wastewater can run downhill into the water supply, we see that flat ground is considered worse and requires a greater difference between the installations: only 5 armspans for sloping ground, but 7 armspans for flat ground. This is presumably due to the greater potential for pooling and groundwater seepage on flat ground, and can therefore be seen to parallel the question to Ibn Ḥanbal in the first report.

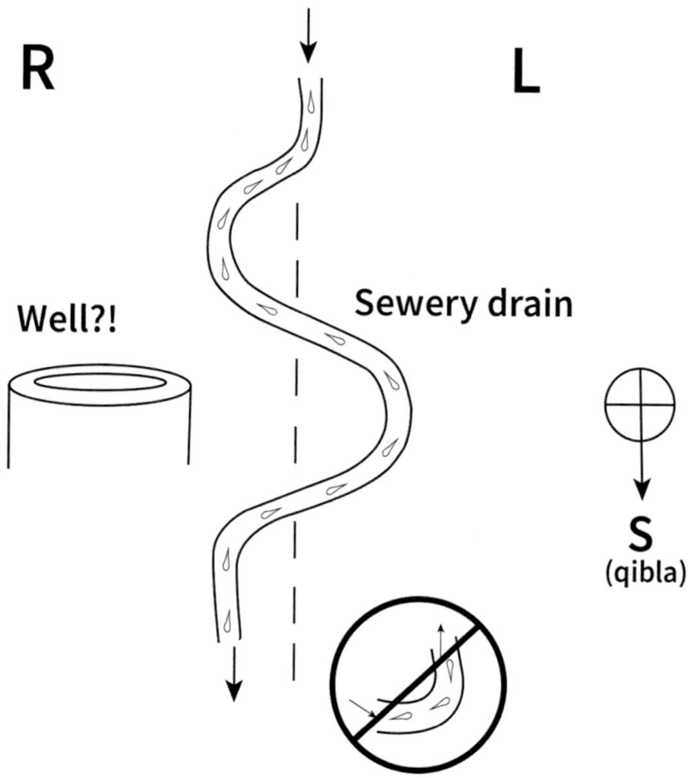

Figure 5.5: A sewer pipe flowing towards the *qibla*. Image created by Dr Aila Santi

[33] al-Kulaynī, 3:8.

The final specification of the direction of flow of the fluid is mysterious, and the reconstruction of the context in Figure 5.5 is speculative, but it seems to indicate that the dirty water flows roughly in the direction of the *qibla*, but meanders to the left and the right of the main direction of travel. Even the voice of the speaker is ambiguous: perhaps it is the questioner giving further details about the topography of the site, or perhaps it is the imam who is giving us a general principle for understanding fluid flows. It is unclear how this relates to what has preceded, except that it might indicate that the ground is somewhat flat, therefore creating a meander. It seems likely that this report was excerpted from a longer report in which further detail was discussed, but which is now lost.

Report 5: Abū al-Ḥasan on Abluting from a Well near Latrines

The next report is attributed to one of the imams, named here Abū al-Ḥasan. This could be one of two men: Mūsā al-Kāẓim (d. 183/799) or ʿAlī al-Hādī (d. 254/868). Both began their lives in Medina, but were brought to Iraq (Baghdad and Samarra, respectively) by Abbasids who were constantly aware of the potential political threat that imams from the holy lineage of ʿAlī and Fāṭima potentially posed to the caliphs. Here we have a short question and answer regarding the case of a well near a latrine:

> From Abū al-Ḥasan [Kāẓim or Hādī] (AS), regarding a well with 5 armspans, or less than that, or more, between it and a latrine. Should one ablute from it? He said: it is not reprehensible to ablute from it or do the major ablution (*ghusl*) from it due to proximity or distance, if the water has not changed.[34]

This is very comparable to what we have seen. There is less contextual detail, which makes it appear a little more abstracted from practical reality. The wordings of both question and answer are generalised, not dealing with one specific distance, but also longer and shorter distances, as if the questioner wanted to get a fuller juristic picture. Again, the central focus is on the change of state, though we are not told of colour or smell: apparently this knowledge is already assumed. In contrast to the above cases then, this may be a casuistic variation on a theme rather than a clear case drawn from experience.

[34] al-Kulaynī, 3:8.

Conclusions

So, what does this little collection of oddly specific cases show us? I think we can draw a few conclusions. These reports are remarkable in the unusual specificity of their description of infrastructural solutions to domestic plumbing arrangements, and also in that they imply the planning of infrastructure to prevent the occurrence of instances of pollution, rather than being *ex post facto* responses to pollution that has already occurred, as is the case in the bulk of rulings on water purity. As such, they suggest that Islamic attitudes to purity generated infrastructures of purity that were designed to prevent and treat relevant issues. This is perhaps not a shocking discovery, but it is an issue that has not really been carefully traced or historicised. This evidence clearly indicates that some people were taking Islamic legal thinking as a guide, not just to rulings on cases of ritual purity, but also for planning, building and maintaining infrastructure. It is not clear how the behaviour of ordinary Muslims might have related to the rigorous expectation of the piety-minded scholars who managed hadith and *fiqh*. And yet in certain ways they have important implications for the ways in which water infrastructure was managed, or at least the ways in which it should have been managed, according to religious thinkers. As guardians of moral behaviour in Muslim society, piety-minded scholars enshrined instructions for behaviour regarding cleanliness and ritual purity in their books of hadith and *fiqh*, and a key subset of these instructions dealt with the nature and effects of water as a purifying agent. Many of these instructions were theoretically binding upon Muslims, or at least highly recommended. The reports gathered here suggest that, at least in planning the domestic infrastructure of purity, some Muslims were thinking in ritual-legal terms to address practical problems.

It is very hard to generalise from these cases or to measure the overall impact of these ways of thinking on urban planning at large, but I think we cannot discount such logic. In thinking about water management, the use cases of drinking and for ablution were considered as distinct but connected, as can be seen in my first quotation from the Prophet, which links ablution and drinking in giving advice to avoid standing water with urine in. When we consider water management in Muslim majority spaces, then we should expect that issues relating to drinking and ablution might have been treated together, not only in terms of intellectual speculation, but also in dealing with on-the-ground problems. This would have become increasingly relevant in Muslim settlements, as Islamic institutions inserted themselves into decision-making for urban planning via institutions like the local mosque and the *waqf*. We might think of ritual purity laws, then, as a kind of proxy for guidelines as to potability, at least within a particular intellectual-social milieu which took hadith and *fiqh* seriously as a guide for appropriate praxis. As an example of the way in which we can see water supply and religion as interacting in concrete locations, we might think of the huge vaulted cisterns beneath the White Mosque in Ramla, which, with such a volume of water stored, appear to have been designed to supply a whole segment of the city, and were probably supplied by the aqueduct built by the Umayyads to supply the city.[35]

The specific cases that I have looked at here focus on the idea of water pollution from underground seepage, showing that this was a recognized issue, and the distancing of water installations was a key strategy to avoid it. This is perhaps not a surprising finding: we could have guessed that this would be the case, but it is interesting to see it articulated explicitly, not in a manual for engineers but in a compilation of authoritative statements on Islamic law, and to compare slightly different cases of the same issue. The location of water channels dug into or resting above the ground was also seen as having implications for groundwater pollution, and might be expected therefore to have generated practical strategies for avoiding this kind of seepage.

Can the reports studied here be trusted as providing truly historical information about actual practice? Perhaps it is just casuistic speculation? It is right to doubt. However, there is a certain level of granular detail in these reports which is both highly unusual, and somewhat encouraging. While the last of the reports quoted here is indeed short and context-poor, like many *fiqhī* hadith in our sources, several of these reports give a greater sense of the contexts in which they might have arisen. In particular, we have hints here and there that very specific cases are being brought from specific locations, often in the Hijaz, where the earlier Shiʿi imams lived, or in Iraq, where many early legal specialists lived.

The *ad hoc* nature of such examples, focused on problems arising from specific environmental situations, is also evident in the way in which different reports had to be ironed out and harmonised in later *fiqhī* discussions. Ultimately, systematising minds among the legal scholars wished to subordinate all concern to some basic principles: that water is intrinsically pure, as suggested in the Quran;[36] that water that is in large quantities or flowing is immune to pollution; that the only case in which these principles are to be set aside is if the water changes state, as indicated by a change in colour, smell or taste. However, these broad principles, while present in early hadith and juristic opinions, do not appear in these early sources in a systematic hierarchy. Yanagihashi shows how attempts were made to harmonise different early reports regarding ablution water. However, these attempts could not fully obscure the intrinsic variation over the details of the earliest reports.[37]

[35] Rosen-Ayalon, "The White Mosque of Ramla," 67–83.

[36] Q 25:48.

[37] See, for example Yanagihashi's analysis of al-Shāfiʿī's attempt to rationalise a hadith report in which it is suggested that the Zamzam well is made impure by a dead body, in spite of its large volume. Yanagihashi, "Water to Be Used for Ablution," 69.

What we have in the material presented, is a set of rules of thumb that vary because of the lack of systematization in the early period. In general, then, I would suggest that we can indeed treat such sources as evidence for the social and environmental history of the early Islamic world, though with caution. We know that these reports are circulated and written down by people largely residing in Iraq, though they may well originate with earlier authorities, and in other locations, in particular the Hijaz. This means that we need to preserve a sense of the ambiguity of their location in mind as we read. In the context of this volume, it is important to note that purity regulations emerging from religious discourses did take into account in a real way the interaction with and planning of infrastructural solutions, in a manner that can be expected to have had influence among those pious Muslims who cared about these issues. This might have been a relatively small number of spiritual athletes in the $2^{nd}/8^{th}$ and $3^{rd}/9^{th}$ centuries, when these narratives purport to date to, but increasingly such concerns may have filtered out into broader society, resulting in, for example, architectural complexes that included bathhouses next to mosques and cemeteries to allow for ablution or the washing of corpses. While it is harder to track archaeologically, these ideas may also, as these reports suggest, have influenced the way people planned their domestic spaces and managed their daily practice. What we see in all of these accounts is that a broad purity-mindedness among early Muslims did result in an interest in managing infrastructure so as to ensure purity in Muslim built environments.

Bibliography

Primary sources

al-Bukhārī, Muḥammad ibn Ismāʿīl. *Ṣaḥīḥ al-Bukhārī*. Edited by Muṣṭafā Dīb Bughā. 9 vols. Dār Ibn Kathīr, 1414.

Ibn Ḥanbal, Aḥmad. *Masāʾil Aḥmad ibn Ḥanbal: Riwāyat ibnihi ʿAbd Allāh*. Edited by Zuhayr al-Shāwīsh. Dār al-Mawadda, 1429.

al-Kulaynī, Muḥammad ibn Yaʿqūb. *Al-Kāfī*. Dār al-Kutub al-Islāmiyya, 1388.

al-Nābulusī, ʿUthmān ibn Ibrāhīm. *The Villages of the Fayyum: A Thirteenth-Century Register of Rural, Islamic Egypt*. Edited by Yossef Rapoport and Ido Shahar. Medieval Countryside. Brepols, 2018.

al-Najāshī, Abū al-ʿAbbās Aḥmad ibn ʿAlī ibn Aḥmad ibn al-ʿAbbās al-Asadī al-Kūfī. *Rijāl*. Edited by Mūsā al-Shubayrī Zanjānī. Muʾassasat al-Nashr al-Islāmī, 1407.

Secondary sources

Abu-Lughod, Janet L. "The Islamic City: Historic Myth, Islamic Essence, and Contemporary Relevance." *International Journal of Middle East Studies* 19, no. 2 (1987), 155–76.

Al Sarhan, Saud. "The Responsa of Aḥmad Ibn Ḥanbal and the Formation of Ḥanbalism." *Islamic Law and Society* 22, no. 1–2 (February 20, 2015): 1–44. https://doi.org/10.1163/15685195-02212p01.

Benhima, Yassir. "Usages de bienséance et règles de pureté dans la littérature juridique de l'Occident musulman médiéval." Médiévales 70 (2016), 13–23.

Blanke, Louise. "A Farewell to *Foricae*: Changing Attitudes to Public Latrines in the Late Antique Near East and Asia Minor." *Antiquité Tardive* 28, no. 1 (2020): 177–92. https://doi.org/10.1484/J.AT.5.122362.

Blyn, George. "The Green Revolution Revisited." *Economic Development and Cultural Change* 31, no. 4 (1983): 705–25.

Cressier, Patrice, Sophie Gilotte, and Marie-Odile Rousset, eds. "Lieux d'hygiène et lieux d'aisance en terre d'Islam (VIIe-XVe siècle)." Special issue, *Médiévales* 70 (2016).

Douglas, Mary. *Purity and Danger. An Analysis of Concepts of Pollution and Taboo*. Routledge and Kegan Paul, 1966.

Garosi, Eugenio. *Projecting a New Empire: Formats, Social Meaning, and Mediality of Imperial Arabic in the Umayyad and Early Abbasid Periods*. De Gruyter, 2022. https://doi.org/10.1515/9783110740820.

Gauvain, Richard. "Ritual Rewards: A Consideration of Three Recent Approaches to Sunni Purity Law." *Islamic Law and Society* 12, no. 3 (2005): 333–93.

Hayes, Edmund. "Encountering the Internal Other: Non-Shiʿi Family Members among the Imami Shiʿa." *Medieval Encounters* 30, nos 2–3 (2024), 280–301. https://doi.org/10.1163/15700674-12340188.

Kennedy, Hugh. "Landholding and Law in the Early Islamic State." In *Diverging Paths?: The Shapes of Power and Institutions in Medieval Christendom and Islam*, edited by John Hudson and Ana Rodríguez, 159–81. Brill, 2014.

Lucas, Noémie. "Le pouvoir de la terre: essai d'histoire de la société bas-iraqienne au IIe/VIIIe siècle." PhD diss., Université Paris 1 Panthéon-Sorbonne, 2020.

Reinhart, A. Kevin. "Impurity/No Danger." *History of Religions* 30, no. 1 (1990): 1–24.

Reklaityte, Ieva. "Les latrines en al-Andalus: leurs principales caractéristiques et les conditions sanitaires urbaines." *Médiévales* 70 (2016): 59–75.

Rosen-Ayalon, Myriam. "The White Mosque of Ramla: Retracing Its History." *Israel Exploration Journal* 56, no. 1 (2006): 67–83.

Yanagihashi, Hiroyuki. "Water to Be Used for Ablution." In *Studies in Legal Hadith*, 61–124. Brill, 2019. https://doi.org/10.1163/9789004391093_004.

CHAPTER 6

Baghdad's System of Water Provision: Infrastructure, Maintenance, and Decline in the City's First Two Centuries*

Josephine van den Bent

Abstract

In the first two centuries after its foundation, the city of Baghdad relied on different systems of water management to provide water to its population of roughly half a million people. This paper first gives a general overview of the system of water provision, showing that it was a hybrid one. It then turns to normative ideas on maintenance responsibilities for different parts of this system, and finally analyses a number of historical examples of this maintenance, in order to analyse these processes in practice. Through this analysis, this paper shows that both forces from "above" (the authorities) and from "below" (local inhabitants) played significant roles in water provision as well as maintenance, and that the hybridity of the water system as well as the efforts "from below" made Baghdad's water provision resilient in times of crisis.

Keywords: Baghdad; Iraq; Abbasid caliphate; water provision; *ḥisba* manuals; maintenance

The city of Baghdad was founded in 145/762 by the Abbasid caliph al-Manṣūr (r. 136–58/754–75) as the capital of the newly-formed Abbasid caliphate. Shortly after its foundation it rapidly expanded into a vast metropolis. Although historians have struggled to calculate the exact size of the city, which was also in flux throughout the centuries as the city grew and declined following the vicissitudes of history, Abbasid Baghdad in its 3rd/9th- and 4th/10th-century heyday, according to modern calculations, may well have been home to between 280,000 and some half a million people.[1] This paper focuses on the period from the foundation of the city in 145/762 until the Buyid turn, i.e., roughly the first two centuries of Baghdad's existence.

* This publication is part of the project "Source of Life: Water Management in the Premodern Middle East (c. 7th–15th centuries)" with file number VI.C.181.016 of the research programme Vici SGW, which is financed by the Dutch Research Council (NWO).

[1] Lassner, *Topography*, 160; Micheau, "Baghdad in the Abbasid Era," 232–35; Kennedy, "The Feeding," 177.

The foundation of a new capital and its ensuing rapid expansion makes for an interesting case when it comes to investigating what happens when a large city is built, what is needed in terms of water supply, and how that supply operates. Baghdad had significant water needs, both for provision and drainage, and it relied on different systems of water management to fulfil these. This paper will first give a general overview of the organisation of the system of water provision in Baghdad, showing that it was in fact a hybrid one. Moreover, all of its parts were reliant on some sort of maintenance and management. Both in the construction and in the maintenance of these systems, forces from "above"—the authorities—and "below"—local citizens—played a role.

Due to the nature of the sources, the dynamics of the forces from above are more accessible to us. Historians under the Abbasid caliphate were primarily interested in the lives and deeds of elite individuals, and much less in the development of institutions and affairs of everyday life.[2] Nor do we have administrative records for the early Abbasid period that could shed light on daily matters regarding water management and the everyday involvement of Baghdadis therein. The role played by "forces from below" is therefore not always easily visible. Nevertheless, a careful reading of a wide variety of sources can still provide significant insights into the way different actors and stakeholders, both from "above" and "below", were involved in Baghdad's water provision. After a general overview of Baghdad's water system, I will therefore combine information from a wide range of sources—including *ḥisba* and *kharāj* manuals, histories, and geographies—in order to analyse normative ideas about this maintenance: who is responsible for what, who pays, and how is that to be organised? However, everyday reality tends to be more unruly than the propagated norm. I will therefore also look at evidence on how maintenance took place in practice, as well as at responses in times of crises when the system was under threat. Although the focus is on maintenance of water provision systems, information on these is scarce. I will therefore also take into consideration maintenance of some water-adjacent infrastructure, such as bridges, as developments in the two areas may have been comparable. Taken together, this shows that the Baghdadi water infrastructure was hybrid in nature, both in its sources of water provision and in the way these were managed, and that it was that hybridity that made the city's water provision resilient in times of crisis.

2 See also Kennedy, *Armies*, xi–xii; Kennedy, *The Early Abbasid Caliphate*, 214.

The Infrastructure of Water Provision in Baghdad

The system of water provision in Baghdad in the first two centuries after its foundation was a diverse one. Not only was water available from the Tigris, but canals and aqueducts constructed by the authorities were used alongside cisterns and wells, and at least some water infrastructure was built through charitable initiative. The services of water carriers also played a key role. The city's founder, caliph al-Manṣūr, ordered the construction of the famous Round City on the west bank of the Tigris, in a location where various existing pre-Islamic canals were found—a key reason why this site was so attractive.[3] He immediately gave orders to dig wells and to construct a *qanāt* from the Karkhāyā Canal, an apparently pre-existing stream to the west of the city that was derived from the Nahr ʿĪsā,[4] so that the water could be used for drinking as well as building activities.[5] At a later stage, he also ordered the construction of another water conduit, this time built out of teak and filled from the Tigris by a waterwheel, which ran directly to his palace. It does not appear that this conduit or the *qanāt* were functional for very long, however, and more importantly, they do not appear to have played a key role in the general provision of water to the city.[6]

Quite the opposite was true for the city's canals. The west of the city was watered by numerous waterways running east from the Euphrates, lying at a higher elevation, to the Tigris. In his description of Baghdad, the 3rd/9th-century author al-Yaʿqūbī describes the importance of the canals:

> For the people of al-Karkh and its surroundings, [water is drawn from] a canal called the Nahr Dajāj […] and from a canal called Nahr Ṭābiq ibn al-Ṣamīh, and they have the great Nahr ʿĪsā which takes water from the main part of the Euphrates […]. They also have cisterns (*ābār*) into which water from these canals enters, and this is sweet water. All the people drink from them and these canals were needed because of the large size and extent of the town.[7]

[3] E.g., al-Ṭabarī, *Taʾrīkh*, 3.1: 272–75; al-Yaʿqūbī, *Kitāb al-buldān*, 234. See also, e.g., Kennedy, "The Feeding," 189–90; Scheiner and Toral, "Baghdād's Topography," 1–3.

[4] On the pre-Islamic credentials of the Nahr ʿĪsā, see Verkinderen, *Waterways*, 29–32, 292–93.

[5] al-Yaʿqūbī, *Kitāb al-buldān*, 238.

[6] Ibn al-Faqīh, *Kitāb al-buldān*, 289; al-Khaṭīb al-Baghdādī, *Taʾrīkh Madīnat al-Salām*, 1:388. See also Van den Bent and Brown, "Constructing," 128–35.

[7] al-Yaʿqūbī, *Kitāb al-buldān*, 250.

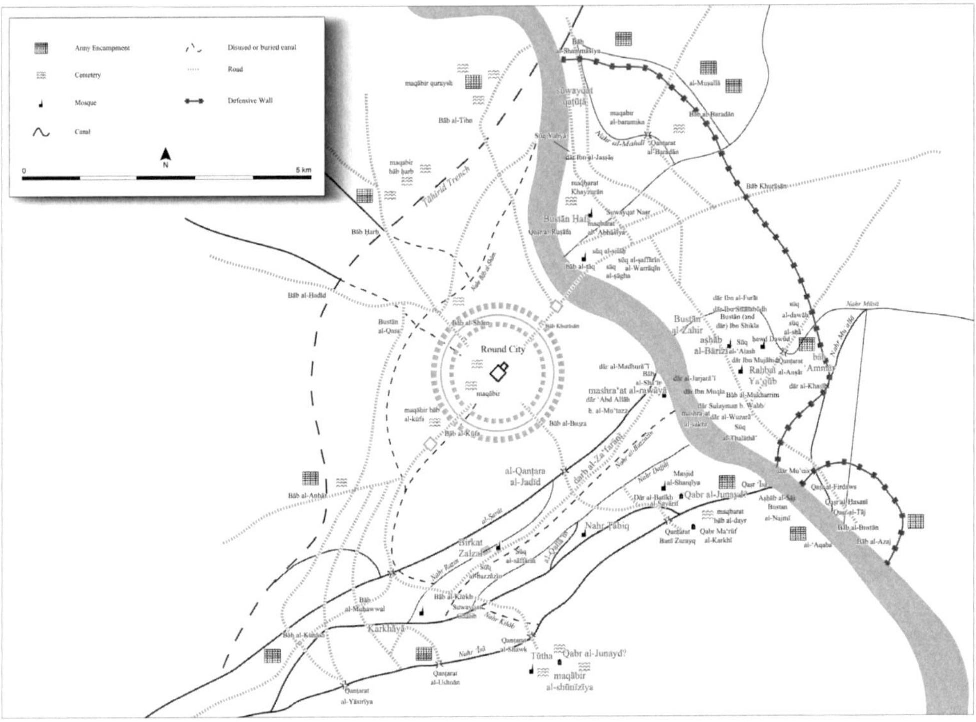

Figure 6.1: Map of Baghdad during the reign of al-Muqtadir (r. 295–320/908–32), based on descriptive sources. This map was made by Judith Ahola and Letizia Osti, and was originally published in Van Berkel et al., *Crisis and Continuity*. Canals that Ahola and Osti definitely identified as active waterways are indicated by a continuous line, while those about which they were in doubt are indicated by a dotted line. Reproduced with permission from Brill.

In his geographical treatise, Suhrāb (4th/10th century) similarly states that the canals flowing through the city provide drinking water for those living close to them.[8] Besides the canals named above, significant canals Suhrāb and other authors mention for the western part of the city also include the Ṣārāt and the Nahr Razīn to the south of the Round City, and the Khandaq Ṭāhir (Trench of Ṭāhir). For the east bank, to where the city rapidly expanded, important canals included the Nahr Mūsā, the Nahr al-Mahdī, and Nahr al-Muʿallā (Fig. 6.1).[9]

[8] Suhrāb, *Kitāb ʿajāʾib*, 129. Suhrāb appears to have written his text during the caliphate of al-Muqtadir (r. 295–320/908–32). Lassner, *Topography*, 29.

[9] E.g., Suhrāb, *Kitāb ʿajāʾib*, 129–34; al-Yaʿqūbī, *Kitāb al-buldān*, 243–51. On these canals, see also the elaborate notes in Le Strange, "Description of Mesopotamia and Baghdād," as well as Van Berkel et al., *Crisis and Continuity*, 223–26.

Both from these canals and from the Tigris—also an important water source, of course—water could be drawn using waterwheels (Ar: *dūlāb*/pl. *dawālīb*), such as in the case of the teak conduit to the caliphal palace in the Round City. They are not commonly mentioned in the sources, but there are some instances. The caliphal Tāj Palace on the east side had waterwheels,[10] but they were also present elsewhere in the city. At least until 251/865, when Turkish troops set fire to the area, there was a "Waterwheel Place" (*mawḍiʿ al-dawālīb*) close to the Bāb al-Ḥadīd in the west of the city;[11] and an area known as Dūlāb Mubārak existed on the east side.[12] They were also found in individuals' gardens.[13]

Regular canals were not the only way in which water was conveyed into the city. Suhrāb also describes the neighbourhood of Ḥarbiyya, to the north of the Round City. He states that "the canals that are in Ḥarbiyya are *quniyy* (sg. *qanāt*) [running] under the ground,"[14] and he mentions a number of small aqueducts that enabled one water course to cross another. Suhrāb describes three such crossings over the Khandaq Ṭāhir, with water from the Nahr Baṭāṭiyā. Two of these he describes as running through something called a *qūraj*, associated with a bridge: "it passes by the aqueduct of the *qūraj* of the Bāb al-Anbār Bridge (*fa-yamurru ʿalā ʿabbārat qūraj qanṭarat Bāb al-Anbār*)" and "it passes by the aqueduct of the *qūraj* of the Bāb Ḥarb Bridge (*wa-yamurru ʿalā ʿabbārat qūraj ʿalā qanṭarat Bāb Ḥarb*)," respectively. The middle crossing seems to not have been associated with a masonry bridge, and is said to have been called ʿAbbārat al-Kūkh.[15] The exact meaning of the term *qūraj* remains unclear, although the implication is clearly that of an aqueduct of sorts.[16] As Le Strange also points out, where Suhrāb uses the term *qūraj*, the later author al-Khaṭīb al-Baghdādī (392–463/1002–71) uses only *ʿabbāra*.[17] Given the fact that

10 Miskawayh, *Tajārib*, 1:196.

11 Miskawayh, *Tajārib*, 4:180. On the conflict between caliph al-Mustaʿīn (r. 248–52/862–66) and the Turkish troops, see for instance Waines, "Internal Crisis," 299–301.

12 The hadith transmitter Abū Bakr ibn Shayba (d. 331/941-2) lived there (al-Khaṭīb al-Baghdādī, *Taʾrīkh Madīnat al-Salām*, 2:248).

13 The hadith transmitter Abū Muslim Ibrahīm ibn ʿAbd Allāh al-Kajjī (d. 292/904) had one that appears to have been operated by hand (al-Khaṭīb al-Baghdādī, *Taʾrīkh Madīnat al-Salām*, 7:37). The author al-Ṣūlī (d. 335/947) described how waterwheels in his garden were destroyed by the Daylamites in 330/941 (*Akhbār*, 218). On this event, see Osti, "Remuneration," 97–98. For more detail on waterwheels in Baghdad, see also Van Renterghem's contribution to this volume.

14 Suhrāb, *Kitāb ʿajāʾib*, 134, and following him al-Khaṭīb al-Baghdādī, *Taʾrīkh Madīnat al-Salām*, 1:435. On this reliance—al-Khaṭīb al-Baghdādī derived this material from one ʿAbdallāh ibn Muḥammad ibn ʿAlī al-Baghdādī—see also Lassner, *Topography*, 33–34.

15 Suhrāb, 134, and following him al-Khaṭīb al-Baghdādī, *Taʾrīkh Madīnat al-Salām*, 1:434. See also Le Strange, "Description of Mesopotamia and Baghdād," 295.

16 See also De Goeje, *Glossarium*, CDXXXVI-CDXXXVII; Verkinderen, *Waterways*, 190.

17 Le Strange, "Description of Mesopotamia and Baghdād," 295.

every *qūraj* mentioned by Suhrāb appears to be associated with a masonry bridge (*qanṭara*), while ʿAbbārat al-Kūkh does not appear to be associated with a bridge and is not called a *qūraj*, it is possible that this indicates a difference in terminology that had been lost by the time of al-Khatīb al-Baghdādī, who otherwise follows Suhrāb's text quite closely in his phrasing here. Suhrāb also reports a similar construction at the Old Bridge (*al-Qanṭara al-ʿAtīqa*) across the Ṣarāt to the south of the Round City, where another *qūraj* is said to have carried water from the Nahr Razīn to the Kufa Gate and to some remains of the Round City of al-Manṣūr.[18]

Unfortunately, for many of these larger hydraulic systems—canals, underground *qanāt*s, aqueducts, etc.—it is unclear (or the evidence is very thin) who took the initiative to construct them. There are some exceptions, such as in the case of the Nahr al-Mahdī, dug on the orders of the caliph al-Mahdī (r. 158–69/775–85).[19] Nevertheless, as the individual instances mentioned above show, major infrastructure—especially that built in the context of the city's initial foundation—seems to generally have been initiated and financed by the authorities.

There are, however, a number of problems with the descriptions of these canals and other conduits. As there is so little archaeological evidence for Abbasid and Buyid Baghdad, historians are largely dependent on the descriptions in these geographical (and other) works. Although these are quite rich, they are not without their challenges and often present the city as it was laid out in its foundational period rather than as it was during their own time, describing it in a rather idealised manner.[20] For the reign of al-Muqtadir (r. 295–320/908–32), Judith Ahola and Letizia Osti have shown that many references to canals appear to indicate neighbourhoods of the same name rather than actual waterways, and that the Nahr Ṭābiq in Suhrāb's text appears to refer to a location rather than to a canal. They also point out that the Khandaq Ṭāhir, which appears to have been dug in the context of the civil war between al-Amīn and al-Maʾmūn in 198/813, is also difficult to identify definitively as a canal. Perhaps, they suggest, it was only filled with water occasionally (or it was little mentioned in anecdotes because it ran through a predominantly agricultural area.[21] Ibn al-Faqīh (d. 290/903), who wrote a little

[18] Suhrāb, *Kitāb ʿajāʾib*, 132, and following him al-Khaṭīb al-Baghdādī, *Taʾrīkh Madīnat al-Salām*, 1:433.

[19] Van den Bent, "Caliphal Involvement," 13–14, 21–22.

[20] Van Berkel et al., *Crisis and Continuity*, 221–22; Micheau, "Baghdad in the Abbasid Era," 221–24. See for instance the chapter by al-Khaṭīb al-Baghdādī, writing in the 5th/11th century, on the Baghdad canals. He elaborately described the city's waterways as if they existed in the present, relying heavily on the information in Suhrāb (al-Khaṭīb al-Baghdādī, *Taʾrīkh Madīnat al-Salām*, 2001, 1:431–36), only to mention somewhere else that many of those canals were not only derelict but that even the trace of them could no longer be found (1:389). On this contrast, see also Lassner, *Topography*, 33–34.

[21] Van Berkel et al., *Crisis and Continuity*, 223–26.

earlier than Suhrāb, does mention a description of Baghdad by "one of the men of letters (*baʿḍ al-udabāʾ*)" who makes mention of a "Nahr al-Khandaq" that may have been the Khandaq Ṭāhir.[22] He also, incidentally, is one of the authors who described the Nahr Ṭābiq as being dug by a Persian called Bābak.[23] Similar to the problem described by Ahola and Osti, I have been unable to identify more references to the conduits described by Suhrāb (and al-Khaṭīb al-Baghdādī).[24] It is clear that not all *nahr*s can be identified as actual canals throughout this entire time period with certainty, and that we lack detailed information about other pieces of water infrastructure. In any case, many of these canals became derelict over time,[25] and we usually do not know when exactly this happened. On the one hand, the sources make clear that these canals were very important for water provision in Baghdad, while on the other hand much about them remains unclear; many of them fell out of use by the 4th/10th century, as will be discussed in more detail below.

Another element of the city's water system was provision through charitable initiative, although here too the lack of evidence hampers research. For the early Abbasid period, I have found evidence for the existence of at least two examples. In the western neighbourhood of al-Karkh, the famous lute-player Zalzal, active in the last quarter of the 2nd/8th century, endowed a pond which was fed by the Nahr Razīn.[26] However, much as was the case for many canals in Baghdad, later mentions in the sources of the so-called Birkat Zalzal refer to the eponymous neighbourhood rather than to the pond itself, making it difficult to determine until when the pond remained in use.[27] The second example is the Ḥawḍ Haylāna, a tank on the east bank that was fed by an offshoot of the Nahr Mūsā. Although the exact identification of the Haylāna this tank was named after remains somewhat unsure, this too seems to have been a charitable enterprise dating to the second half of the

[22] Ibn al-Faqīh, *Kitāb al-buldān*, 310–11. The city and its waterways are described as follows (facing south): "The Tigris is in the middle of it [Baghdad], the Ṣarāt is on its right and the Nahr al-Malik in front of it and the Nahr ʿĪsā crosses through it, and the Nahr Karkhāyā passes through its streets, and the Nahr al-Khandaq circles around it (*wa-nahr al-khandaq dāʾir bihā*)." This Nahr al-Khandaq does not appear to indicate the moat al-Yaʿqūbī describes around the Round City, filled with water from the Karkhāyā Qanāt (al-Yaʿqūbī, *Kitāb al-buldān*, 239), as the identification of the city in this quote as solely the Round City would contradict the description of the Tigris and the Nahr ʿĪsā going through it.

[23] Ibn al-Faqīh, *Kitāb al-buldān*, 294. See Van Berkel et al., *Crisis and Continuity*, 224.

[24] See also Van Berkel et al., *Crisis and Continuity*, 227, where it is pointed out that the corresponding bridges are only mentioned by Suhrāb.

[25] See for instance Miskawayh, *Tajārib*, 2:406.

[26] Ibn al-Faqīh, *Kitāb al-buldān*; al-Khaṭīb al-Baghdādī, *Taʾrīkh Madīnat al-Salām*, 1:401. See also Van den Bent and Isoldi, "Pleasing God," 208–10.

[27] See for instance al-Khaṭīb al-Baghdādī, *Taʾrīkh Madīnat al-Salām*, 5:301; 6:61; 7:408.

2nd/8th century.[28] It is of note that both reservoirs were connected to the city's canal systems, but dug on the initiative and at the cost of the respective donors. These charitable initiatives are slightly difficult to classify as being from either "above" or "below". On the one hand, at least these two examples were clearly elite initiatives, endowed by people who had very close ties to the Abbasid authorities. On the other hand, they were initiated and carried out by individuals as private actors rather than in the role of an official authority.

As al-Manṣūr's order upon the foundation of Baghdad already indicates, the city also had wells. These, however, were not a prime choice for drinking water. Reports on the unpalatability of well water are found for various cities on the Tigris. According to the 4th/10th-century geographer al-Muqaddasī, Mosul's population took its drinking water from the Tigris and the Zubayda river, because the wells were saline.[29] For Samarra, al-Yaʿqūbī similarly reports that the wells not only had to be dug very deep but that they were unpalatably salty. Drinking water there came from the Tigris as well, brought in by the many available water carriers.[30] Although Ibn al-Faqīh, in contrasting the water situation in Samarra to that in Baghdad, reports that the Baghdadis had wells with agreeable drinking water,[31] this assessment appears to be an outlier. When al-Amīn (r. 193–98/809–13) was under siege in the Round City in 198/813, and was blocked from receiving food and water, he grew thirsty and asked for water to drink. There was none available, so when he reached the Basra Gate on his way out of the city to surrender, he asked for water from the guards' wells. Once he received it, however, he refused to drink it: he was too disgusted by its foul smell.[32] Of course, it was too foul for the caliphal taste, but perhaps others in the Round City did consume well water during this siege. However, other sources also report the unpalatability of Baghdad's well water,[33] and although it was certainly drunk in case of dire necessity—as we shall see below—well water was normally used for other purposes.

Like in Samarra, water carriers also played a significant role in Baghdad's water provision. These workers were a "force from below" *par excellence*. They are not, however, easy to trace. As mentioned above, Baghdad as a whole suffers from a lack of archaeological evidence from the Abbasid period. But even if such evidence were

[28] Ibn al-Faqīh, *Kitāb al-buldān*; al-Khaṭīb al-Baghdādī, *Taʾrīkh Madīnat al-Salām*, 1:414–15. See also Van den Bent and Isoldi, "Pleasing God," 210–12.

[29] al-Muqaddasī, *Aḥsan al-taqāsīm*, 138.

[30] al-Yaʿqūbī, *Kitāb al-buldān*, 263.

[31] Ibn al-Faqīh, *Kitāb al-buldān*, 329.

[32] al-Ṭabarī, *Taʾrīkh*, 3.2:908, 917. On the conflict between al-Amīn and his brother al-Maʾmūn (r. 198–218/813–33), see for instance Kennedy, *The Early Abbasid Caliphate*, 135–63. On the relationship between water purity, smell, and taste in *fiqh* texts, see Edmund Hayes' contribution to this volume.

[33] E.g., Miskawayh, *Tajārib*, 2:406.

there, the water carriers' activity would of course not be visible in it. Nor are they particularly present in the textual sources: those deal overwhelmingly with courtly circles, and pay little attention to the general populace as a whole, or to the humble profession of water carriers in particular. Nevertheless, they are occasionally visible in anecdotes and stories: some reports relate that al-Manṣūr gave the orders for the construction of the abovementioned teak conduits in response to his discovery that water-carrying animals came into his courtyard.[34] There are also occasional other bits and pieces of historical information, and we can deduce their presence at a larger scale, and their importance to the city's water provision, in some historical episodes.

Several streets in Baghdad appear to have been named after them, presumably because a large number of water carriers lived there at some point. Al-Yaʿqūbī relates the existence of a Water Carriers Street (*Sikkat al-Saqqāʾīn*) in the Round City, between the Basra and Khurasan Gates.[35] On the East bank, at least in the 4th/10th century, there was a Darb al-Saqqāʾīn, near the Sūq al-Silāḥ. At least at this point in time, this street may not have been exclusively populated by water carriers, as the hadith transmitters Abū Aḥmad al-Bazzāz ibn Nāfiʿ (d. 376/986) and al-Ḥusayn ibn ʿUmar al-ʿAllāf (341–426/953–1035) reportedly also resided there and no mention is made of them working as water carriers.[36] There was also the Mashraʿat al-Rawāyā, "the Wharf of the Water Carriers", on the west bank of the Tigris: an access point to the water, clearly used by water carriers.[37] Sources also allude, between the lines, to the presence of water carriers: for instance, the 3rd/9th-century Ṣūfī leader Abū Jaʿfar al-Ḥaddād was said to have worked in the market everyday but not to have spent any of his income on visits to the bathhouse or on drinking water, spending it

[34] al-Khaṭīb al-Baghdādī, *Taʾrīkh Madīnat al-Salām*, 1:388. A slightly different version, including water-carrying animals but without the conduits is found in al-Ṭabarī, *Taʾrīkh*, 3.1:322–23. See also Van den Bent and Brown, "Constructing," 128–29.

[35] al-Yaʿqūbī, *Kitāb al-buldān*, 240.

[36] al-Khaṭīb al-Baghdādī, *Taʾrīkh Madīnat al-Salām*, 8:640; 12:255–56. Al-Khaṭīb al-Baghdādī also relates an interesting report on Wakīʿ ibn al-Jarrāḥ (129–97/746–812), who was said to go, after the Ẓuhr prayer, to the road to the watering place used by the water carriers (*aṣḥāb al-rawāyā*) where they would rest their animals and he would teach them about their religious duties until the ʿAṣr prayer. This, however, appears to have taken place in Kufa rather than in Baghdad (al-Khaṭīb al-Baghdādī, *Taʾrīkh Madīnat al-Salām*, 15:654). On the hadith transmitter Abū ʿAbdallāh Muḥammad the Water Carrier, see the contribution by Van Renterghem in this volume.

[37] The earliest reference I have found is to the early 330s/940s (al-Khaṭīb al-Baghdādī, *Taʾrīkh Madīnat al-Salām*, 13:261), but as he is writing in the 5th/11th century, he may have used the name known to him. The area may have been known by a different name a century earlier, but the name is clearly current in the 5th/11th century (e.g., al-Khaṭīb al-Baghdādī, *Taʾrīkh Madīnat al-Salām*, 1:437; 8:678; 12:191). On this Mashraʿat al-Rawāya as well as the terminology of *mashraʿa*/wharf, see also the contribution by Van Renterghem in this volume. For similar topographic names in Cairo, see the contribution by Bakhoum in this volume.

all on the poor instead.[38] Seeing as he must have hydrated himself in some manner, the expenditure on drinking water likely referred to the fees for water carriers' services.[39] The baths that Abū Jaʿfar al-Ḥaddād purposefully avoided, appear to also have had water carriers among their staff, who were normally employed alongside a bath attendant, a manager, a sweeper, and a stoker.[40] We should perhaps take the claim, which al-Khaṭīb al-Baghdādī based on al-Ṣūlī, that these five were the minimum number of staff per bathhouse as doubtful, especially in light of his exaggerated claim that there were no fewer than 60,000 bathhouses at the time. We can conclude, however, that water carriers were a common type of employee in bathhouses, although whether these water carriers worked exclusively for these bathhouses or also worked for themselves is impossible to say.[41]

Responsibilities

Baghdad in its first two centuries therefore had a differentiated system of drinking water provision. But who were the parties responsible for its construction and its upkeep? As discussed above, some of the canals predated the Islamic period, and other water infrastructure was constructed in the context of the city's foundation. The physical elements of large water infrastructure required not only a significant financial investment during construction, but also needed to be maintained. For the charitable initiatives by Zalzal, Haylāna, and others, that responsibility resided with the underlying *waqf* foundation. That this did not always go as the founders had planned may be illustrated by the traveller and geographer Ibn Ḥawqal (d. after 367/978), who claimed that the reservoirs collecting rainwater in Mecca were in ruins because the overseers (*al-mutawallūn*) of the *waqf*s supposedly funding these reservoirs had appropriated the funds.[42] Canals also required upkeep. They could become silted up or otherwise blocked, and breaches upstream could also result in

38 al-Khaṭīb al-Baghdādī, *Taʾrīkh Madīnat al-Salām*, 16:593.

39 The later Egyptian author Ibn al-Ukhuwwa (d. 729/1329), in his *ḥisba* manual, makes mention of shops where drinking water is sold (*Maʿālim*, 97–98 (Eng.)/239 (Ar.)). I have not encountered direct evidence for similar shops in Abbasid Baghdad, but they may well have existed. Even so, they likely would have been furnished by water carriers.

40 al-Khaṭīb al-Baghdādī, *Taʾrīkh Madīnat al-Salām*, 1:439. On the economics of bathhouses, see also the contribution by De Haan in this volume.

41 On the water carriers, see also the contributions by Van Renterghem, Isoldi, and Bakhoum in this volume, as well as—more generally—Brown and Van Berkel, "Water Provision," 92–93. Water carriers may also have been involved in procuring water for the Friday Mosques, where water was offered as well (al-Ṭabarī, *Taʾrīkh*, 3.4: 2165).

42 Ibn Ḥawqal, *Ṣūrat al-arḍ*, 1:30.

canals drying up. Water carriers, as private labourers, were a fairly independent group and would have been personally responsible for the upkeep of their vessels (waterskins) and beasts of burden. They did, however, need access to the water, and that included the access places, the *mashraʿāt*, on the Tigris and the canals.

Early authors of land tax (*kharāj*) manuals demonstrate the existence of a tradition that argues that the management of the main waterways is the responsibility of the authorities, which includes things like the construction of dams and bridges, but also elements of upkeep, such as excavation and the maintenance of river banks.[43] This certainly applied to the Tigris, but at least according to Abū Yūsuf (113–82/731–98) it also applied to the major canals running from the Tigris and the Euphrates. If those needed to be excavated, the taxpayers who were drawing the canal water and the Treasury had to bear those costs together.[44] These manuals were focused on the agricultural context—as the success of water management translated directly into tax revenues for the Treasury—but cities located on these main rivers and canals, including Baghdad, would still have benefited from these norms enjoining the authorities' responsibility for waterways, if indeed they were adhered to.[45] However, these texts say nothing about the authorities' responsibilities for water provision to cities. Overall, little is to be found in the sources about this. Some indications and general ideas about maintenance responsibilities appear in the so-called *ḥisba* manuals, however.

Ḥisba manuals prescribe the responsibilities of the *muḥtasib*. The *muḥtasib* was appointed by the caliph or his representative and had an important role in a city's markets, including the inspection of weights and measures (and thereby fraud prevention), but also supervising other domains such as hygiene. Over time, from the 3rd/9th century onwards, the *muḥtasib*'s job description also came to include more general moral and religious concerns.[46] Many of the well-known manuals are

[43] Ben Shemesh, *Taxation in Islam, Vol. II*, 99v-100r (Ar.), 62 (Eng.); Abū Yūsuf, *Kitāb al-kharāj*, 97, 110; Ben Shemesh, *Taxation in Islam, Vol. III*, 106, 127 (Eng.). See also Campopiano, "Cooperation," 400–401; Van den Bent, "Caliphal Involvement," 4–5.

[44] Abū Yūsuf, *Kitāb al-kharāj*, 110; Ben Shemesh, *Taxation in Islam, Vol. III*, 106 (Eng.). See also Campopiano, "State," 22–23.

[45] Abū Yūsuf, *Kitāb al-kharāj*, 110; Ben Shemesh, *Taxation in Islam, Vol. III*, 106 (Eng.). See also Campopiano, "Cooperation," 400; Van den Bent, "Caliphal Involvement," 5–6.

[46] For the history of *ḥisba* and the *muḥtasib*, see, e.g., Buckley, *Book of the Islamic Market Inspector*, 1–12; Ghabin, "Muḥtasib"; Lange, "Changes." A distinction is often made between more theoretical literature on *ḥisba*, dealing with the notion of *al-amr bi-l-maʿrūf wa-l-nahy ʿan al-munkar* ("commanding right and forbidding wrong"), and the more practical advice texts, known as "*ḥisba* manuals" in modern scholarship, dealing with day-to-day situations *muḥtasib*s might encounter (Klein, "Between Public and Private," 42). On the former, see for instance Cook, *Commanding*.

from later time periods and/or from geographical areas far removed from Iraq.[47] However, there are two relatively early texts, dating to the two centuries under investigation here, that provide interesting insights and may suggest something about prevailing ideas in the first two Abbasid centuries.

In 1979, J. Sadan described and partially edited an anonymous manual, known as *Siyāsat al-mulūk*, a single copy of which is extant and is kept at the Süleymaniye Library in Istanbul. Despite its title, it focuses on various government offices in Abbasid-era Iraq. Sadan and Adam Silverstein argued that it should be dated to the early Buyid period, to the middle of the 10th century CE (ca. 329–338/940–950).[48] Recently, however, Eugénie Rébillard revisited this issue and argued for an earlier date of composition, between the late 3rd/9th century and the early 4th/10th century, before the Buyids came to power in 334/945.[49] Rébillard's argument appears convincing, but even without that it is clear that the practices the anonymous author describes go back to the pre-Buyid period. The informants he mentions were all active in the 3rd/9th century, and the last Abbasid political figure mentioned is al-Muwaffaq (d. 278/891), who acted as regent for his brother al-Muʿtamid (r. 256–79/870–92).[50]

The manual consists of thirteen sections (*abwāb*) on a wide range of positions and government institutions containing all sorts of practical advice, and appears to be aimed at Abbasid officials within these various institutions. The offices and institutions discussed include the ruler and the vizier, but also, among others, the police (*al-shurṭa*), the royal physician (*al-ṭabīb*), and the chef (*al-ṭabbākh*). It also contains a section on the *muḥtasib*. The *ḥisba* section in this anonymous Iraqi source is understandably shorter than its (mostly later) relatives devoted exclusively to the topic, at a bit over five folios in the manuscript or about four pages in the 1979 partial edition. It contains a few water-related injunctions. The order that shop-owners should keep a bucket of water handy in case of fire seems a prudent one for a city struck by fires on a regular basis.[51] The author also states that the *muḥtasib* should make sure that water from drainpipes (*al-mayāzīb*)—presumably from the roof—doesn't run onto the roads, as it is bad for the roads and hinders passers-by.

47 E.g., al-Shayzarī's *Nihāyat al-rutba fī ṭalab al-ḥisba* from the 6th/12th century, relied upon heavily by the Egyptian authors Ibn al-Ukhuwwa (*Maʿālim al-qurba fī aḥkām al-ḥisba*) and Ibn Bassām (*Nihāyat al-rutba fī ṭalab al-ḥisba*) (Buckley, *Book of the Islamic Market Inspector*, 13–14).

48 Sadan, "A New Source"; Sadan and Silverstein, "Ornate Manuals." The article "A New Source" contains a partial edition, and this partial edition as well as a facsimile of the rest of the text can also be found in Sadan, *Nouvelle source*.

49 Rébillard, *Imposer l'ordre*. Préambule: les sources, par. 57-87.

50 Rébillard, *Imposer l'ordre*, Préambule: les sources, par. 85-86. See also Silverstein, *Postal Systems*, 97n27, which still assumes an early Buyid-era composition but states that the text summarises Abbasid practices from the late 3rd/late 9th and early 10th centuries for the benefit of the Buyid amirs.

51 Edition: Sadan, "A New Source," 375.

Instead, he should command that a slit be made in the wall for the water to flow through.[52] Being inconvenienced by outward-draining pipes seems to have been a fairly common issue. The famous vizier Abū al-Ḥasan ibn al-Furāt (241–312/855–924) fell victim to one, which dirtied the vizier himself as well as his riding animal to the extent that he had to go clean up in a nearby perfumer's shop.[53] And Ibn al-Jawzī's (510–97/1126–1200) life of the famous theologian Aḥmad ibn Ḥanbal (164–241/780–855) underlines the latter's scrupulosity with an anecdote about such drainpipes. Ibn Ḥanbal reportedly had work done on the roof of a house he owned, and the water outlet (*masīl al-mā'*) was made to empty into the street. The next day, he had the carpenter come to redirect the drainpipe (*al-mīzāb*) toward the house.[54] The suggestion is that Ibn Ḥanbal was uncommonly conscientious here, implying that from the perspective of our anonymous author, the *muḥtasib* likely had his work cut out for him in this regard.

The *muḥtasib* was apparently in the position to order inhabitants of the city to make changes to their drainage systems. This would have involved primarily the inhabitants' personal residences. However, *Siyāsat al-mulūk* also contains a section discussing the general upkeep of the city's quarters. The anonymous author states that if the government refrains from cleaning the streets, the *muḥtasib* should have the inhabitants (*ahl kull nāḥiya*) clean the streets in their neighbourhoods. Moreover, the *muḥtasib* is also the one who should order the owners of houses and shops to repair the local streets as well as "the access places (*mashāri'*) on the Tigris and the canals".[55] According to the text, the *muḥtasib* was a more appropriate person

52 Edition: Sadan, "A New Source," 373–74. Similar passages can be found in various later *ḥisba* manuals. Al-Ghazālī's *Iḥyā 'ulūm al-dīn* lists water coming out of drainpipes from walls onto narrow streets as a "common wrong", as it dirties clothes and obstructs roads (al-Ghazālī, *Iḥyā 'ulūm al-dīn*, 2:339). Al-Shayzarī also describes that drainpipes emptying from walls into the road in winter (and waste coming out of houses in summer) are a hinder to passers-by, and states that the *muḥtasib* should order the owners of these drainpipes to excavate a drain plastered with lime in the wall instead (al-Shayzarī, *Nihāyat al-rutba*, 14). Al-Ghazālī's continuation that this is less of problem on wider streets if people can walk around it may offer some elucidation on the difficult to read sentence that follows the injunction in the anonymous source (see Sadan, "A New Source," 374n143).

53 Hilāl al-Ṣābi', *Tuḥfat al-umarā'*, 73. On Ibn al-Furāt, see for instance Van Berkel, "Ibn Al-Furāt."

54 Ibn al-Jawzī, *Virtues (Manāqib)*, 1:500 (Ar.), 501 (Eng.), par. 49.28. As Michael Cooperson has pointed out, the accidental nature of the presence of such "realia" in the text make them fairly reliable (Cooperson, "Excavation," 56–57). But even if Ibn Ḥanbal did not in reality have the direction of his drainpipes corrected, the fact that this could be used as an example at all underlines the prevalence of this issue in Abbasid Baghdad. See also the contribution of Hayes in this volume, on Ibn Ḥanbal dealing with a problem created by waste runoff into the street.

55 *Wa-ya'khudhuhum bi-iṣlāḥ al-durūb wa-l-sikak wa-shirā al-aqfāl wa-iṣlāḥ al-mashāri' 'alā dijla wa-l-anhār* (Sadan, "A New Source," 374). I interpret the *-hum* as referring back to the "owners of houses and shops" in the previous sentence, but it could also refer back further to either the guards or even the aforementioned *ahl kull nāḥiya*. The house and shop owners or all local residents appear

for such tasks, as well as more reliable, than the police's district- and *maṣāliḥ* chiefs (*aṣḥāb al-arbāʿ wa-l-maṣāliḥ*).[56] It is not entirely clear whether the responsibility for the maintenance of access places to water and other infrastructure resided with the inhabitants of these neighbourhoods on principle, or that—as in the case of street cleaning—the primary responsibility for this lay with the authorities (*al-sulṭān*), only to devolve onto the locals if the government failed to do so. Reading the text as straightforwardly as possible, it would seem that the former is the case, but this is far from certain. In either case, there was apparently a shared responsibility of sorts: the *muḥtasib* as the authorities' representative had a commanding role, possibly even a managing one, while the inhabitants of the city were responsible for the realisation of necessary repairs in their own neighbourhoods. Moreover, the fact that the likely alternatives to the *muḥtasib* for this role were the local district chief, or the chief of the even smaller *maṣlaḥa*, underlines the strong local character of the organisation of these repairs. This normative text thus clearly places water-related responsibilities on "forces from below".

An even older and significantly longer *ḥisba* manual is the treatise known as *Kitāb al-iḥtisāb*. The treatise's editor R.B. Serjeant has argued that the text's author was al-Nāṣir li-l-Ḥaqq al-Ḥasan ibn ʿAlī al-Uṭrūsh (d. 304/917), the imam of the Zaydi state in Ṭabaristān, south of the Caspian Sea.[57] This text should probably be interpreted as situated in the realities of the Zaydi state's capital city of Āmul.[58] The treatise is significantly longer than the *ḥisba* section in *Siyāsat al-mulūk*, but as it was produced in the Zaydī context it cannot simply be extrapolated to the Abbasid territories. Nevertheless, it is worth looking at how it discusses water and infrastructure maintenance to see what kind of basic principles underlie its approach. Where water and waterways are discussed, it is clear that one concern is hygiene: people are forbidden from urinating in the water and throwing rubbish and filth

to make more sense than the guards in this respect. Other responsibilities of the *muḥtasib* also appear to result in involvement in construction works of sorts. According to the manual, the *muḥtasib* is also responsible for the control over incoming and outgoing ships, he is to construct a raft (*ṭāʾif*) on the river in order to block anything from travelling along it at night, except with a permit (Sadan, 374). See also Bosworth, "Some Remarks," 84–85.

56 Sadan, "A New Source," 374. The district chiefs (*aṣḥāb al-arbāʿ*) acted under the authority of the head of police (*ṣāḥib al-shurṭa*). Under the authority of the district chiefs were the *aṣḥāb al-maṣāliḥ*, who were responsible for the smallest unit within the urban police force and commanded the guards (*ḥāris, ḥurrās/aḥrās*) and other subordinates (Rébillard, *Imposer l'ordre*, chapter 5, par. 22–32; chapter 7, par. 15–22).

57 Serjeant, "A Zaidī Manual," 1–6. For an overview of the history of the Zaydis in the Caspian region, see for instance Daftary, *A History of Shiʿi Islam*, 151–62.

58 Lev, *Administration*, 185.

into it;[59] fullers and dyers may not pour wastewater into canals (or the street).[60] The *muḥtasib* also has the responsibility to make sure that people do not dig wells or build waterwheels (or mosques) in places where they would hinder passers-by; the same principle is at work here as in the case of the drainpipes. And if there is any damage to roads, mosques or bridges, this needs to be addressed. The *muḥtasib* is to bring the issue to the *qāḍī*. Presumably this is to check for the presence of a *waqf*, for the text continues to say that if the structure has a *waqf* to fund it, those funds should be used for the repairs. Otherwise, the issue should be raised with the Zaydī imam (i.e., al-Nāṣir li-l-Ḥaqq himself or his successors) and paid for from the Treasury if there are sufficient funds. If there is no money in the Treasury, the responsibility for funding the repairs lies with the Muslim community.[61]

This text was written in a different context from that of the central Abbasid cities that are the focus of this chapter, and the Zaydī situation cannot simply be extrapolated to the Abbasid caliphate. Nevertheless, a few elements from the Zaydī text are noteworthy. One is that private citizens clearly took the initiative to construct wells and waterwheels. Second, the authorities have a responsibility for the upkeep of key infrastructure, for which the Treasury should pay, if at all possible. Although waterways such as canals are not mentioned in this text, bridges are, which suggests that this responsibility may extend to other elements of public infrastructure as well. In itself, the text does not tell us anything about the situation in Baghdad. At the same time, however, as Sadan and Silverstein point out, the *ḥisba* section of *Siyāsat al-mulūk* appears to draw on older practices and texts: the anonymous author writes that "there is a book about [the *ḥisba*] that [the *muḥtasib*] must obtain and study always."[62] There are also parallels in themes and principles (e.g., weights and measures; the obligation to wear waist-coverings in the bath-house; keeping thoroughfares clear) between the anonymous and Zaydī texts that point to possible underlying shared traditions.[63] These might include a general idea of a government responsibility for the upkeep of infrastructure, including elements of water management.

This idea of the authorities being responsible for the maintenance of hydraulic infrastructure is certainly present in the work of the well-known jurist al-Māwardī (d. 450/1058), as I have discussed elsewhere. Not only should one of the ruler's first concerns upon founding a new city be proper water provision,[64] but the

[59] *Kitāb al-iḥtisāb* in Serjeant, "A Zaidī Manual," 18. On water pollution, including by urine, see also Hayes' contribution to this volume.

[60] *Kitāb al-iḥtisāb* in Serjeant, "A Zaidī Manual," 31.

[61] *Kitāb al-iḥtisāb* in Serjeant, "A Zaidī Manual," 28.

[62] Edition: Sadan, "A New Source," 376. See Sadan and Silverstein, "Ornate Manuals," 344, 351.

[63] See also Sadan and Silverstein, "Ornate Manuals," 344; Sadan, "A New Source," 359.

[64] al-Māwardī, *Tashīl*, 210–11. See also Van den Bent, "Caliphal Involvement," 6.

ḥisba-section of al-Māwardī's *al-Aḥkām al-sulṭāniyya* also offers some interesting insights into his ideas about responsibility for urban maintenance. In a discussion of "commanding good" (*al-amr bi-l-maʿrūf*) with regard to the rights of individuals, he mentions various examples concerning infrastructure: in case of failure of a town's water supply, or the collapse of a town's wall, mosques or Friday mosques, these all should be repaired using money from the Treasury (*bayt al-māl*), on the condition that sufficient resources are available. In case of insufficient funds, however, the responsibility for repairs "is directed to all those amongst the inhabitants who have the means," but not imposed upon one of them individually.[65] Al-Māwardī continues to specify what should happen if these inhabitants should fail to take care of these repairs. If a town remains inhabitable, the *muḥtasib* should just leave the inhabitants to their own devices. If not, the course of action rests on the town's location. If it is a frontier town (*thaghr*) and its abandonment would imperil the Dār al-Islām, it is obligatory for the inhabitants with means to take up the work, and the *muḥtasib* should encourage them as well as inform the ruler (*al-sulṭān*). If it is not a frontier town, however, the *muḥtasib* cannot force the inhabitants to take on the responsibility of rebuilding, "because the ruler (*al-sulṭān*) is more obligated to do so". If the sultan, however, does not have the funds required for reconstruction, the inhabitants may choose between leaving the town or repairing it under the supervision of the *muḥtasib* (who is also to seek permission from the sultan).[66]

Like the Zaydī manual, we cannot simply extrapolate these norms from 5th/11th-century texts to an earlier period in Baghdad. However, al-Māwardī regularly refers to texts by earlier authors and builds on existing traditions, meaning that it is probable that these notions on responsibility for upkeep were shared in the earlier Abbasid period.[67] Based on these texts, it would seem that there was a general idea that the (central) authorities were responsible for the upkeep of important infrastructure and major waterways. It is noteworthy that according to the anonymous *ḥisba* text, responsibilities for street cleaning reside with authorities and only devolve onto neighbourhood residents when the former neglect to do so, while the residents seem to have been liable for the more strenuous duty of structural repairs to streets and *mashāriʿ* (in contrast to the Zaydī text!). Perhaps responsibilities for larger elements of water provision and infrastructure—canals, bridges—were considered to belong to the authorities and to be financed by the Treasury, while

65 al-Māwardī, 357 (see also 316–17). Needy travellers who are not provided for by the inhabitants are also listed as an example. See also Van den Bent, "Caliphal Involvement," 6–7.

66 al-Māwardī, *al-Aḥkām*, 358. In his discussion on the *bayt al-māl* and spending money on the public interest (*al-maṣlaḥa*), al-Māwardī states something similar (*al-Aḥkām*, 316–17). See also Van den Bent, "Caliphal Involvement," 6–7.

67 See, e.g., Melchert, "Māwardī's Legal Thinking," 68–86. See also Van den Bent, "Caliphal Involvement," 6.

the *mashāriʿ*—as smaller but no less pivotal elements of water provision infrastructure—were a communal duty for the neighbourhood. Or perhaps this too should be understood to be a duty that only devolves onto residents when the authorities default. In either case, whether from the outset or only in case of government neglect and/or insolvency, what is clear is that responsibilities for maintenance were considered to be shared between the authorities and the residents.

Failures in Maintenance and Decline

The *ḥisba* manuals are of a prescriptive nature and it is difficult if not impossible to determine to what extent their advice was carried out in practice. That the central authorities were far from able and/or willing to adhere to the advice given in the *kharāj* manuals with regard to the upkeep of irrigation infrastructure in the countryside is visible in the decline in tax revenues in the 3rd/9th and 4th/10th centuries (although other factors were at play as well, including occasional deliberate damage done during wars).[68] However, the collection of potential problems described in these texts give good insight in the issues a *muḥtasib* and other inhabitants of Abbasid Baghdad and other cities might encounter.[69] The intimate contact between these texts and real-life practices, to paraphrase Yaron Klein,[70] is visible in the staunchly realist approach they take to the financial aspects of upkeep and repair. These texts all operate on the assumption that the authorities are likely to default at one point or other, and that locals would have to step up both practically and financially, perhaps under the supervision of the *muḥtasib*. As Sadan points out, the chapters in the *Siyāsat al-mulūk* reveal an "endemic budget deficit",[71] and within that context investments in water management—whether in the fiscally important agricultural areas or in the city—do not seem to have been anyone's priority.[72] These financial problems also contributed to the political struggles in the Abbasid caliphate, which eventually resulted in the appointment of Aḥmad ibn Buwayh in 334/945 as the *amīr al-umarāʾ*. This new position of "chief commander" had been introduced in 324/936 and its holder effectively had more power than the caliph.

68 Waines, "Internal Crisis"; Campopiano, "State," 8–12, 22–24. With regard to intentional damage, the breach of the Nahrawan Canal by Ibn Rāʾiq in 938 is probably the most famous example. It would take until the reign of Muʿizz al-Dawla almost twenty years later until it was repaired, with devastating consequences to the area. (See e.g. Christensen, *Decline*, 94.)

69 Klein, "Between Public and Private," 43.

70 Klein, "Between Public and Private," 43.

71 Sadan, "A New Source," 358.

72 On these financial problems, see for instance Kennedy, *Prophet*, 169–78.

The appointment of Ibn Buwayh, also known by his title Muʿizz al-Dawla, marked the start of the Būyid period (334–447/945–1055).[73]

There is not much information in the historical sources about the maintenance of water infrastructure in Baghdad's first two hundred years. Perhaps, to an extent, that is because the sources were mostly focused on the court and, at least for the early decades of the Abbasid caliphate, things ran smoothly. However, it is clear that maintenance of water infrastructure was frequently inadequate. Ibn al-Faqīh reports that the teak conduits in the Round City fell out of use by the middle of the 3rd/9th century.[74] He does not mention who would have been responsible for their upkeep, but it was presumably the caliphal court or the caliphal administration—the caliphs themselves had relocated to the East Bank, and the caliphal administration followed with the reign of al-Maʾmūn, r. 197–218/813–33), whereupon the conduits were abandoned.[75] The fact that the Abbasid court moved to Samarra shortly afterwards, and remained there until 279/892, surely did not help. Given that the conduits seem to have only been used by the palace, this was not necessarily a problem, even if the Friday Mosque remained in use. A much bigger problem, one with dire consequences, was the neglect of the city's canals. By the second half of the 4th/10th century, authors describe the silting up and disappearance of these canals. For the most part, the exact causes and timelines are obscure—as discussed above, already for the reign of al-Muqtadir (295–320/908–32) various canals are difficult to trace.[76] There are a few instances, however, where the sources offer a glimpse of problems with these canals as well as occasional, albeit largely unsuccessful, attempts by the authorities to address the issues.

In his description of the reign of al-Rāḍī bi-llāh (r. 934–40), the scholar and courtier Abū Bakr al-Ṣūlī (d. 335/947) recollects an anecdote from the year 327 (938–39). He reports that al-Rāḍī gave the order to distribute alms worth 1,000 dinars to the people inhabiting the area between the Sarāt and the Nahr ʿĪsā, "because of the lack of water due to the breach" (*li-ʿawaz al-māʾ min ajl al-bathq*). Al-Ṣūlī then offers an alternative solution, which he says would be better than distributing alms: "Add another thousand to the thousand, and fix the breach (*yaʿmal al-bathq*)."[77]

[73] For these struggles, see for instance Bonner, "The Waning of Empire."

[74] Ibn al-Faqīh, *Kitāb al-buldān*, 289–90. See also Van den Bent and Brown, "Constructing," 132.

[75] On these movements, see for instance Micheau, "Imperial Foundation," 411–12; Scheiner and Janos, "Baghdād," 8–9.

[76] Van Berkel et al., *Crisis and Continuity*, 223–26.

[77] al-Ṣūlī, *Akhbār*, 137. A *bathq* or *bithq* (pl. *buthūq*) is the place where water flows out of an opening in the river bank or dam (Lane, *An Arabic-English Lexicon*, 1:152. The term, however, is used in various ways. It can be on purpose, in order to divert water from a river into a canal (Bosworth, "Some Remarks," 82), like a sluice gate (for that translation see Morony, *Iraq*, 178), e.g., the Bathq al-Ḥīrī, the Bathq Sayyār, and the Bathq Shīrīn mentioned by al-Balādhurī, *Futūḥ*, 357–58, 366–69.

Unfortunately, no further background information on the precise nature of the problem is given—al-Ṣūlī is primarily interested in the political aspect of the story, which is his manoeuvring of the eunuch Rāghib into the role of overseer of this project. Consequently, we don't know where this breach took place, what happened exactly, and why this resulted in a lack of water in a significant part of the western part of Baghdad. However, we do learn that the caliph allotted another two thousand dinars to the project, bringing the total up to three thousand dinars. According to al-Ṣūlī, one of the reasons Rāghib would be a good overseer is that he would not allow the workers to prolong the work in order to receive more pay. This purportedly efficient pace led to the repair of the breach in a little over 50 days.[78]

A number of years later, during the reign of al-Muttaqī li-llāh (r. 329–33/940–44), there was another attempt to deal with a breach. Seeing as it is described as the "breach (*bathq*) at the Nahr ʿĪsā", it is possible that it was in the same or a nearby location.[79] In the year 333/944–45, al-Ṣūlī reports, the Turkish *amīr al-umarāʾ* Tūzūn

Such *buthūq* are also listed by Abū Yūsuf (*Kitāb al-kharāj*, 110; Wiedemann, "Beiträge," 314–15). The term is also sometimes interpreted as a dam of sorts (Wiedemann, "Beiträge," 314-315n4; Saliba, *The History of Al-Ṭabarī, Volume XXXV. The Crisis of the ʿAbbāsid Caliphate*, 78). The term *bathq* can also, however, mean an involuntary bursting of the river bank (which is also how the English translation interprets the *buthūq* in Abū Yūsuf: Ben Shemesh, *Taxation in Islam, Vol. III*, 106). This use of the term can be found, for instance, in the description of the breaches at the Nahr Būq and the Nahr Rufayl in 329/940-1 by Miskawayh. These two breaches (*hādhayn al-bathqayn*) were not fixed at the time, leading to the destruction of the Bādūrayā district (southwest of Baghdad) for more than a decade, the author reports. Only in 345/956-7, under Muʿizz al-Dawla—who reportedly took a very active role in the repairs—were these breaches repaired by damming them up (*sadd bathq Nahr al-Rufayl wa-sadd bathq Bādūriyā*) (Miskawayh, *Tajārib*, 2:9, 165). The Nahr Rufayl refers to (part of) the Nahr ʿĪsā (see Verkinderen, *Waterways*, 30, 292–93), and this breach might be related to the one described by al-Ṣūlī. The Nahr Būq however was on the east side of Baghdad, which is rather confusing in light of Miskawayh's phrasing which makes clear that *both* breaches affect Bādūrayā. With regard to al-Ṣūlī's report on the water problems in southwestern Baghdad: I am interpreting it here as an involuntary breach rather than repairs to a sluice gate/dam. A breach upstream in the Nahr ʿĪsā, especially before Muhawwal as that would also affect the Ṣarāt, whether in a retaining dam or in the riverbank, may well have caused the drying up of the canals in the city of Baghdad leading to a lack of water. See also the translation by Marius Canard, who renders it "la rupture du barrage" (al-Ṣūlī, *Akhbâr*, 1:210).

[78] al-Ṣūlī, *Akhbār*, 137–38. The cost of the material is unknown, but the workers intended here were likely unskilled labourers carrying out the hard work (although some people involved in the project may well have been experts). An unskilled labourer in Baghdad in 338-9/950 made c. 1.6 dinars per month (Pamuk and Shatzmiller, "Plagues," 202). If we round up the 50-plus-some days to two full lunar months, each labourer should have earned about 3.2 dinars.

[79] It may also refer to the Nahr Rufayl breach described by Miskawayh (Miskawayh, *Tajārib*, 2:9, 165), see note 77. Overall, it is difficult to identify these floods exactly. As Waines points out, authors tended to only record those natural disasters that had especially severe consequences (Waines, "Internal Crisis," 289n10). See also Christensen, *Decline*, 96, who appears to consider the breaches mentioned by al-Ṣūlī and Miskawayh for 327 and 329 respectively as one and the same.

and his secretary Ibn Shīrzād attempted to fix the breach. Tūzūn travelled to the site, and spent a lot of his own money in an attempt to dam it. However, al-Ṣūlī writes: "God Almighty did not allow it, and the water carried away most of the work. The *amīr* was very distressed."[80] Incidentally, just after relating al-Rāḍī's repair of the breach, al-Ṣūlī mentions another report that suggests that the caliphal authorities could not quite keep up with the maintenance that was their responsibility according to the manuals. When Abū al-Walīd ibn Ḥamdān died in 327/939, he bequeathed money to repair the upper bridge (*al-jisr al-fawqānī*) across the Tigris.[81]

At the very beginning of this chapter, I discussed al-Yaʿqūbī's description of the canals in the neighbourhood of al-Karkh, which he reported were so important for its water provision.[82] It was precisely this neighbourhood that was hit hard by the Nahr ʿĪsā breach. It is noteworthy that al-Rāḍī's initial solution of distributing alms appears to have been intended to shift water supply in these neighbourhoods from a reliance on the canals to one of the other pillars in the water management system: the water carriers. These alms were most likely intended to help people pay for their services.

The shift from a reliance on the canals to other water sources is also visible in the description of the situation in Baghdad in the early Buyid period by Miskawayh (ca. 320–421/ca. 932–1030), a historian and philosopher who also worked as a librarian and secretary at court. Moreover, it also suggests that, when the authorities fail to deliver, the local communities did attempt to address at least some of the issues, as the authors of the *ḥisba* manuals advised. Like his predecessors al-Yaʿqūbī and Suhrāb, Miskawayh expressed the importance of Baghdad's canals for the city's drinking water provision:

> There were many canals in Baghdad, like Nahr al-ʿAbbāra, Nahr Masjid al-Anbār, Nahr al-Bazzāzīn, Nahr al-Dajāj, Nahr al-Qallāʾīn, Nahr Ṭābiq and its aqueduct (*mīzāb*) to the Tigris, the Ṣarāt, Nahr ʿĪsā, and a canal in the Ḥarbiyya neighbourhood, which takes from the Dujayl. From these were conduits (*murāfiq*) for the people to water their gardens and for drinking in the areas farther removed from the Tigris. Their channels have become filled in and their traces have been effaced. Generations have grown up without knowing of them. The poor (*al-ḍuʿafāʾ*) were forced to drink the unpalatable well water (*miyāh al-ābār al-thaqīla*) or to take the trouble of fetching water from the Tigris at a long distance.[83]

80 al-Ṣūlī, *Akhbār*, 278.
81 al-Ṣūlī, *Akhbār*, 138.
82 al-Yaʿqūbī, *Kitāb al-buldān*, 250.
83 Miskawayh, *Tajārib*, 2:406.

This passage shows a number of things. First, it points again to the importance of these canals for Baghdad's water provision. It is true that the city's water supply did, at least at some points in time, consist of a variety of elements. There were aqueducts and canals constructed primarily by the authorities, but also cisterns, and at least some water reservoirs that were donated as charity. However, these container facilities appear to have been pretty much entirely reliant on the canal system to fill them up. Neither of the canals feeding the Birkat Zalzal and the Ḥawḍ Haylāna that I mentioned earlier are listed by Miskawayh here as being silted up, so perhaps those were still available for those living closer to them. But the cisterns in al-Karkh, mentioned by al-Yaʿqūbī, likely suffered significantly.

Nevertheless, despite Miskawayh's dramatic depiction here, if anything, his account shows that the system was resilient, precisely because it was built on more than one pillar: besides the canals and related infrastructure, there were the water carriers and—in dire circumstances—the wells. The latter were clearly a last resort: when there were riots in Baghdad in 361/971-2, most people were unable to reach the Tigris for water and had to drink from wells.[84] Despite it not being the "sweet water" championed by Ibn al-Faqīh, the water must have been potable enough for people to survive on and they could fall back on it in emergencies.

The water carriers are not in fact mentioned by Miskawayh, yet as with al-Rāḍī's charity, we can read them in his silence. Miskawayh only mentions the struggles of the poor: they either have to drink the unpleasant well-water, or have to haul water home from the far-away Tigris. But what about the more affluent inhabitants? Those who were rich and lived on the Tigris could of course bring in water directly using waterwheels. Those who did not have that option could perhaps make use of their own servants and other resources, such as beasts of burden,[85] or solicit the services of water carriers. Although their services were too expensive for the poor, it shows that the fact that the system of water provision consisted of different components made it resilient.

Miskawayh follows this passage with another interesting bit of information on failing maintenance, that also underlines the involvement of local residents in upkeep and repairs. He describes how the Nahr ʿĪsā, the Ṣarāt, and the Khandaq Ṭāhir used to have bridges (*qanāṭir*) that had fallen into disrepair. In some cases crossing them was no longer possible, while in others the general populace

[84] Miskawayh, *Tajārib*, 2:308.

[85] For example—although I suspect that this took place in al-Anbār rather than in Baghdad, albeit the latter cannot be ruled out—it is reported as an illustration of the generosity of the legal scholar and hadith transmitter Isḥāq ibn al-Bahlūl al-Tanūkhī (164–252/780-1–866-7) that he had a slave or servant (*ghulām*) and a mule to draw water and provide it to his relatives (al-Khaṭīb al-Baghdādī, *Taʾrīkh Madīnat al-Salām*, 7:393).

(*al-raʿiyya*) had shoddily repaired them, in accordance to their means (*ʿalā ḥasab al-iqtiṣād*). According to Miskawayh, anyone who was not a healthy adult man—he lists cattle, women, children and weaklings—was at risk of falling off.[86] The main point he wishes to convey in what follows, is that the Buyid ruler ʿAḍūd al-Dawla (r. 367–72/978–83) had them all rebuilt as sturdy structures, and similarly repaired the pontoon bridge (*al-jisr*) across the Tigris. ʿAḍūd al-Dawla also ordered the digging out of the canals, restoring their water-flow.[87] This underscores, again, a general notion that this is the responsibility of those in power, and consequently has a legitimizing function here. However, it is also noteworthy that apparently, when the authorities dropped the ball, the locals did band together to do what they could. They were constrained by their circumstances and their finances, but they did attempt to repair this infrastructure. It also suggests that locals may in fact have taken care of the *mashāriʿ* in their neighbourhoods, as the *Siyāsat al-mulūk* instructs. The ones on the canals, of course, are only useful as long as there is water in them, but the *mashāriʿ* on the Tigris and their maintenance were now of even more importance to the city's water provision system than before.

Conclusions

From the foundation of Baghdad in 145/762, the city rapidly became one of the biggest in the world with some 300,000 to 500,000 inhabitants in the 3rd/8th and 4th/9th centuries.[88] Because of this size, as al-Yaʿqūbī also pointed out, its elaborate system of canals was no superfluous luxury. Nevertheless, the system did not rest exclusively on these canals, and that was a good thing.

Both in the establishment of elements of the system of water provision and in its maintenance, forces from "above" and "below" played a role. The Muslim authorities were widely viewed as having a significant responsibility when it came to providing water in cities like Baghdad. The primary responsibilities for the upkeep of canals and other major infrastructure appears to have resided with them. But other parties played a role as well. There were citizens involved in charitable initiatives, for instance, and individuals also had a variety of ways of procuring water

86 Miskawayh, *Tajārib*, 2:406. The somewhat mysterious conduits (*qūrāj*) described by Suhrāb (and al-Khaṭīb al-Baghdādī) as attached to the bridges crossing the Khandaq Ṭāhir and to the Old Bridge are unlikely to have survived this dilapidation. As I mentioned above, I have found no other mentions of these conduits, and Miskawayh similarly makes no mention of them breaking down nor of them being reconstructed under ʿAḍūd al-Dawla.

87 Miskawayh, *Tajārib*, 2:406. See also Christensen, *Decline*, 97.

88 Lassner, *Topography*, 160; Micheau, "Baghdad in the Abbasid Era," 232–35; Kennedy, "The Feeding," 177.

for their personal use (waterwheels, wells). However, while contemporary histories and geographies emphasise the importance of the canals for the city's water provision, if we read between the lines of the historical anecdotes, we perceive another all-important pillar of Baghdadi water provision: the water carriers.

Even the prescriptive sources acknowledge that the state is not always able (or possibly: not always willing) to live up to its maintenance responsibilities. Consequently, authors of the earliest known *ḥisba* manuals agree that if the governmental authorities fail to uphold their responsibilities, that responsibility devolves upon the community as a whole—after all, clean drinking water is a primary necessity. Little mention is made in the sources of upkeep of the important canals. Excavation was needed to prevent silting up, but maintenance in the rural areas upstream was also important: breaches there could have a disastrous effect on the city as well, through floods or a lack of water. While for the early Abbasid period, one could make the argument from silence that the system appears to have run smoothly, this was certainly not the case in later years. Yet even then, not a great deal of information is offered in the sources. However, as the few examples mentioned in this paper suggest, interventions by the authorities appear to have been haphazard and largely unsuccessful.

When, in the 4th/10th century, many of the canals failed, there were two elements of key importance to the system of water provision: its hybridity, and the role played by "forces from below". Despite its unpalatability, Baghdadis could fall back on well water in emergencies. But more importantly, the water carriers could stop part of the gap left by the canals. This is suggested by al-Rāḍī's attempt to solve the breach-induced water crisis in the southwest of the city by means of alms, as well as by Miskawayh's text: despite the dire picture he paints, he inadvertently shows the importance of the water carriers and the mixed nature of the system. That is not to say that the system worked equally well for all Baghdadis: for many people water would have become harder to come by and significantly more expensive, with this development—as crises always do—having a disproportionately severe effect on the poor.

The "forces from below" also played a significant role in maintenance and repairs. According to the *Siyāsat al-mulūk*, locals were responsible for maintenance of local streets as well as the *mashāriʿ*, the access places, on the canals as well as the Tigris. Whether they actually did so is impossible to say—the continued and growing role played by the water carriers is insufficient evidence, and this task could have been taken on by other parties—although one can probably safely assume that few people would bother to maintain a *mashraʿa* on a silted-up canal. But when circumstances required it, local Baghdadis also took on responsibilities that traditionally and according to the manuals lay with the authorities, much like the manuals advised—although perhaps not in the well-coordinated manner al-Māwardī and his peers had envisioned. Despite being limited in funds and perhaps in other ways, ordinary people attempted to repair bridges (and perhaps other infrastructure) to the best of their ability. The early geographies of Baghdad boast of its wonderful canals, their sweet water, and its impressive architecture and infrastructure. But it is the elements that we find the least in the sources—the wells, generally maligned, the water carriers, and the forces "from below" in general—that made the city resilient in times of crisis.

Bibliography

Primary Sources

Abū Yūsuf. *Kitāb al-kharāj*. Dār al-Maʿrifa, 1979.

al-Balādhurī. *Kitāb futūḥ al-buldān*. Edited by M.J. de Goeje. E.J. Brill, 1866.

al-Ghazālī. *Iḥyā ʿulūm al-dīn*. Vol. 2. 4 vols. Dār al-Maʿārif, 1982.

Ben Shemesh, A., trans. *Taxation in Islam, Vol. II. Qudāma b. Jaʿfar's Kitāb al-Kharāj*. Brill/Luzac & Co. Ltd, 1965.

Ben Shemesh, A., trans. *Taxation in Islam, Vol. III. Abū Yūsuf's Kitāb al-Kharāj*. Brill, 1969.

Ibn al-Faqīh, Abū Bakr al-Hamadhānī. *Kitāb al-buldān*. Edited by Y. al-Hādī. ʿĀlam al-Kutub, 1996.

Ibn al-Jawzī. *Virtues of the Imam Ahmad Ibn Ḥanbal (Manāqib Abī ʿAbd Allāh Aḥmad ibn Muḥammad ibn Ḥanbal)*. Edited by Michael Cooperson. Vol. 1. 2 vols. New York University Press, 2013.

Ibn Ḥawqal. *Kitāb ṣūrat al-arḍ*. Edited by J.H. Kramers. 2nd ed. Vol. 1. 2 vols. Brill, 1938.

Ibn al-Ukhuwwa. *The Maʿālim al-qurba fī aḥkām al-ḥisba of Ḍiyāʾ al-Dīn Muḥammad ibn Muḥammad al-Qurashī al-Shāfiʿī, known as Ibn al-Ukhuwwa*. Edited by Reuben Levy. Luzac & Co, 1938.

al-Khaṭīb al-Baghdādī. *Taʾrīkh Madīnat al-Salām*. Edited by Bashshār ʿAwwād Maʿrūf. Vol. 1, 2, 5, 6, 7, 8, 12, 13, 15, and 16. 17 vols. Dār al-Gharb al-Islāmī, 2001.

al-Māwardī. *al-Aḥkām al-sulṭaniyya*. Edited by Aḥmad Jād. Dār al-Ḥadīth, 2006.

al-Māwardī. *Tashīl al-naẓar wa-taʿjīl al-ẓafar fī akhlāq al-malik wa-siyāsat al-mulk*. Edited by Riḍwān al-Sayyid. Dār al-ʿUlūm al-ʿArabiyya, 1987.

Miskawayh. *Kitāb tajārib al-umam*. Edited by H.F. Amedroz. Vol. 1. 2 vols. Basil Blackwell, 1920.

Miskawayh. *Kitāb tajārib al-umam*. Edited by H.F. Amedroz. Vol. 2. 2 vols. Basil Blackwell, 1921.

Miskawayh. *Kitāb tajārib al-umam*. Edited by Sayyid Kusrawī Ḥasan. Vol. 4. 7 vols. Dār al-Kutub al-ʿIlmiyya, 2003.

al-Muqaddasī, Muḥammad ibn Aḥmad Shams al-Dīn. *Kitāb aḥsan al-taqāsīm fī maʿrifat al-aqālīm*. Edited by M.J. de Goeje. 2nd ed. Brill, 1906.

al-Ṣābiʾ, Hilāl. *Tuḥfat al-umarāʾ fī taʾrīkh al-wuzarāʾ: The Historical Remains of Hilāl al-Ṣābiʾ. First Part of his Kitab al-Wuzara (Gotha Ms. 1756) and Fragment of his History 389–393 A.H. (B.M. Ms, add. 19360)*. Edited by H.F. Amedroz. Brill, 1904.

Siyāsat al-mulūk, edition in Sadan, J. "A New Source of the Būyid Period." In *Israel Oriental Studies IX*, edited by Gideon Goldenberg, 355–76. Tel-Aviv University, 1979.

al-Shayzarī, ʿAbd al-Raḥmān ibn Naṣr. *Nihāyat al-rutba fī ṭalab al-ḥisba*. Edited by Al-Sayyid al-Bāz al-ʿArīnī. Maṭbaʿat Lajnat al-Taʾlīf wa-l-Tarjama wa-l-Nashr, 1946.

Suhrāb. *Kitāb ʿajāʾib al-aqālīm al-sabʿa ilā nihāyat al-ʿimāra*. Edited by Hans von Mžik. Otto Harrassowitz, 1929.

al-Ṣūlī, Abū Bakr Muḥammad ibn Yaḥyā. *Akhbār al-Rāḍī bi-Llāh wa-l-Muttaqī Lillāh*. Edited by J. Heyworth Dunne. Luzac & Co., 1935.

al-Ṣūlī, Abū Bakr Muḥammad ibn Yaḥyā. *Akhbâr Ar-Râdî Billâh Wa'l-Muttaqî Billâh. Histoire de La Dynastie Abbaside de 322 à 333/934 à 944*. Translated by Marius Canard. Vol. 1. 2 vols. Imprimeries La Typo-litho et Jules Carbonel Réunies, 1946.

al-Ṭabarī. *Taʾrīkh al-rusūl wa-l-mulūk*. Edited by M.J. de Goeje. Vol. 3.1. E.J. Brill, 1879.

al-Ṭabarī. *Taʾrīkh al-rusūl wa-l-mulūk*. Edited by M.J. de Goeje. Vol. 3.2. E.J. Brill, 1881.

al-Ṭabarī. *Taʾrīkh al-rusūl wa-l-mulūk*. Edited by M.J. de Goeje. Vol. 3.4. E.J. Brill, 1890.

al-Yaʿqūbī. *Kitāb al-buldān*. Edited by M. J. de Goeje. E.J. Brill, 1892.

Secondary sources

Bent, Josephine van den. "Caliphal Involvement in Water Provision in the Cities of the Early Abbasid Period." *Journal of Abbasid Studies* 13 (2025): 1–30.

Bent, Josephine van den, and Peter J. Brown. "Constructing Hydraulic Infrastructure in the ʿAbbāsid and Ṭūlūnid Capitals: Water Conduits in Baghdad, Samarra and Cairo between the Eighth and Ninth Centuries." *Al-Masāq* 37, no. 1 (2025): 126–45.

Bent, Josephine van den, and Angela Isoldi. "Pleasing God, Serving the Citizens. Charity and Water Supply in Cairo and Baghdad." In *City, Citizen, Citizenship, 400–1500. A Comparative Approach*, edited by Els Rose, Robert Flierman, and Merel de Bruin-van de Beek, 199–229. Palgrave Macmillan, 2024.

Berkel, Maaike van. "Ibn al-Furāt." In *Encyclopaedia of Islam, THREE Online*, edited by Kate Fleet, Gudrun Krämer, Dennis Matringe, John Nawas, and Devin J. Stewart. Brill, 2020. https://doi.org/10.1163/1573-3912_ei3_COM_30778.

Berkel, Maaike van, Nadia Maria El Cheikh, Hugh Kennedy, and Letizia Osti. *Crisis and Continuity at the Abbasid Court*. Brill, 2013.

Bonner, Michael. "The Waning of Empire, 861-945." In *New Cambridge History of Islam*, edited by Chase F. Robinson, 305–59. Cambridge University Press, 2010.

Bosworth, C. Edmund. "Some Remarks on the Terminology of Irrigation Practices and Hydraulic Construction in the Eastern Arab and Iranian Worlds in the Third-Fifth Centuries A.H." *Journal of Islamic Studies* 2, no. 1 (1991): 78–85.

Brown, Peter J. "Supplying a Medieval Metropolis: Water Management and Agriculture in the Hinterland of Early Islamic Basra." *Water History* 14 (2022): 379–98.

Brown, Peter J., and Maaike van Berkel. "Water Provision in Early Islamic Cities: A Multidisciplinary Approach to Urban Water Governance." In *City, Citizen, Citizenship, 400–1500. A Comparative Approach*, edited by Els Rose, Robert Flierman, and Merel de Bruin-van de Beek, 75–103. Palgrave Macmillan, 2024.

Buckley, R.P., trans. *The Book of the Islamic Market Inspector. Nihāyat al-Rutba Fī Ṭalab al-Ḥisba (The Utmost Authority in the Pursuit of Ḥisba) by ʿAbd al-Raḥmān b. Naṣr al-Shayzarī*. Oxford University Press, 1999.

Campopiano, Michele. "Cooperation and Private Enterprise in Water Management in Iraq: Continuity and Change between the Sasanian and Early Islamic Periods (Sixth to Tenth Centuries)." *Environment and History*, no. 23 (2017): 385–407.

Campopiano, Michele. "State, Land Tax and Agriculture in Iraq from the Arab Conquest to the Crisis of the Abbasid Caliphate (Seventh-Tenth Centuries)." *Studia Islamica* 1, no. 107 (2012): 1–37.

Christensen, Peter. *The Decline of Iranshahr: Irrigation and Environments in the History of the Middle East, 500 B.C. to A.D. 1500*. Museum Tusculanum Press, 1993.

Cook, Michael. *Commanding Right and Forbidding Wrong in Islamic Thought*. Cambridge University Press, 2001.

Cooperson, Michael. "The Abbasid 'Golden Age': An Excavation." *Al-ʿUṣūr al-Wusṭā* 25 (2017): 41–65.

Daftary, Farhad. *A History of Shiʿi Islam*. I.B. Tauris, 2013.

Ghabin, Ahmad. "Muḥtasib." In *Encyclopaedia of Islam, THREE*, edited by Kate Fleet, Gudrun Krämer, Dennis Matringe, John Nawas, and Devin J. Stewart. Brill Online, 2008. http://dx.doi.org/10.1163/1573-3912_ei3_COM_40360.

Goeje, M.J. de, ed. *Annales quos scripsit Abu Djafar Mohammad ibn Djarir at-Tabari. Introductio, Glossarium, Addenda et Emendanda*. E.J. Brill, 1901.

Kennedy, Hugh. *The Early Abbasid Caliphate. A Political History*. Croom Helm, 1981.

Kennedy, Hugh. *The Armies of the Caliph: Military and Society in the Early Islamic State*. Routledge, 2001.

Kennedy, Hugh. "The Feeding of the Five Hundred Thousand: Cities and Agriculture in Early Islamic Mesopotamia." *Iraq* 73 (2011): 177–99.

Kennedy, Hugh. *The Prophet and the Age of the Caliphates*. Fourth Edition. Routledge, 2022.

Klein, Yaron. "Between Public and Private: An Examination of *Ḥisba* Literature." *Harvard Middle Eastern and Islamic Review* 7 (2006): 41–62.

Lane, Edward William. *An Arabic-English Lexicon*. Librairie du Liban, 1968.

Lange, Christian. "Changes in the Office of *Ḥisba* under the Seljuqs." In *The Seljuqs. Politics, Society and Culture*, edited by Christian Lange and Songül Mecit, 157–81. Edinburgh University Press, 2011.

Lassner, Jacob. *The Topography of Baghdad in the Early Middle Ages: Texts and Studies*. Wayne State University Press, 1970.

Le Strange, Guy. "Description of Mesopotamia and Baghdād, Written about the Year 900 A.D. by Ibn Serapion. The Arabic Text Edited from a MS. in the British Museum Library, with Translation and Notes (Part I)." *Journal of the Royal Asiatic Society of Great Britain & Ireland* 27, no. 2 (1895): 255–315.

Lev, Yaacov. *The Administration of Justice in Medieval Egypt: From the 7th to the 12th Century*. Edinburgh University Press, 2020.

Melchert, Christopher. "Māwardī's Legal Thinking." *Al-ʿUṣūr al-Wusṭā* 23, no. 1 (2015): 68–86.

Micheau, Françoise. "Baghdad in the Abbasid Era: A Cosmopolitan and Multi-Confessional Capital." In *The City in the Islamic World*, vol. 1, edited by Salma Khadra Jayyusi, Renata Holod, Antillio Petruccioli, and André Raymond, 221–45. Brill, 2008.

Micheau, Françoise. "Baghdad, an Imperial Foundation (762–836 CE)." In *The Cambridge World History*, edited by Norman Yoffee, 397–415. Cambridge University Press, 2015.

Morony, Michael. *Iraq after the Muslim Conquest*. Princeton University Press, 1984.

Osti, Letizia. "The Remuneration of a Court Companion in Theory and Practice: A Case Study." *Journal of Abbasid Studies* 1 (2014): 85–107.

Pamuk, Şevket, and Maya Shatzmiller. "Plagues, Wages, and Economic Change in the Islamic Middle East, 700–1500." *Journal of Economic History* 74, no. 1 (2014): 196–229.

Rébillard, Eugénie. *Imposer l'ordre. La police dans les villes et les campagnes de l'Iraq abbasside (132-334/750-945)*. Beirut: Presses de l'Ifpo, 2024. https://books.openedition.org/ifpo/26737.

Sadan, J. "A New Source of the Būyid Period." In *Israel Oriental Studies IX*, edited by Gideon Goldenberg, 355–76. Tel Aviv University, 1979.

Sadan, J. *Nouvelle source sur l'époque būyide*. Tel Aviv University, n.d.

Sadan, J., and A. Silverstein. "Ornate Manuals or Practical Adab? Some Reflections on a Unique Work by an Anonymous Author of the 10th Century CE." *Al-Qanṭara* 25, no. 2 (2004): 339–55.

Saliba, George, trans. *The History of Al-Ṭabarī, Volume XXXV. The Crisis of the ʿAbbāsid Caliphate*. State University of New York Press, 1985.

Scheiner, Jens, and Damien Janos. "Baghdād: Political Metropolis and Intellectual Center." In *The Place to Go: Contexts of Learning in Baghdad, 750-1000 C.E.*, edited by Jens Scheiner and Damien Janos, 1–45. Gerlach Press, 2021.

Scheiner, Jens, and Isabel Toral. "Baghdād's Topography and Social Composition." In *Baghdād. From Its Beginnings to the 14th Century*, edited by Jens Scheiner and Isabel Toral, 1–43. Brill, 2022.

Serjeant, R. B. "A Zaidī Manual of Ḥisbah of the 3rd Century (H)." *Rivista Degli Studi Orientali* 28, no. 1 (1953): 1–34.

Silverstein, Adam J. *Postal Systems in the Pre-Modern Islamic World*. Cambridge University Press, 2007.

Verkinderen, Peter. *Waterways of Iraq and Iran in the Early Islamic Period. Changing Rivers and Landscapes of the Mesopotamian Plain*. I.B. Tauris, 2015.

Waines, David. "The Third Century Internal Crisis of the Abbasids." *Journal of the Economic and Social History of the Orient* 20, no. 3 (1977): 282–306.

Wiedemann, Eilhard. "Beiträge Zur Geschichte Der Naturwissenschaften. X. Zur Technik Bei Den Arabern." *Sitzungsberichte Der Physikalisch-Medizinischen Sozietät Zu Erlangen* 38 (1906): 307–57.

CHAPTER 7

Fighting over Water: Everyday Water Issues in Late Buyid and Early Seljuq Baghdad ($5^{th}/11^{th}$ Century)

Vanessa Van Renterghem

Abstract

This paper discusses the relations between civil population and rulers in the middle period of Abbasid Baghdad ($5^{th}/11^{th}$ century, under Buyid then Seljuq rule), as seen through the prism of water management. Drawing on a corpus of Arabic historiographical sources, mainly chronicles and biographical collections, it examines the actors and infrastructures providing access to water, and particularly drinking water, during this period, and tries to assess the role of rulers and administration in this task. It then investigates some cases of social conflicts involving the access to water during the first half of the $5^{th}/11^{th}$ century, in the context of the religious and political fragmentation of Baghdad's population. In doing so, it investigates the role of access to water in shaping the social geography of late Abbasid Baghdad.

Keywords: Baghdad; Iraq; Abbasids; Buyids; Seljuqs; water management; waterwheels; water carriers; Tigris; canals; *fitna*; Shiʿi-Sunni strife; *ʿayyārūn*; *khafāra* ("protection" tax)

Baghdad, as an imperial foundation, was built in a location where water was easily available. As demonstrated by Josephine van den Bent in the previous chapter, during the two first Abbasid centuries, Baghdadis had at their disposal a mixed system assuring water provision, for drinking but also for hygienic, agricultural and industrial needs. This system rested on the network of canals bringing water from the Euphrates to the Tigris in West Baghdad and, in the eastern part of the city, from the Nahrawān, itself a derivation of the Tigris.[1] From these canals, as well as from the Tigris itself, water was drawn and made accessible to the population by way of waterwheels, individually-funded open-air tanks (*birka*, *ḥawḍ*), but also through the work of water carriers (*saqqāʾūn*, pl. of *saqqāʾ*) who sold water carried by animals. Van den Bent discussed to what extent the Abbasids constructed and took care of the main pieces of infrastructure, the canals, and showed how the

[1] See the contribution of Van den Bent in this volume; see also Le Strange, *Baghdad*, 47–56, 174–77.

local population might participate in their maintenance. The period here under study is usually presented as a period of urban decay, in parallel with the loss of Abbasid authority under Buyid (mid-4th/10th to mid-5th/11th centuries) then Seljuq rule (mid-5th/11th to mid-6th/12th centuries), even though the city was still an active and lively metropolis.[2] Exploring access to drinking water and other water-related topics[3] is a way to investigate the more general issue of urban decline, as well as the relations between the civil population and local rulers and administration in the context of the overall weakness of Abbasid authority. From the thirties of the 4th/forties of the 10th century, the political, fiscal and military powers of the caliphs were confiscated by the occupant of the highest military position, that of Great Emir (*amīr al-umarāʾ*). For a decade, elite leaders from the Abbasid entourage were appointed to this position. It was then taken over by Iranian military rulers, the Buyids, from Daylam (south of the Caspian Sea), who established their firm tutelage over the Abbasid caliphate (334–447/945–1055). The Buyids were Shiʿi Muslims and their rule was more favourable to the Baghdadi Shiʿi population than the rule of the early Abbasids had been. The Buyids themselves were divided, different members of the family ruling regional emirates in Iranian lands, and their rule in Iraq progressively disintegrated in the course of the first half of the 5th/11th century. They were eventually swept away by the Seljuqs, themselves military rulers, and relying on Turkmen troops from Central Asia. The Seljuqs, despite the fact that they were Sunni Muslims, exerted a similar political, military, fiscal and administrative tutelage over the Abbasid caliphate. They ruled for a century over Baghdad and Iraq (447–547/1055–1152), generally favouring Sunni groups.[4]

The 5th/11th century is rarely studied for its own sake, although it witnessed a meaningful change for the local population: the substitution of the Shiʿi Buyids by the Sunni Seljuqs. For this very reason, studying this period can be taken as an opportunity to compare the situation under both dynasties. The choice of studying the 5th/11th century is also justified by the historiographical rupture that occurred at the end of the 4th/10th century: the sources available to study the end of the Buyid period were mostly written in the 6th/12th century or later, and also cover the Seljuq period.

[2] Van Renterghem, *Élites*. On Baghdad during Buyid and Seljuq times, see Alshaar, "Baghdād under Būyid rule;" Van Renterghem, "Baghdād under the Saljūqs."

[3] Except for the topography of the canals and the floods of the Tigris, the specific issue of drinking water supply in medieval and Ottoman Baghdad has only been investigated by Iraqi scholar ʿAbd al-Salām Raʾūf, *Mashāriʿ*. For a synthesis on the state of the canals at the beginning of the 4th/10th century, see Ahola and Osti, "Baghdad," 223–26.

[4] On the political context: Bowen, "Last Buwayhids;" Busse, *Chalif und Grosskonig*; Kabir, *Buwayhid Dynasty*; Donohue, *Buwayhid Dynasty*; Hanne, *Putting the Caliph*; Kennedy, "Late ʿAbbasid pattern;" Peacock, *Great Seljuk Empire*; El-Hibri, *Abbasid Caliphate*.

The first half of the 5th/11th century, which witnessed the decline of Buyid rule over Iraq, and particularly over Baghdad, was an extremely troubled time. Both Buyid and Abbasid representatives experienced the utmost difficulties in controlling Baghdad, a city then plagued by the exactions of civil militias and brigands called *ʿayyārūn*,[5] various military groups, and also economic difficulties. These troubles culminated with the Fatimid seizure of Baghdad, which lasted one full year (450–51/1058–60) during which the Abbasid caliph al-Qāʾim (r. 422–67/1031–75) was exiled to the Jazīra. The second half of the century, under the rule of the Seljuq sultans who swept the Buyids as well as the Fatimids out of Baghdad and central Iraq, saw the return of al-Qāʾim to Baghdad, but also saw the city become the theatre of many sectarian riots and fights. Throughout the 5th/11th century, civilians were also struck by floods, epidemics, and famines.[6] In these troubled, mainly unprosperous times, access to water was a daily concern, rarely taken care of by the rulers. It could also be instrumentalised during conflicts between neighbouring districts.

This chapter first studies the actors and infrastructures providing access to water, and particularly drinking water, during the 5th/11th century, and investigates the role of rulers and the public administration in this task. It then focuses on three cases of social conflict involving access to water during the first half of the century, and assesses the social, political and topographical consequences of water issues in Baghdad. It relies on a corpus of Arabic historiographical sources mainly produced in the 6th-7th/12th-13th centuries: chronicles such as the *Kitāb al-muntaẓam fī taʾrīkh al-mulūk wa-l-umam* by the Ḥanbali Baghdadi Ibn al-Jawzī (d. 597/1201); the *Mirʾāt al-zamān wa-taʾrīkh al-aʿyān* of his grandson Sibṭ ibn al-Jawzī (d. 654/1256); and the *Kitāb al-kāmil fī al-taʾrīkh* of Ibn al-Athīr (d. 630/1233); biographical collections of Baghdadis or people who visited Baghdad; travelogues, and other literary productions. Although none of these texts is specifically dedicated to the questions of water and water management, they do include anecdotes about everyday life in Baghdad that provide precious, albeit unsystematic, information on these topics. It should be made clear that each author had his own positioning towards the political and military authorities: for example, Ibn al-Jawzī, a strongly Sunni-orientated scholar living in Baghdad who sought the patronage of the Abbasid entourage, always took care to give a positive image of the Abbasid caliph, never emphasising or even portraying his weakness or lack of power, while Ibn al-Athīr, who lived in Mosul under the local rule of the Zangids, vassals of the Seljuqs, was

[5] On the *ʿayyārūn* in Abbasid and Buyid times, see Sabari, *Mouvements populaires*; Cahen and Hanaway, "*ʿAyyār*;"; Cheikh-Moussa, "L'historien;" Tholib, "al-Qādir billāh", 84–94; Donohue, *Buwayhid Dynasty*, 338–43; Tor, *Violent Order*; Tor, "*ʿAyyār*;" Van Renterghem, *Élites*, 1:450–58; Amabe, *Urban Autonomy*, 24–25; Rébillard, *Imposer l'ordre*, 451–56.

[6] Floods of the Tigris have been studied by al-Sūsa, *Fayḍānāt*.

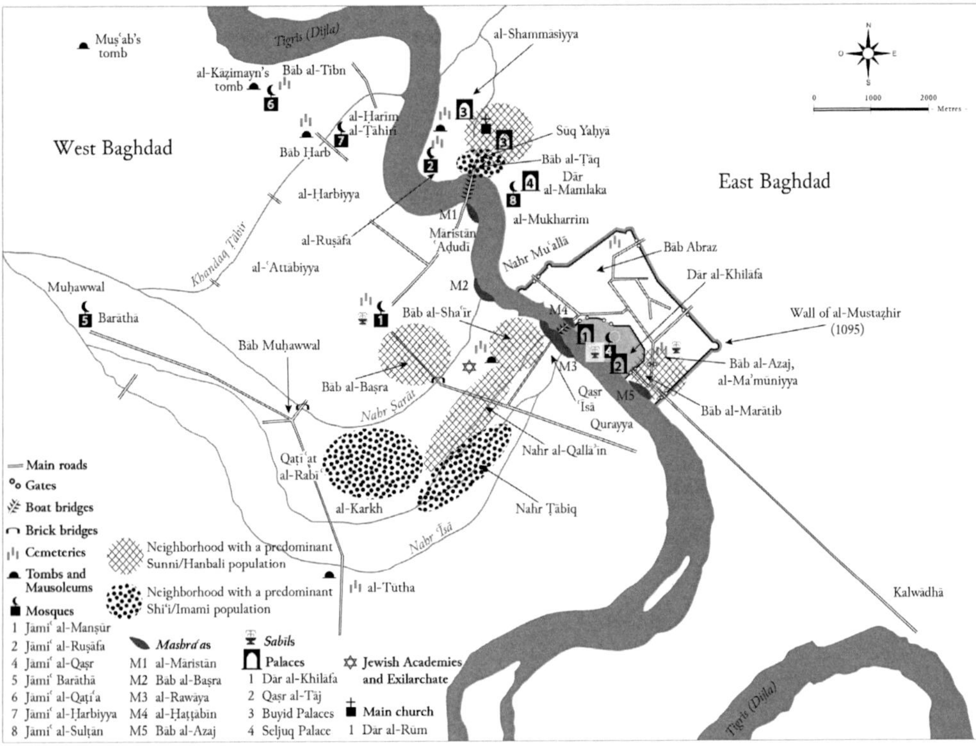

Figure 7.1: General Map of Baghdad during the Buyid and Seljuq periods (mid-4th/10th to mid-6th/12th centuries), with an emphasis on water and localisation of confessional groups. Created by the author.

more detached from this necessity to praise and defend the caliphs—but on the other hand, his knowledge of the urban context was less precise than the Baghdadi Ibn al-Jawzī's. These medieval authors also had different sectarian perspectives: Ibn al-Jawzī takes some strong anti-Shiʿi stances, while Ibn al-Athīr adopts more nuanced perspectives. These standpoints may influence these medieval narratives, and must be taken into consideration when reading and using the sources.

Water Facilities in 5th/11th-Century Baghdad

Like Cairo, Baghdad was built besides a large river. Water was abundantly provided by the Tigris, which crossed the city and divided it into West and East Baghdad, and by the system of canals, mainly derived from the Euphrates on the west side, but from the Tigris on the east side. Water was used for drinking, washing, watering the gardens, orchards and palm groves inside and outside the city, for hygienic purposes in hammams and mosques, and for numerous industrial activities like paper or textile production. Water was as necessary to animals as to humans, and could also be used as a source of energy for water-mills. But water was also a cause of death and epidemics, when polluted by dead human or animal corpses; it might lack in some areas (for example when canals were dried up by drought or breaches), or become too abundant in others, and destroy entire districts when the Tigris flooded, a kind of catastrophic event that was recurrent in medieval Baghdad. River and waterways were also used as transport routes, and wharves on the Tigris were important elements of the urban topography and structure. The Tigris played a crucial role in dividing the city into two parts, East and West Baghdad, connected by one, two or three boat-bridges depending on the period; only one or two functioned during the 5th/11th century.[7] It must also be noted that the question of wastewater disposal is very difficult to deal with, since the sources do not inform us about the way it was put back into the Tigris or into waterways in other areas of the city.

The infrastructure ensuring access to water in 5th/11th-century Baghdad was still the same as that designed for the imperial foundation of the city. As shown in the previous chapter, some facilities—such as the canals—seem to have deteriorated in the 3rd/9th or 4th/10th century, in comparison to the first Abbasid period, but the Buyid ruler ʿAḍud al-Dawla (r. 338–72/949–83) restored at least some of them.[8] The main source for water were the canals and the Tigris, which was accessible in particular places, the *mashraʿāt*. The term *mashraʿa* (pl. *mashraʿāt*) is often translated by "wharf",[9] but this translation can be misleading since the Arabic term does not refer to a structure built over the water, but to a place where the river was accessible to humans, animals, and possibly light boats. Boat bridges were also placed and tied from one *mashraʿa* to another. In times of siege, *mashraʿāt* were strategic points whose control was vital. When the rebel *amīr* al-Basāsīrī, supported by Fatimid subsidies, besieged Baghdad in Dhū al-qaʿda 450/December 1058, he settled with his troops in Mashraʿat al-Rawāyā ("the *mashraʿa* of the waterskins"), outside of al-Karkh;[10] his ally, the ʿUqaylid Quraysh ibn Badrān, established his own camp in Mashraʿat Bāb al-Baṣra.[11] Both places were situated on the west bank of the city, since al-Basāsīrī, Quraysh ibn Badrān and their troops came from the west,

7 On the bridges of Baghdad, see Le Strange, *Baghdad*: 177–86; Ahola and Osti, "Baghdad," 226–27. For the 5th/11th century, see Makdisi, "Topography," 186n2; Van Renterghem, *Élites*, 1:404.

8 See the contribution of Van den Bent in this volume.

9 See, for example, Le Strange, *Baghdad*, 85n1.

10 This localisation is given by Yāqūt, *al-Buldān*, 4:421.

11 Ibn al-Jawzī, *al-Muntaẓam*, 16:32; Ibn al-Athīr, *al-Kāmil*, 8:153; Sibṭ ibn al-Jawzī, *Mirʾāt al-zamān*, 19:65.

from Syria and Jazīra. Establishing a military camp in a place where access to the Tigris was possible ensured the provision of drinking water for the troops, but also had a strategic aspect, since the troops had to cross the Tigris to reach the caliph's palaces, in East Baghdad, while the boat bridges had been dismantled to prevent the pro-Fatimid armies from reaching it easily.

As for drinking water during peace time, *mashraʿāt* were the places where the water carriers, the *saqqāʾūn* (sing. *saqqāʾ*), would collect the water that they then sold in the districts located farther away from the Tigris. Water was poured into waterskins, then placed on beasts of burden, either camels or mules, called *rawāyā* (sing. *rāwiya*).[12] Drinking water could be stocked in jars (*jarra*, pl. *jirār*; *ḥubb*, pl. *ḥibāb*; *muzammala*, pl. *muzammalāt*), and drunk using a mug or drinking cup (*kūz*, pl. *kizān*, *akwāz*).

Mashraʿāt played an important role as an interface between the Tigris and the population. They were the destination of the main roads perpendicular to the river, and as such, they were structuring elements of the local road network. Overall, they were significant landmarks of Baghdad's topography, and may deserve greater attention than what they have received so far.[13]

Water could also be drawn from the Tigris and the smaller canals, for irrigation or for industrial use, by the use of waterwheels. Some are mentioned for 5th/11th-century Baghdad and its surroundings. The main denomination for waterwheels was the word *dūlāb* (pl. *dawālīb*), of Persian origin. Moved by animals (camels or cows), or sometimes by hand, such waterwheels were made of wood.[14] They are occasionally mentioned by the chronicles when destroyed or damaged by a fire. In 431/1039 for example, Bedouin Arabs plundered some villages to the west of Baghdad, encouraged by the ʿUqaylid ruler of Mosul, Qirwāsh ibn al-Muqallad, then opposed to the Buyid ruler Jalāl al-Dawla. To make sure that the rural economy would be lastingly affected, the Arabs let the cattle loose and set fire to the

12 The term *rāwiya*, pl. *rawāyā*, could also be used to indicate the water carrier himself, as specified by Yāqūt: "the *rāwiya* is the camel [or beast of burden] used to draw [and carry and sell] water; the water carrier is also [called] a *rāwiya*" (*al-rāwiya huwa al-baʿīr alladhī yustaqā ʿalayhi al-māʾ, wa-l-rajūl al-mustaqī ayḍan rāwiya*; Yāqūt, *al-Buldān*, 3:116). In Baghdad, the term *rāwiya* or *rawāyā* also referred figuratively to the load of water itself, when carried and sold by the *saqqāʾūn*.

13 To my knowledge, no specific study has been devoted to the ports and wharves of Baghdad or even to fluvial transportation on the Tigris or Euphrates towards the Abbasid capital. Interestingly, no tentative map of medieval Baghdad, including Le Strange's, systematically indicates the localisation of these *mashraʿāt*, though they were crucial interfaces between the city and the river, and only the pontoon bridges are depicted.

14 The technical aspects of waterwheels and water-lifting apparatuses have been discussed based on textual, iconographical and archaeological evidence. See for example Hill, *Engineering*, 156 sq; Ruggles, *Islamicate Environments*, 28–38.

waterwheels (*dawālīb*) of this area.[15] At the end of the same year (431/1040), Turkish soldiers rebelled against Jalāl al-Dawla and perpetrated similar actions, including the burning of waterwheels, in Baghdad and its surroundings.[16]

Terminology for waterwheels fluctuated. They were sometimes named *nāʿūra* (pl. *nawāʿīr*, norias). Ibn al-Jawzī signals one that got stuck in the great cold during the winter of 418/1027: the water had frozen on both sides of the Tigris, and icicles hung down from the blocked noria.[17] Another term was *sāniya* (pl. *sawānī*), which may allude to a smaller wheel or to wheels used to take water from wells.[18] The construction of such equipment is mentioned in 502/1108, among other important public works undertaken by Behrūz, the popular Seljuq *shiḥna* (civil governor) of Baghdad.[19]

The functioning of such waterwheels is rarely detailed for Baghdad for the period, but some information is given about similar wheels in Basra that burnt in a fire in 483/1090. On this occasion, Ibn al-Jawzī describes the system of water abduction in Basra:

> [This system] was destroyed [by the fire]. It rested on waterwheels (*dawālīb*) that turn and carry [up] the water, which is thrown into lead pipes (*qanāt al-raṣāṣ*) that go to the cisterns (*maṣāniʿ*), distant one *parsanj*[20] from the water.[21]

Waterwheels, moved by animal force in order to elevate the water that was then poured into a canal system to reach more or less distant areas of the city, should be distinguished from watermills (*raḥā*, pl. *arḥāʾ*), some of which were reported on the Euphrates when a terrible drought prevented them from functioning during the summer of 420/1029, along with the (grain) mill (*ṭaḥūn*, pl. *ṭawāḥīn*) next to them.[22]

15 Ibn al-Jawzī, *al-Muntaẓam*, 15:274; Sibṭ ibn al-Jawzī, *Mirʾāt al-zamān*, 18:433.

16 Sibṭ ibn al-Jawzī, *Mirʾāt al-zamān*, 18:436.

17 Ibn al-Jawzī, *al-Muntaẓam*, 15:184.

18 See Lane, *Lexicon*, 1450, root < SNY >.

19 Ibn al-Jawzī, *al-Muntaẓam*, 17:112. On the *shiḥna* Behrūz al-Ghiyāthī, who bore the laudatory *laqab* Mujāhid al-Dīn, see Van Renterghem, *Élites*, 1:466.

20 Measure of distance, equivalent to 6 to 8 km.

21 Ibn al-Jawzī, *al-Muntaẓam*, 16:289. Unless otherwise specified, all translations are mine.

22 Ibn al-Jawzī, *al-Muntaẓam*, 15:197. The presence of a grain mill (confirmed by the fact that the chronicler gives, after mentioning the mill itself, the price of the flour that was produced by this mill) next to other watermills implicates that in this case the grain mill was also moved by water and not by animal or human force. Donald Hill states that "very powerful mills driven by paddle-wheels were used on the Tigris and Euphrates" in medieval times, and describes the technology of such watermills, drawing from Arabic sources (*Engineering*: 145, 164–66). Medieval and Ottoman watermills from the area of Mosul have been studied from an archaeological perspective by Tonghini and Usta, "Mesopotamian Waterscapes" and Tonghini, "Flour for the Caliph," showing a continuity of activity of these mills from the Umayyad period until the 20th century. The wheels they describe were horizontal.

A district of Baghdad bore the name of "Gate of the (Water) Mills" (Bāb al-Arḥā'), but the mills themselves are not alluded to in the sources used for this research. The last point that should be documented is that aristocratic houses in Baghdad had their own waterwheels on the Tigris, as described by the Baghdadi scholar Ibn ʿAqīl (d. 513/1119):

> [...] On the west bank is al-Karkh, on the shore of which are palaces, in orderly disposition, all with waterwheels, gardens, and balconies facing [the palaces across the Tigris]. [...] And the ducks playfully swim together on the wharf of the riverside palace. Many a time would the singing voices of this quarter mix with the sound of its waterwheels, the quacking of its ducks, the clamour of its soldiers and servants, while the Tigris gently streamed along between the two rows of its riverside palaces.[23]

Ibn ʿAqīl's description raises the attention to the sound made by the wooden waterwheels, certainly an important element of the Baghdadi soundscape; the sound he alludes to, a tinkling (*ranīm*), seems to have been pleasant to his ears. Private waterwheels ensured that the rich owners of these riverside palaces could get sufficient water to drink, cook and water their gardens, for themselves and their households. This is one of the elements that explains how the banks of the Tigris were favoured as aristocratic neighbourhoods on both sides of the river.[24]

While residents of the aristocratic palaces had easy access to water, modest Baghdadis living areas more distant from the Tigris had either to buy their water from the professional water carriers or to go to the river and bring the water they needed themselves. There is a clear opposition in the sources between the water sold by the *saqqā'ūn*—some chronicles recorded the price of a skin of water, especially when it was rising—and the free water delivered by some facilities, in this case called *subul* (sing. *sabīl*). Free drinking water could in particular be obtained from fountains (*siqāya*) situated in the mosques or in some public places. Unfortunately, little information is available about these: one is mentioned in Bāb al-Azaj in 448/1056,[25] and another one, called Siqāyat al-Raḍī, was situated inside al-Manṣūr's mosque in West Baghdad.[26] No more details are available regarding the number of people to whom these *subul* provided water.

As for the previous period,[27] the Arabic chronicles give very little information about the water carriers, despite the fact that it is clear that their role in providing

[23] Quoted in Ibn al-Jawzī (attributed), *Manāqib Baghdād*, 26–27. English translation by Makdisi, "Topography", 189–90.

[24] Van Renterghem, *Élites*, 1:415–28.

[25] By Sibṭ ibn al-Jawzī, *Mir'āt al-zamān*, 18:496.

[26] Ibn al-Jawzī, *al-Muntaẓam*, 18:205. Based on the chronicles, Ra'ūf describes and localises seven *siqāyāt* in Baghdad for the Abbasid period. Ra'ūf, *Mashāriʿ*, 44–55

[27] As shown in the contribution of Van den Bent in this volume.

Baghdadis with water was important. During the summer of 485/1092, *saqqāʾūn* were recruited by the Abbasid vizier Ibn Jahīr to fight a fire that was devastating the Nahr Muʿallā area and its adjacent markets, to the north of the caliphal precinct in East Baghdad. The details of their intervention are not specified, but their importance as professionals water bearers was clear in this case of emergency.[28]

Saqqāʾūn were obviously a modest fraction of all Baghdadi workers. For the 6th/12th century, Ibn al-Dubaythī (d. 637/1239) alludes to a hadith transmitter and Quran reader, Abū ʿAbd Allāh Muḥammad the water carrier (*al-saqqāʾ*), who died in 572/1176, and who had only this humble occupation: "He drew water from the Tigris and carried it to people's houses, and made a living only from what he gained [from this work]."[29] The same expression is used about the Sufi master Abū al-Najīb al-Suhrawardī (d. 553/1168) to emphasise his ascetic character and his modest way of life.[30] Even so, *saqqāʾūn* were not among the poorest inhabitants of the city (such as day labourers), because they usually owned the beasts of burden they used for carrying the waterskins. To lose their animal meant that they were no longer able to work. One *saqqāʾ* is mentioned in 461/1068 as having foolishly slaughtered the cow with which he was working, and distributing the meat in charity, as a manifestation of his tremendous happiness caused by the nomination of Ibn Jahīr to the vizierate after he had been in disgrace.[31] In 426/1034 and 428/1035, among other exactions, unruly *ʿayyārūn* stole camels and mules belonging to the *saqqāʾūn*.[32] Both anecdotes emphasise the cost of beasts of burden and prove that they were the property of the *saqqāʾūn* themselves. The water carriers thus appear to have been private labourers who used their own assets as the tools of their trade.[33]

Were the *saqqāʾūn* all men? Both Ibn al-Jawzī and Ibn al-Athīr report that in 481/1088, the people of Bāb al-Azaj, a neighbourhood of East Baghdad, attacked a woman who was pouring water for people on the Tigris bank.[34] They took her water for free, shouting "*al-sabīl!*" ("free water"), smashing the jars (*muzammalāt*) she owned. The woman had to seek protection from the Seljuq military governor

[28] Ibn al-Jawzī, *al-Muntaẓam*, 16:299.

[29] *Kāna yastaqī al-māʾ min Dijla wa-yaḥmiluhu ilā buyūt al-nās, wa-lā yaʾkul illā min kasbihi*; Ibn al-Dubaythī, *Dhayl Taʾrīkh*, 1:491, #348.

[30] *Ishtaghala bi-l-zuhd wa-l-mujāhada mudda, ḥattā annahu kāna yastaqī al-māʾ bi-Baghdād wa-yaʾkul min kasbihi.* Curiously, this fact is reported by Ibn ʿAsākir (d. 571/1176) in his *History of Damascus*, and not by the Iraqi biographers of Abū al-Najīb al-Suhrawardī. This leads us to consider this *saqqāʾ*-narrative as a trope used to emphasise poverty or renouncement. Ibn ʿAsākir, *Taʾrīkh madīnat Dimashq*, 36:412. The same sentence is used later by Yāqūt, about the same person: *al-Buldān*, 3:290.

[31] Sibṭ ibn al-Jawzī, *Mirʾāt al-zamān*, 19:209.

[32] Ibn al-Jawzī, *al-Muntaẓam*, 15:245, 257.

[33] For the previous period, see the contribution of Van den Bent in this volume.

[34] Ibn al-Jawzī, *al-Muntaẓam*, 16:277; Ibn al-Athīr, *al-Kāmil*, 8:319.

(*shiḥna*) who was nearby, on his boat. This woman was clearly selling the water she was drawing from the Tigris. Ibn al-Athīr specifies that this behaviour (attacking this woman, or maybe other women working in *siqāya*) was customary to this group of men, which leads us to characterise it as a gendered assault.[35] Though the anecdote is not entirely clear, it indicates that some *saqqāʾūn* were women—although this particular woman does not seem to have been working with beasts of burden who carried waterskins, but rather in all likelihood drew the water from the river and kept it cool in her jars. It also alludes to the existence of conflicts over selling water instead of giving it for free (*sabīl*). This event happened in the context of conflicts between the Sunni and Shiʿi neighbourhoods of the west bank, and will be investigated further in the third part of this chapter. There is no other reference to women *saqqāʾāt* in the sources for the period.

In all cases, the *saqqāʾūn*'s role was crucial in everyday life, in particular for the inhabitants of modest means who lived far away from the Tigris and had to purchase their own water. To get fresh water to his palace every day, the caliph himself used *saqqāʾūn* who worked exclusively for him. In 530/1136, to make the Seljuq sultan Masʿūd realise how much money he needed just to run his palaces, the newly enthroned caliph al-Muqtafī (r. 530–55/1136–60) simply informed him that he had eighty mules carry water to the palace every day.[36] It may be assumed that contrary to the preceding example, the mules belonged to the caliph and the *saqqāʾūn* were his employees. It also suggests that the caliph's palaces were probably not endowed with waterwheels at this period. Later on, some time before the end of the 6th/12th century, a remarkable waterwheel (*dūlāb*) was built to supply the Abbasid palaces. It is described by Ibn al-Sāʿī as bordering on the palace of Banafshā (d. 598/1201), the concubine of caliph al-Mustaḍīʾ (r. 566–75/1170–80):

> The palace was adjacent to four waterwheels that brought water from the Tigris to the magnificent Caliphal Palace. Each wheel was positioned higher than the previous one, so that the first would take water from the Tigris, the second from the first, the third from the second, and the fourth from the third.[37]

The building of this impressive work, however, is not recorded in the chronicles of the time.

Waterwheels and water carriers were the main suppliers of drinking water in 5th/11th-century Baghdad. We also know that wells (*biʾr*, pl. *ābār*) were in use in the city, but as is made obvious from the medieval sources and as analysed in the

35 Ibn al-Athīr uses the expression "as was their custom" (*ʿalā ʿāda lahum*).

36 Ibn al-Athīr, *al-Kāmil*, 9:78.

37 Ibn al-Sāʿī, *Nisāʾ al-khulafāʾ*, 110–111.

previous chapter,[38] the water from the wells was not meant to be drunk or at least was not a first choice for drinking water. It might have been used to water orchards and gardens (*basātīn*), or maybe given to the cattle, but no information is available on this subject. If wells are mentioned in chronicles, it is either to signal them as places where things were hidden or where corpses were thrown. Drinking the water from the wells is only reported during sieges, as a sign of the great extremity endured by the besieged population.[39] We may nevertheless understand that their role in providing (non-drinking) water in Baghdad was important, as indicated by the fact that wells were reported empty in times of drought.[40] Conversely, water would overflow from the wells when the Tigris was flooding, damaging the surrounding buildings,[41] or at times of heavy rain.[42] Canals also played a role in supplying water, but the sources do not specify if their water was only used for crafts and industry, for watering gardens and orchards, for hygienic purposes (such as feeding the hammams), or if it was also drunk by people and animals.

Caring for Water: State Involvement or Private Regulation?

During the entire 5th/11th century, the role of rulers, be they Abbasid, Buyid or Seljuq, in guaranteeing and regulating the distribution of water seems to have been a far from onerous one. Rulers sometimes played a role in having major water infrastructure built, but do not appear to have deeply cared about the question of the civil population's access to potable water.

Information about the building or repair of water infrastructures is scarce for the period. In 475/1083, on the occasion of the reconstruction of the caliphal palace mosque, Jāmiʿ al-Qaṣr, the Abbasid vizier Fakhr al-Dawla ibn Jahīr[43] endowed it with a fountain ending with a jet of water (*fawwāra*), thus providing free water (*siqāya*) to people. The water was drawn from the vizier's own house,[44] and reached the mosque through an underground canal (*qanāt*). Ibn al-Jawzī emphasises the

38 See the contribution of Van den Bent in this volume.

39 This happened in 419/1028 to people besieged by rebel soldiers inside the Dār al-Mamlaka, the Buyid palaces (Ibn al-Jawzī, *al-Muntaẓam*, 15:191).

40 As noted by Ibn al-Jawzī for year 458/1066 (*al-Muntaẓam*, 16:96).

41 Ibn al-Jawzī, *al-Muntaẓam*, 16:155 (year 466/1074).

42 Ibn al-Jawzī, *al-Muntaẓam*, 17:204 (year 516/1122).

43 On the family of the Banū Jahīr, see Hanne, "Banū Jahīr;" Van Renterghem, *Élites*, 1:370–75.

44 The sources do not give the location of his palace (*dār*), but it is highly probable that, like most of the aristocratic residences of the time, it was situated on the bank of the Tigris, and possessed its own waterwheel(s). On the areas of residence of political elites in Seljuq Baghdad, see Van Renterghem, *Élites*, 1:417–21.

public usefulness of this equipment, although it is difficult for modern eyes to evaluate to what extent this construction can be considered private charity, and to what extent it stemmed from a state duty. Nothing is said about the funds with which the fountain was founded, but the source of the water must have been private waterwheels owned by the vizier.[45] This may guide us to conclude that the notions of "public" and "private" are not relevant for analysing the politics of these endowments in the period under consideration.

By contrast, chroniclers report that in 502/1108 the Seljuq sultan Muḥammad Tapar (r. 498–511/1105–18) entrusted his *shiḥna* Behrūz with the building of the Seljuq palaces and mosque (Dār al-Mamlaka, Jāmiʿ al-Sulṭān) in East Baghdad. Behrūz was also responsible for the public works (*aʿmāl*) in Iraq. The first work mentioned was the cleaning and building of waterwheels; Sibṭ ibn al-Jawzī adds that he also dug some canals, which could mean that he had new canals constructed or that he had old ones maintained and cleaned.[46] Interestingly, Ibn al-Athīr allows us to understand where the money necessary to these important works was taken from: the sultan had beforehand arrested, and heavily taxed, the Abbasid state treasurer (*ṣāḥib al-makhzan*) and a member of the prestigious Banū al-Muslima family.[47]

Building water infrastructure was a demonstration of political as well as economic power, meaning that the authority to undertake such works could be contested. In 427/1035 or 428/1036, in a context of political conflict between the Abbasid caliph, the Buyid ruler, and the army, the greatest rival of the Abbasid dynasty, the Fatimid caliph, sent 5,000 dinars for fixing a canal bringing water from the Euphrates to Kufa. The inhabitants of Kufa asked the Abbasid caliph for permission to accept this considerable amount of money, coming as it did from the principal enemy of the Abbasids. As reported by the Sunni sources, the initially reticent caliph authorised the use of this sum when jurisconsults testified that it was sound Muslim money—despite his political opposition to the Fatimids, the Abbasid caliph then had no good reason to reject the offer.[48] Along the same lines, one of the first actions undertaken by the rebel amir al-Basāsīrī, three months after he had taken Baghdad in the name of the Fatimids in 451/1059, was to enrol and pay workers to dig a canal in the vicinity of Karbala, the main destination of Shiʿi pilgrims in Iraq. Ibn al-Jawzī states that he undertook this work in order to fulfil a vow that he had made, and that he used an entire boatload of cereals to

[45] Ibn al-Jawzī, *al-Muntaẓam*, 16:224.

[46] Ibn al-Jawzī, *al-Muntaẓam*, 17:112; Sibṭ ibn al-Jawzī, *Mirʾāt al-zamān*, 20:27: *kirāʾ al-sawāqī*.

[47] Ibn al-Athīr, *al-Kāmil*, 8:574. The link between the two events is not clearly made by Ibn al-Athīr, but they are presented one after the other, as is often the case with related events in the chronicles. On the Banū al-Muslima, see Van Renterghem, *Élites*, 1:367–70.

[48] Ibn al-Jawzī, *al-Muntaẓam*, 15:257 (year 428/1036); Sibṭ ibn al-Jawzī, *Mirʾāt al-zamān*, 18:402 (year 427/1035).

remunerate the workers.[49] In both cases, the Shiʿi Fatimids paid for the extension of water infrastructure in a mainly Shiʿi area, linked to the holy place of Karbala, both fulfilling the needs of the local population in terms of access to water, and at the same time asserting their authority over the south of Iraq. The amounts of money or goods necessary to pay for these works were considerable and necessitated the engagement of states —whether the Abbasids or the Fatimids or their representative.

The Seljuq sultans understood the importance of such infrastructures, and in 479/1087, just after his first coming to Baghdad, sultan Malik Shāh (r. 465–85/1072–92) ordered the construction of a canal linking the Euphrates to Najaf[50]—in all likelihood, the same canal that had been planned by al-Basāsīrī, on which the sources give no further information; in fact, we don't know if his project was even completed, since al-Basāsīrī was killed at the end of 451/beginning of 1060. This waterway must have had great potential, since the Mongol governor of Baghdad had it dug (again) in 661/1263.[51] This demonstrates that building major water infrastructure in southern Iraq was, for Sunni as much as for Shiʿi powers ruling from Baghdad, a way to affirm their sovereignty over this area.

Only central or regional powers had the economic capacity to build important water facilities such as canals. Arabic sources do not provide much information about the funding and control of the maintenance of such structures. Chronicles occasionally report works being carried out to fix or clean the canals in Baghdad and its vicinity, especially when breaches (*bathq*, pl. *buthūq*) caused entire areas to be flooded, and canals themselves dried up. Such work required an abundant workforce. In 466/1074, when caliph al-Qāʾim (r. 422–467/1031–1075) ordered the Seljuq amir Aytakīn al-Sulaymānī to fix the breaches of the Qūraj, the main canal in north-east Baghdad,[52] he commandeered the Baghdadi populace (*ʿawāmm*) for this task. Aytakīn, who had briefly been *shiḥna* of Baghdad before being revoked by the caliph, did not perform the duty that he had been assigned to, and immediately fled Baghdad. A great flood then occurred that devastated Baghdad; the Qūraj

49 Ibn al-Jawzī, *al-Muntaẓam*, 16:45.

50 Ibn al-Jawzī, *al-Muntaẓam*, 16:259.

51 Honigmann and Bosworth, "al-Nadjaf."

52 As described by Yāqūt, who indicates the precise vocalisation of the term: "al-Qūraj is a canal (*nahr*) between al-Qāṭūl and Baghdad. Every time it floods, Baghdad is flooded from it. [...] Today, it is a calamity to the people of Baghdad. They work very hard to obstruct it and consolidate it, but when the water grows, its breaches open and [the water] submerges their houses and their city, and destroys them." Yāqūt, *al-Buldān*, 4:412. On the term *qūraj*, see also the contribution of Van den Bent in this volume.

was eventually fixed by an Abbasid official, the chief *naqīb* (*naqīb al-nuqabāʾ*).[53] Maintenance of this canal seems to have been an Abbasid responsibility, because on different occasions throughout the 6th/12th century, we see Abbasid viziers, judges, or other officials visiting the Qūraj to check its condition and trying to fix it if necessary.[54]

The caliphs also played a role in the maintenance of certain waterwheels. The unique indication we have of this fact comes from Sibṭ ibn al-Jawzī who mentions, in his obituaries, a man entrusted by caliph al-Mustaẓhir (r. 487–512/1094–1118) to watch over the fodder given to the cows who turned the wheels —implying that the man, a renowned hadith transmitter, received a stipend for this function.[55] But no more detailed information is given about the waterwheels themselves: were they for public use, or were they private wheels belonging to the Abbasid palaces?

Another anecdote underlines the fact that the caliphs, through the *muḥtasib* they appointed, sometimes tried to control the quality of the water, and to avoid its pollution. In 467/1075, just after the death of caliph al-Qāʾim and the enthronement of his grandson al-Muqtadī (r. 467–87/1075–94), the *muḥtasib* of the Abbasid precincts (al-Ḥarīm) forbade hammams from discharging their wastewater into the Tigris—wells should be dug for this—and proclaimed that fish and salted goods had to be washed in al-Najmī, a place on the west bank of the Tigris, and not anywhere else.[56] The other measures proclaimed the same day were of a moral kind: bans on prostitutes, pigeon racing, and nudity in hammams, and enjoining separation of sexes in boats crossing the Tigris.[57] Both sets of measures appear to proclaim the morality, hence good government, of the new caliph. No other intervention of the *muḥtasib* in the domain of the public use of water is recorded in the chronicles for the period.

To sum up, we have some (scarce) indication that there was a certain investment of Abbasid and Seljuq rulers and dignitaries in the building and maintenance of water amenities during the second part of the 5th/11th century. The first half of this century, in contrast, appears to have been a time of crisis and low investment in this field on the part of both Buyid and Abbasid authorities, certainly because of the diminution in their power and financial resources. Investment in building and management of important facilities can be considered a way of asserting control over given areas, and of proclaiming one's good government. Nonetheless, the scarcity of information about the public management of water, and in particular drinking water, also

53 Ibn al-Jawzī, *al-Muntaẓam*, 16:154–155, 157. See also the slightly different version of Sibṭ ibn al-Jawzī, *Mirʾāt al-zamān*, 19:276 (sic. al-Fwrḥ for al-Qūraj).

54 See the events of years 516/1122, 554/1159, 568/1172 and 573/1178 in the chronicles.

55 Naṣr ibn Aḥmad ibn al-Baṭar (d. 494/1101): *jaʿalahu [al-khalīfa] ʿalā al-dawālīb, mushrifan ʿalā ʿulūfāt al-baqar*; Sibṭ ibn al-Jawzī, *Mirʾāt al-zamān*, 19:520.

56 Ibn al-Jawzī, *al-Muntaẓam*, 16:166; Ibn al-Athīr, *al-Kāmil*, 8:378. Ibn al-Jawzī has "fish and salted goods" (*al-samak wa-l-māliḥ*), Ibn al-Athīr "salted fish" (*al-samak al-māliḥ*).

57 These facts are also noted by Ghabin, *Ḥisba*, 49–50.

demonstrates that a large part of this management was let to private or communal initiative, starting with the work of the *saqqāʾūn*. This point may be confirmed by the considering water use during episodes of sectarian strife in 5th/11th-century Baghdad.

Water and the *Fitna*: Three Cases of Conflicts over Access to Water

The 5th/11th century was particularly rich in *fitan* (pl. of *fitna*), a term that can be translated by "riot" or "rebellion" as well as by "protest" or "social unrest". *Fitan* have been investigated from the point of view of urban control, violence, sectarianism, and popular agency.[58] In Baghdad, they were mostly of a sectarian kind. Under Buyid rule, the Shiʿi festivals and public processions were allowed and officially supported by the Great Emirs, in contrast to the early Abbasid period. The Buyid Muʿizz al-Dawla even imposed a day of mourning for ʿĀshūrāʾ in 352/963,[59] a decision that was not to the taste of sections of the Sunni population of Baghdad. Generally speaking, the Buyid period witnessed the multiplication of sectarian fights between civilians, even if such clashes had already broken out in early Abbasid times. On these occasions, groups of men from different neighbourhoods with predominantly Shiʿi or Sunni populations confronted each other; larger categories of civilians, such as women and children, could be harmed, as the conflicts sometimes degenerated into fires or general looting. No area of Baghdad had a "single confession" population, but some districts had a higher concentration of Sunni or Shiʿi inhabitants. The districts of Bāb al-Ṭāq and al-Ruṣāfa, north of East Baghdad, or Nahr al-Qallāʾīn, south of West Baghdad, were strongly Sunni, whilst the neighbourhoods of Bāb al-Marātib, south of the caliphal precincts on the east bank, and Bāb al-Baṣra, on the west bank, were well-known Ḥanbali strongholds, rich in Sunni traditionalists and activists. The tension with some mainly Shiʿi districts neighbouring these areas was considerable: the markets of Sūq Yaḥyā, known to have an important Shiʿi population, was situated inside the Sunni Bāb al-Ṭāq district, and the Shiʿi populations of al-Karkh,[60] south of the former Round City, were direct neighbours to the markedly Sunni districts of Bāb al-Baṣra and Nahr al-Qallāʾīn (Fig. 7.1).

58 On *fitan* in general, see Neggaz, "Revolts." On *fitan* in Baghdad during the Abbasid, Buyid and Seljuq periods, see Cahen, "Mouvements populaires;" Laoust, "Agitations religieuses;" Sabari, *Mouvements populaires*; Van Renterghem, *Élites*, 1:447–450. For *fitan* in other periods and contexts, and in a comparative view, see Elbendary, *Crowds and Sultans*; Lantschner, "Fragmented Cities;" *idem*, "Invoking legitimacy."

59 See Neggaz, "Al-Karkh," especially 299 n174.

60 For an analysis of the processes by which al-Karkh progressively became a Shiʿi neighbourhood, see Neggaz, "Al-Karkh."

On some occasions, chronicles indicate that access to water was used by one faction as a weapon to weaken the other. It was mainly inhabitants of Sunni districts located next to the Tigris, in West Baghdad, who prevented their (mainly) Shiʿi neighbours, who did not have a direct access to the river, from obtaining drinking water from it. The first half of the 5th/11th century, a time of political weakness for both the Buyids and the Abbasids in Iraq, and of low urban control over Baghdad, was particularly rich in *fitan*,[61] but the first half-century of Seljuq control over the Abbasid capital (second half of the 5th/11th century) also saw much more urban strife than the first half of the 6th/12th century.[62]

The late Buyid period was a time of great difficulty for Baghdadis. The political context was highly confusing, since different Buyid pretenders were fighting over the title of Great Emir, harassing the caliph to obtain his formal recognition. They used all the means of pressure that were at their disposal, from political interventions in the caliph's entourage to the use of military force. Another factor of unrest were the soldiers themselves, namely the troops that were supposed to be under Buyid control and were garrisoned in Baghdad, who regularly rebelled against the Great Emir when he failed to pay their wages. Such rebellions could lead them to loot Baghdad, at the expense of the civilian population. Local actors, the *ʿayyārūn*, meanwhile took advantage of these episodes of chaos to add their own looting and depredations to those of the soldiery. During this period of low urban control, Shiʿi and Sunni groups often fought each other, especially in West Baghdad, where the authority of the political powers was weaker than in East Baghdad, where both the Buyid and Abbasid palaces and administrations were located.

The first mention of water being used as a weapon during a sectarian *fitna* dates back to the beginning of Ramadan 425/July 1034.[63] At that time, the *ʿayyārūn* had repeatedly been attacking Baghdad under the leadership of a chief named al-Burjumī. The inhabitants of commercial districts in Baghdad were forced to organise themselves, forming militias in order to defend their markets, homes and caravanserais. The *ʿayyārūn* eventually took over the commercial taxes and other city revenues, including both the taxes on goods that came by boat on the Tigris[64] and the taxes (*jibāya*) that the police (*aṣḥāb al-masāliḥ*) used to collect from the

61 More than 30 *fitan* are mentioned by the chronicles during the first half of the 5th/11th century; the year 422/1031 in itself witnessed around ten events of this kind. See Van Renterghem, "Un an de *fitna*-s."

62 More than 30 *fitan* occurred between the first arrival of Ṭughril Beg in Baghdad (in 447/1055) and the end of the 5th/11th century.

63 On these events, see Ibn al-Jawzī, *al-Muntaẓam*, 15:239–41; Sibṭ ibn al-Jawzī, *Mirʾāt al-zamān*, 18:392.

64 Ibn al-Jawzī, *al-Muntaẓam*, 15:239.

markets (*aswāq*),[65] and promised in exchange to take charge of police duties and assure the security of specific districts. Under the title of *qāʾid* (pl. *quwwād*), they plundered both sides of the city, extorting the whole population. Turkish soldiers and Bedouins added their own attacks to the general disorder.

In this troubled context, two brothers, "the sons of al-Iṣbahānī"—both "repentant" *ʿayyārūn* who had joined the Buyid military troops—prepared processional banners (in Arabic *munjūq*, pl. *majānīq*)[66] to undertake an ostentatious pilgrimage to the tomb of Muṣʿab ibn al-Zubayr (d. 72/691), the brother of the anti-Umayyad rebel ʿAbd Allāh ibn al-Zubayr, famous for his repression of Alid groups. Visitation of Muṣʿab's tomb, north-west of Baghdad, had been established at the end of the 4th/10th century as a Sunni counterpart to the Shiʿi procession for ʿAshūrāʾ, openly permitted by the Buyids.[67] The Sunni militants paraded noisily in the markets, and stopped facing the Buyid palaces, where they chanted slogans in favour of the Abbasid caliph. This led to a major *fitna* between Shiʿi and Sunni neighbourhoods, especially in West Baghdad, where the Shiʿis of al-Karkh fought their neighbours from al-Qallāʾīn and al-Ṣaffārīn, two areas separated from al-Karkh by canals that could be crossed on footbridges (*qanāṭir*, sing. *qanṭara*) (Fig. 7.1). Each faction prevented people from the other camp from going across the footbridges; a number of people were killed, and looting took place. The Iṣbahānī brothers then took a radical action against the Shiʿis of al-Karkh:

> The two sons of al-Iṣbahānī forbade the carrying of water (*manaʿa* [...] *man ḥamala al-māʾ*) from the Tigris to al-Karkh and its canals,[68] to the extent that the inhabitants [of al-Karkh] suffered and were stricken with hardship. A waterskin (*al-rāwiya*) was sold for 2 or 3 dirhams. Then the matter was mediated between the two parties, and they came to reconciliation.[69]

Preventing water carriers from reaching al-Karkh with their drinking water was a severe retaliatory measure directed towards the entire population of that area, and not only towards the demonstrators. The rise in the price of a waterskin[70] must have been a powerful instrument with which to end the fighting, weakening

[65] Ibn al-Jawzī, *al-Muntaẓam*, 15:240.

[66] On this term, see Van Renterghem, "Un an de *fitna*-s," 105n25.

[67] On the Sunni counterparts of Shiʿi celebrations, see Neggaz, "Al-Karkh," 300.

[68] The Arabic term is *rawāḍiʿ* (sing. *rāḍiʿ*), interpreted by Raʾūf, *Mashāriʿ*, as little canals or waterways built to carry water to the neighbourhoods, presumably from waterwheels, or as derivations of the main canals.

[69] Ibn al-Jawzī, *al-Muntaẓam*, 15:241.

[70] We do not know what the price of a skin of water was in normal times, but the fact that the price was specified here indicates that it rose notably.

the support of the residents of al-Karkh for the *fitna*. Indeed, the next sentence in Ibn al-Jawzī's strongly Sunni-orientated chronicle signals that a settlement had been reached by the two parties, though it gives no details about the terms of this reconciliation. Some days after the end of the *fitna*, the *ʿayyār* leader al-Burjumī was arrested and drowned at the mouth of the Dujayl, a canal of West Baghdad.[71]

The next year, in 426/1035, a similar story is reported by Ibn al-Jawzī, Sibṭ ibn al-Jawzī and Ibn al-Athīr. The *ʿayyārūn* were again strong in Baghdad, and groups of Kurds and Bedouin Arabs were plundering the surrounding areas.[72] This time, the *fitna* was caused by a young soldier, a *ghulām*[73] in the service of the Buyid *malik*. The *ghulām* is reported to have stolen fruits from an orchard belonging to caliph al-Qāʾim. The caliph, offended by this trespass on his property, demanded that the Buyid ruler punish the offender, but the *malik* refused to deliver him up to justice. The angry caliph used the only means at his disposal, blocking the administrative system: "He ordered the judges (*quḍāt*) not to dispense justice, the *fuqahāʾ* not to issue fatwas, and the *khuṭabāʾ* not to conclude a [single] wedding[74] and not to write any contract, and to have the doors of the congregational mosques closed to make prayer impossible."[75] The *ghulām* was briefly imprisoned in the Abbasid palace, then quickly released. This struggle between the Sunni caliph and the Shiʿi *malik* triggered new sectarian riots that left many dead, and in which drinking water was used again as a weapon, this time in East Baghdad:

> The people of Sūq Yaḥyā prevented [the water carriers] from carrying water (*manaʿa* [...] *ḥaml al-māʾ*) from the Tigris to the people of Bāb al-Ṭāq and al-Ruṣāfa.[76] The Turks and the caliph (*al-sulṭān*) did not intervene, knowing that even if they had tried to stop the exactions, they would have increased. [Then] the *ʿayyārūn* became the masters of the country.[77]

Once again, in a context of sectarian strife, a faction situated near the Tigris (here the Shiʿi inhabitants of Sūq Yaḥyā) was able to prevent the entire population of a neighbouring area (here the Sunnis of Bāb al-Ṭāq and al-Ruṣāfa) from accessing affordable

71 Ibn al-Jawzī, *al-Muntaẓam*, 15:241; Sibṭ ibn al-Jawzī, *Mirʾāt al-zamān*, 18:392; Ibn al-Athīr, *al-Kāmil*, 7:765–66.

72 Ibn al-Jawzī, *al-Muntaẓam*, 15:245; Sibṭ ibn al-Jawzī, *Mirʾāt al-zamān*, 18:397; Ibn al-Athīr, *al-Kāmil*, 7:768.

73 "One of the mamluks" according to Sibṭ ibn al-Jawzī, *Mirʾāt al-zamān*, 18:397.

74 Sic *amlāk*; I chose to read *imlāk*, following Sibṭ ibn al-Jawzī who has *nikāḥ*.

75 Ibn al-Jawzī, *al-Muntaẓam*, 15:245–46; see the similar versions in Sibṭ ibn al-Jawzī, *Mirʾāt al-zamān*, 18:397, and Ibn al-Athīr, *al-Kāmil*, 7:768.

76 Bāb al-Ṭāq and al-Ruṣāfa were Sunni districts, when Sūq Yaḥyā had mainly an Imāmī population. All three districts were situated north of East Baghdad. See Fig. 7.1.

77 Ibn al-Jawzī, *al-Muntaẓam*, 15:246.

drinking water (Fig. 7.1). This water war was directly associated with the general lack of control over the situation by the powers in place: the Turkish garrisons of Baghdad, who were supposed to obey the Buyid *malik*, and the Abbasid caliph. As a result, the *ʿayyārūn* again held free rein over Baghdad and its surroundings.

Sibṭ ibn al-Jawzī gives another version of the conflict, emphasising the role of the *ʿayyārūn* instead of the sectarian aspects of the quarrel:

> That year, the *ʿayyārūn* took hold of both sides of Baghdad. Both the caliph and Jalāl al-Dawla (the Buyid ruler) had lost all power. [...] The *ʿayyārūn* resided in the houses of the Turks and the guards; they stayed [there] during the day and went out at night, carrying out attacks. They prevented the people of the neighbourhood (*ahl al-maḥāll*)[78] located far away from the Tigris [from getting] drinking water. [Water] could only be drunk by paying a protection tax (*khafāra*), and they took several dirhams for each waterskin (*rawāyā*).[79]

It is not easy to assess if the conflict over the water was caused by the Shiʿi residents of Sūq Yaḥyā mobilizing against the Sunni neighbourhood of North-East Baghdad, as asserted by Ibn al-Jawzī, or by (usually Sunni) *ʿayyārūn* acting against the (plausibly Shiʿi) inhabitants of the west bank, as stated by his grandson. A careful comparison of the chronicles, with the addition of the version of Ibn al-Athīr, suggests that both events occurred, first through the initiative of the *ʿayyārūn*, then as retaliation on the part of the Shiʿis of East Baghdad. Indeed, the theft of the *saqqāʾūn*'s mules, mentioned at the beginning of the same year by Ibn al-Jawzī,[80] must clearly be related to these events: it then triggered the water blockade in East Baghdad. In any case, the anecdote emphasises the importance of water and its role as a weapon during conflicts opposing neighbouring areas, as well as a source of criminal income for the *ʿayyārūn*.[81]

For almost two decades, upheavals continued to shake Baghdad, provoked by the *ʿayyārūn*'s exactions, by inter-sectarian strife, or by the city's unruly Turkish troops. In 442/1050, to the surprise of all chroniclers, a truce was concluded between the Sunni and Shiʿi populations of the Abbasid capital, allied together against a corrupt chief of police called al-Nasawī. The truce was short-lived, and sectarian quarrels recommenced in 443/1051, lasting for almost one decade. The *fitna* began with a polemic about religious inscriptions on the mosque of al-Karkh. There were fierce fights between inhabitants of al-Karkh and those of neighbouring areas, in

[78] This expression usually refers to the inhabitants of West Baghdad, and especially of al-Karkh.

[79] Sic *zawāyā*, obviously inaccurate. Sibṭ ibn al-Jawzī, *Mirʾāt al-zamān*, 18:397

[80] Ibn al-Jawzī, *al-Muntaẓam*, 15:245.

[81] The *khafāra* phenomenon is still to be explored in urban context. On a similar practice in rural areas, called *ḥimāya* ("protection"), see Paul, "*Ḥimāya* Revisited."

particular the ones mainly populated by Sunnis; access to water was denied again to the Karkhis. Ibn al-Jawzī gives the following description of the conflict over water:

> The people of Bāb al-Shaʿīr prevented the taking of water (*manaʿa* […] *man ḥamala al-māʾ*) from the Tigris for al-Karkh and its canals (*rawāḍiʿ*). Added to this was the fact that there was no more water in the Nahr ʿĪsā. A waterskin (*rāwiya*) was sold for a *qīrāṭ* when it passed under "protection" (*idhā khufirat*), and the poor (*ḍuʿafāʾ*) suffered terribly from this. Souks were closed, and day work was stopped. Some humble people from al-Karkh went at night to draw some loads of water (*rawāyā*) from the Tigris of the Ṣarāt. They poured it into jars and set them up in the markets. They mixed it with rosewater and shouted: "free water [for all]"! (*ṣāḥū al-sabīl*). They [then] took refuge on a small boat (*samāriyya*) in the wharf (*mashraʿa*) of Bāb al-Shaʿīr,[82] took [the jars] and carried them to [Bāb] al-Sammākīn.[83]

Once again, the local population, affected by the restriction on water, had to find ways to get free or cheap water despite the blockade, figuring out a way to bring it into their own district, this time by means of the river. The water was scented with rosewater, probably to conceal its poor quality, or alternatively to give it a festive taste in celebration of the victory over the blockade.

As is often the case, a political explanation is to be found in Ibn al-Athīr's version of the events:

> The representatives of [the Buyid ruler] al-Malik al-Raḥīm refrained from preventing [the *fitna*] because they were angry with [the Abbasid vizier] Raʾīs al-Ruʾasāʾ because of his leaning towards the Hanbalites. These Sunnis prevented water from being transported from the Tigris to al-Karkh; [at the same time], a breach in Nahr ʿĪsā had opened up [and so this canal had dried up], and the situation became very serious for them [the people of al-Karkh]. A group of them were deputised; they went to the Tigris, took water and transferred it into waterskins (*ẓurūf*, sing. *ẓarf*). They poured rosewater into it, and proclaimed: "free water [for all!]" (*al-māʾ li-l-sabīl*). This aroused the Sunnis against them, and Raʾīs al-Ruʾasāʾ hardened his policy against the Shiʿis.[84]

Such comments have their importance, since the anti-Shiʿi policy of the Abbasid vizier Raʾīs al-Ruʾasāʾ was the main reason why the local Shiʿa rallied to the cause of the pro-Fatimid amir al-Basāsīrī when he seized Baghdad some years later,

82 That is, the mouth of the Ṣarāt Canal on the Tigris.

83 "The Fishmongers' (or Fishermen's) Gate", one of the gates of al-Karkh, facing the Sunni neighbourhood of al-Qallāʾīn. See Fig. 7.1.

84 Ibn al-Athīr, *al-Kāmil*, 8:95–96.

in 449/1058. In any case, both versions underline the fact that the rulers did not intervene—be it because they did not have the capacity to put an end to the *fitna*, or because their passivity was a kind of retaliation against civilians suspected of not supporting them—and that the local population had to find their own solution to getting hold of precious drinking water. The Buyid as well as the Abbasid authorities appear here at the best unable, at the worst unwilling, to provide basic water "services"[85] to the population. Civilians had to endure the reign of the *ʿayyārūn* and the *khafāra*, with no possible recourse to the nominal authorities, something that can be interpreted in modern terms as mafia-style extortion. The fact that one of these conflicts was ended by a reconciliation (*istiṣlāḥ*), obtained through unofficial mediation between the two parties, reveals also to what point the rulers were unwilling to intervene, whatever their reasons may have been.

In any case, these three conflicts show that the question of access to water, both for drinking and for economic activities (crops, crafts, industries), was crucial for the population and could easily become an object of pressure in a context of conflict and strife. They also demonstrate that the neighbourhoods that were situated on the banks of the Tigris were clearly at an advantage when it came to water, since they could not be deprived of water themselves, but were in a position to block its distribution to their enemies. This fact must have had an important impact on Baghdad's social topography. Indeed, Raʾūf notes, at the end of the Abbasid era, a gradual shift of Baghdad's population from the districts situated far from the Tigris to those close to the river and therefore to accessible water.[86] The aristocratic character of some of the Tigris' banks, where luxurious palaces were built over the centuries,[87] must also have been influenced by the convenience of being located by the river's waters, especially for rich residences equipped with their own waterwheels. This fact would compensate for the danger also embodied by the Tigris, which regularly flooded and was quite capable of destroying the adjacent buildings.

Conclusion

Despite the scarcity of the materials provided by the sources, the few anecdotes that can be found in the chronicles emphasise the importance of water issues for everyday life in medieval Baghdad, a fact that invites us to revisit the whole of urban history through the prism of water issues. This is obvious for the history of

85 It is not clear that this term can be used here without any risk of anachronism, since the notion of "services" (*khadamāt*) is not present in the sources.

86 Raʾūf, *Mashāriʿ*, 33–34.

87 Van Renterghem, *Élites*, 1:415–28.

Abbasid Baghdad's topography, still to be explored for the 4th–7th/10th–13th centuries: beyond the well-known question of the canals, the topography of residential districts must be reinvestigated, focusing on the *mashraʿāt*, considered as essential points for the circulation of people and goods, an interface between fluvial and terrestrial circulation, as well as crucial spots where drinking water was collected and then sold in districts further from the Tigris.

Access to drinking water must also be considered an essential factor in the social geography of Abbasid (and later) Baghdad. A social and religious geography of the access to water, following the path established by Raʾūf, is still to be written. This first look at the events of the 5th/11th century demonstrates that, at that time, Shiʿi neighbourhoods were less favoured than Sunni districts when it came to access to the Tigris. Karkhis were dependent on canal water, but canals were fragile, subject to breaches and drought. This fact conferred on al-Karkh a vulnerability which the neighbouring Sunni districts could take advantage of in time of conflict. The practice of blocking access to water, or the circulation of water for sale, was employed either by *ʿayyārūn* as a criminal practice for extorting civilians, or by religious factions during the sectarian conflicts that regularly burst out in Baghdad. In neither case could victims of water blockades rely on the ruling authorities. Though investing in important works, such as digging new canals, restoring old ones, or constructing water amenities like public *siqāyāt*, was visibly considered by the rulers as a means to stake their claim to possession and control of territories, such initiatives became rare during the 5th/11th century, obviously because of the political and economic weakness of the local authorities. When criminal or sectarian activities led to water blockades, the inhabitants' collective action was the only way to restore their access to the water necessary to their daily survival.

Bibliography

Primary Sources

Ibn ʿAsākir. *Taʾrīkh madīnat Dimashq*. Edited by Muḥibb al-Dīn al-ʿAmrawī. 80 vols. Dār al-Fikr, 1415/1995.

Ibn al-Athīr. *Kitāb al-kāmil fī al-taʾrīkh*. Edited by ʿUmar ʿAbd al-Salām Tadmūrī. 11 vols. Dār al-Kitāb al-ʿArabī, 2012.

Ibn al-Dubaythī. *Dhayl Taʾrīkh Madīnat al-Salām Baghdād*. Edited by Bashshār ʿAwwād Maʿrūf. 5 vols. Dār al-Gharb al-Islāmī, 1427/2006.

Ibn al-Jawzī. *Kitāb al-muntaẓam fī taʾrīkh al-mulūk wa-l-umam*. Edited by Muḥammad and Muṣṭafā ʿAbd al-Qādir ʿAṭā. 18 vols. Dār al-Kutub al-ʿIlmiyya, 1412/1992.

Ibn al-Jawzī (attributed to). *Manāqib Baghdād*. Edited by Muḥammad Bahjat al-Atharī. Maṭbaʿat Dār al-Salām, 1342/1924.

Ibn al-Sāʿī. *Nisāʾ al-khulafāʾ al-musammā Jihāt al-aʾimma al-khulafāʾ min al-ḥarāʾir wa-l-imāʾ*. Edited by Shawkat M. Toorawa, English translation by the editors of the Library of Arabic Literature. In *Consorts of the Caliphs. Women and the Court of Baghdad.* Edited by Julia Bray. New York University Press, 2015.

Sibṭ ibn al-Jawzī. *Mirʾāt al-zamān fī taʾrīkh al-aʿyān*. Edited by Muḥammad Anis al-Khinn *et al.* 23 vols. Dār al-Risāla al-ʿĀlamiyya, 1423/2013.

Yāqūt al-Ḥamawī. *Kitāb muʿjam al-buldān*. [No editor given], 5 vols. Dār Ṣādir, 1995.

Secondary Sources

Ahola, Judith and Letizia Osti. "Baghdad at the time of al-Muqtadir." In Maaike van Berkel *et al.*, *Crisis and Continuity at the Abbasid Court: Formal and Informal Politics in the Caliphate of al-Muqtadir (295–320/908–32)*, 221–41. Brill, 2013.

Alshaar, Nuha. "Baghdād under Būyid rule." In *Baghdād: From Its Beginnings to the 14th Century*, edited by Jens Scheiner and Isabel Toral, 194–226. Brill, 2022.

Amabe, Fukuzo. *Urban Autonomy in Medieval Islam: Damascus, Aleppo, Cordoba, Toledo, Valencia and Tunis*. Brill, 2016.

Bowen, Harold. "The Last Buwayhids." *Journal of the Royal Asiatic Society* 2 (1929): 225–45.

Busse, Heribert. *Chalif und Grosskönig: Die Buyiden im Iraq (945–1055)*. Franz Steiner Verlag, 1969.

Cahen, Claude. "Mouvements populaires et autonomisme urbain dans l'Asie musulmane du Moyen Âge." *Arabica* 5 (1958): 225–250, and 6 (1959): 25–56, 233–65. Reprint Brill, 1991.

Cahen, Claude, and William L. Jr Hanaway. "*ʿAyyār*." *Encyclopaedia Iranica* III/2 (1987): 159–63.

Cheikh-Moussa, Abdallah. "L'historien et la littérature arabe médiévale." In *L'Œuvre de Claude Cahen: Lectures Critiques, Arabica* 43, no. 1 (1996): 152–88.

Donohue, John J. *The Buwayhid Dynasty in Iraq 334 H./945 to 403 H./1012: Shaping Institutions for the Future*. Brill, 2003.

Elbendary, Amina. *Crowds and Sultans: Urban Protest in Late Medieval Egypt and Syria*. The American University in Cairo Press, 2015.

El-Hibri, Tayeb. *The Abbasid Caliphate: A History*. Cambridge University Press, 2021.

Ghabin, Ahmad. *Ḥisba. Arts and Craft in Islam*. Harrassowitz, 2009.

Hanne, Eric. *Putting the Caliph in His Place: Power, Authority, and the Late Abbasid Caliphate*. Fairleigh Dickinson University Press, 2007.

Hanne, Eric. "The Banū Jahīr and Their Role in the ʿAbbāsid and Saljūq Administration." *Al-Masāq* 20, no. 1 (March 2008): 29–45.

Honigmann, E. and C.E. Bosworth. "al-Na<u>dj</u>af". In *Encyclopaedia of Islam New Edition Online*, edited by P. Bearman *et al.* Brill, 2012.

Hill, Donald. *A History of Engineering in Classical and Medieval Times*. Routledge, 1996 (1st ed. 1986).

Kabir, Mafizullah. *The Buwayhid Dynasty of Baghdad (334/946–447/1055)*. Iran Society, 1964.

Kennedy, Hugh. "The Late ʿAbbasid Pattern, 945–1050." In *The New Cambridge History of Islam*, edited by Chase Robinson, vol. 1, *The Formation of the Islamic World: Sixth to Eleventh Centuries*, 360–93. Cambridge University Press, 2010.

Lane, Edward William. *An Arabic-English lexicon*. 8 vols. Williams and Norgate, 1863–93.

Lantschner, Patrick. "Fragmented Cities in the Later Middle Ages: Italy and the Near East Compared." *The English Historical Review* 130 (544) (June 2015): 546–82.

Lantschner, Patrick. “Invoking and Constructing Legitimacy: Rebels in the Late Medieval European and Islamic Worlds.” In *The Routledge History Handbook of Medieval Revolt*, edited by J. Firnhaber-Baker and D. Schoenaers, 168–88. Routledge, 2017.

Laoust, Henri. “Les agitations religieuses à Baghdad aux IVe et Ve siècles de l’Hégire.” In *Islamic Civilisation 950-1150*, edited by D.S. Richards, 169–86. Cassirer, 1973.

Le Strange, Guy. *Baghdad during the Abbasid Caliphate, from Contemporary Arabic and Persian Sources*. Clarendon Press, 1900.

Makdisi, George. “The Topography of Eleventh Century Baghdād, Materials and Notes, I & II.” *Arabica* 6 (1959): 178–97, 281–306.

Neggaz, Nassima. “Revolts in the Late Medieval Middle East 1200–1500.” In *A Companion to Crime and Deviance in the Later Middle Ages*, edited by Hannah Skoda, 196–222. ARC Humanities, 2023.

Neggaz, Nassima. “Al-Karkh: the Development of an Imāmī-Shīʿī Stronghold in Early Abbasid and Būyid Baghdad (132–447/750–1055).” *Studia Islamica* 114 (2019): 265–315.

Paul, Jürgen. “*Ḥimāya* Revisited.” In *Acts of Protection in Early Islamicate Societies*, edited by Edmund Hayes and Eline Scheerlinck. *Annales islamologiques* 54 (2020): 83–106.

Peacock, Andrew C.S. *The Great Seljuk Empire*. Edinburgh University Press, 2015.

Raʾūf, ʿImād ʿAbd al-Salām. *Tārīkh mashāriʿ miyāh al-sharb al-qadīma fī Baghdād* (*A History of ancient canalisation of drinking water in Baghdad*). Dār al-Shuʾūn al-Thaqāfiyya al-ʿĀmma, 2002.

Rébillard, Eugénie. *Imposer l’ordre: la police dans les villes et les campagnes de l’Iraq abbasside (IIe–IVes./VIIIe–Xe s.)*. PIFD 306. Presses de l’Ifpo, 2024.

Ruggles, D. Fairchild. *Islamicate Environments. Water, Land, Plants, and Society*. Cambridge University Press, 2025.

Sabari, Simha. *Mouvements populaires à Bagdad à l’époque ʿabbasside, IXe-XIe siècles*. Maisonneuve, Librairie d’Amérique et d’Orient, 1981.

al-Sūsa, Aḥmad. *Fayḍānāt Baghdād fī al-taʾrīkh* [*The Floods of Baghdad in History*], 3 vols. Maktabat al-Adīb al-Baghdādiyya, 1963–1965.

Tholib, Udjang. “The Reign of the ʿAbbāsid Caliph al-Qādir billāh (381/991–422/1031): Studies on the political, economic and religious aspects of his caliphate during the Buwayhid rule of Baghdād.” PhD diss., McGill University, 2002.

Tonghini, Cristina and Onur Usta. “Mesopotamian Waterscapes: Watermills in the ‘Land behind Mosul’.” *Mesopotamia* 56 (2021): 91–119.

Tonghini, Christina. “Flour for the Caliph: Watermills in the ‘Land behind Mosul’.” In *The Historian of Islam at Work. Essays in Honor of Hugh N. Kennedy*, edited by Letizia Osti and Maaike van Berkel, 234–53. Brill, 2022.

Tor, Deborah G. *Violent Order. Religious Welfare, Chivalry, and the* ʿAyyār *Phenomenon in the Medieval Islamic World*. Ergon Verlag, 2007.

Tor, Deborah G. “*ʿAyyār*.” In *Encyclopaedia of Islam Three Online*, edited by K. Fleet, G. Krämer, D. Matringe, J. Nawas and D. J. Stewart. Brill, 2014.

Van Renterghem, Vanessa. “Baghdād under the Saljūqs.” In *Baghdād: From Its Beginnings to the 14th Century*, edited by Jens Scheiner and Isabel Toral, 227–64. Brill, 2022.

Van Renterghem, Vanessa. “Un an de *fitna*-s à Bagdad. Les désordres urbains de l’année 422/1031.” *Annales Islamologiques* 56 (2022): 95–138.

Van Renterghem, Vanessa. *Les Élites bagdadiennes au temps des Seldjoukides*. 2 vols. Presses de l’Ifpo, 2015.

CHAPTER 8

Urban Crafts Associated with the Provision of Water in Islamic Syria*

Marcus Milwright

Abstract

The Islamic Middle East presents a diverse assemblage of physical evidence concerning the collection, storage, and transportation of water, both in rural and urban contexts. This chapter focuses on the practices of the specialised artisans who facilitated the transportation and storage of water in urban areas of pre-modern Syria, from the construction of infrastructure to the manufacture of individual components. Evidence on these issues is drawn from archaeology, visual sources, primary texts (6th/12th to the early 14th/20th centuries), and ethnographic observations. The textual analysis considers the detailed evidence contained within the *Qāmūs al-ṣināʿāt al-Shāmiyya* (Dictionary of the Damascene Crafts), written between about 1890 and 1909. This text helps to establish the scope of the specialisms involved in this vital aspect of the everyday life of Syrian cities, the socio-economic significance of these activities within the wider craft sector, and the extent of the continuity of craft practices from the Medieval period to the early 20th century.

Keywords: Syria; Damascus; Islamic; hydraulic infrastructure; crafts; technology; architecture; urbanism

Across the pre-modern Middle East, a wide range of technologies were employed to transport water into and around urban areas. Other types of infrastructure were, of course, required for the storage of fresh water and for the removal of wastewater from buildings. While the Islamic period provides abundant archaeological evidence for the construction and renovation of hydraulic infrastructure, both in urban settlements and the rural environment, there were relatively few technological innovations after the 1st/7th century. Rather, we encounter

* I am most grateful to Maaike van Berkel for the kind invitation to take part in the workshop held at the Netherlands-Flemish Institute in Cairo in February 2024. I benefited from the comments offered by the participants and organizers to the initial draft of this paper. Thanks also to the editors for their insightful critiques of the subsequent drafts. The translations presented were prepared by Hala Qasqas and the author. The research for this chapter was made possible through the support of the British Academy.

considerable areas of continuity with earlier design and practice. There was, however, a transmission of technological knowledge—perhaps also accompanied by the movement of skilled practitioners—between regions, leading to increasingly sophisticated solutions to the challenges of collecting and transporting water. The resulting networks of canals (including those running underground), aqueducts, cisterns, and associated structures, remained sensitive to the local conditions, also generating regional traditions and methods relating to the fair distribution of the water itself.[1]

State investment was required for larger projects, such as aqueducts, with smaller-scale infrastructure often resulting from the patronage of private individuals or religious institutions (using revenue from *waqf*s). Little can generally be gathered about the individual lives and experiences of skilled craftspeople who worked in crafts related to the transport and storage of water, though textual evidence from the 19th and 20th centuries permits some insights into craft practices and the socio-economic context. This chapter explores the data gathered from the late Ottoman phase in Damascus and questions what these findings might contribute to the understanding of craft practice in earlier periods. Importantly, the continuity of the craft itself does not mean that the methods used in the manufacturing processes were necessarily the same. Although crafts were often relatively conservative, artisans remained alive to the potential offered by new materials and technologies.

What was required for all such hydraulic systems was the existence of a diversified craft sector that could manufacture the necessary components and install them. The working practices of these people form the principal subject matter of this chapter. The first section offers some brief comments about the role of archaeology in the study of urban water systems, with a focus on the evidence from Hama. The second section deals with evidence for crafts operating in Damascus at the end of the Ottoman period, with comparative data brought into the discussion from a longer timeframe in the Middle East. The aim is to identify which crafts were involved in the creation of hydraulic infrastructure in and around the Syrian capital, and how those specialisms interacted in larger construction projects. Although the available evidence is rather fragmentary, texts and archival documents can give a sense of the costs involved in the installation and maintenance of water systems.

[1] On the archaeological study of Late Antique and Islamic water systems, see: Milwright, *Introduction*, 64–70, 161–64; contributions in Bienert and Häser, *Men of Dikes*. On legal issues and time measurement, see, for example, Wilkinson, "Muslim Land"; Hehmeyer, "Diurnal Time."

An Archaeological Case Study

It is beyond the scope of the present chapter to review the available archaeological evidence in detail, though some general points can be made about the findings from the excavations of pre-modern urban infrastructure designed for the provision of fresh water and the removal of wastewater. The methods employed for the transport of water into towns and cities varied across the Islamic world, though the most effective was the construction of channels, usually either underground or in the form of aqueducts, that fed into public reservoirs/cisterns.[2] Water could then be piped to other locations, such as public fountains. Urban centres located near to natural springs could employ simple gravity-fed systems to distribute water to different districts (Fig. 8.1),[3] though water-lifting technologies were employed in circumstances where the settlement was above the water table.

For example, the citadel at Hama in Syria, destroyed by the Central Asian conqueror, Tīmūr (d. 1405) in the late 8th/early 15th century, made use of norias (waterwheels) to raise the water needed by the inhabitants of the fortified zone (Fig. 8.2). The excavations in the citadel itself revealed a sophisticated network of ceramic pipes connecting a series of fountains.[4]

In his publication of the water systems on the Hama citadel, Peter Pentz notes the care with which the ceramic pipes making up the network were connected with caulking to reduce water loss. The skill of those who installed the pipes (a craft known as the *qanawātī*; see below) is also evident in the way they negotiated angles, through the use of specially-made ceramic pipes. He notes that caulking was not applied to the downspouts that carried water down from the gutters of building, presumably because this water was carried at a lower pressure.[5] Not all excavated lines of drainage pipes were laid with such care; for example, a water conduit discovered in the Monastery of Apa Jeremias in Saqqara contained ceramic pipes held within stone troughs, that were, in the words of the excavator, "laid,

[2] For example, on the aqueducts of al-Fusṭāṭ, see: Scanlon, "Housing and Sanitation," 188; Kubiak and Scanlon, *Fustat Expedition*, figs. 29, 35; Gayraud, "Fostat", 450–52, figs. 12–15. On the infrastructure of Cairo in the Mamluk period, see Levanoni, "Water Supply." The term *qanāt* is applied in archaeological publications to a variety of water channels, though it is understood in the present context to refer to those which are constructed underground.

[3] For example, an Ottoman-period system of this type is still partially operational in the Greek port town of Nakpaktos. See Milwright and Baboula, "Water on the Ground," 214–27.

[4] Pentz, *Hama*, 37–89. While the archaeological evidence points to limited occupation of the citadel into the Late Mamluk and Ottoman periods, the buildings and infrastructure can be confidently dated to prior to 1402. On the complex water systems of the Crusader and Mamluk eras at Crac des Chevaliers, see: Vágner, Zsolt and Zsófia Csóka."New Research."

[5] Pentz, *Hama*, 89.

Figure 8.1: Ottoman fountain (*çeşme*) in the upper town of Nafpaktos, Greece. Photograph shows the water channels running to and from the fountain. Photograph by Marcus Milwright and Evanthia Baboula, 2006.

Figure 8.2: Norias in Hama, Syria. Photograph by Vyacheslav Argenberg ©, 2008. (Wikimedia Commons.)

in defiance of good practice, with the narrow ends pointing against the flow of water."[6] In general, however, the hydraulic infrastructure recovered through archaeological research appears to have been made to be robust and durable, as can be seen in the vaulted brick sewerage that served the wealthier parts of early Islamic al-Fusṭāṭ. Main sewers ran under the central part of the street into which waste from the houses would be directed through secondary channels.[7]

Crafts and Craft Practice in the 19th and Early 20th Centuries

Where the evidence concerning the pre-modern Islamic crafts associated with the provision of water is rather limited,[8] the situation for the 19th and early 20th centuries is more promising. In particular, there is a greater abundance of detailed written sources that allows for the reconstruction of a reasonably complete picture of the crafts operating in specific localities and of the ways in which artisans with different specialisms collaborated in the creation of urban water systems. This can also be correlated with extant physical evidence preserved in historic houses and the urban areas surrounding them.[9] This section reviews data relating to Greater Syria in this period, with a concentration on the city of Damascus. For this city, the most important text that deals with craft activities is the *Qāmūs al-ṣināʿāt al-Shāmiyya* (Dictionary of Damascene Crafts), composed by three Damascene authors—Muḥammad Saʿīd al-Qāsimī (d. 1900), Jamāl al-Dīn al-Qāsimī (d. 1910), and Khalīl al-ʿAẓm (d. 1926)—between about 1890 and 1909.[10]

I have discussed the history, aims, and character of this text elsewhere,[11] but it is worth reiterating two points. First, this source is constrained in certain respects by the fact that none of the authors were specialists in the crafts (sing. *ṣināʿa* or *ḥirfa*) that they describe in the book. As a result, they do not always display a full grasp of the technical details of the activities they observed,[12] though they remain highly

[6] Quibell, *Excavations at Saqqara*, 12–14, pls XXIX, XXX.

[7] Scanlon, "Housing and Sanitation."

[8] For example, see sources collected in Milwright, "Out of Sight," 198–204.

[9] For example, Dayoub, Kasmo and Mollenhauer, *Culture of Building*.

[10] First published in an Arabic edition with a French index and introduction in 1960. See al-Qāsimī, al-Qāsimī and al-ʿAẓm, *Qāmūs*. The editor, Ẓāfir al-Qāsimī, suggests a timespan of c. 1890–1906 for the research and writing of the book. However, the final chapter assessed in this section (on the *saqqā*ʾ) mentions events in 1909, meaning the final stages of writing occurred in this year or soon after.

[11] On the character, aims and context of the *Qāmūs*, and with references to earlier scholarship, see: Milwright, "Manual Crafts."

[12] They do, however, make use of interviews they conducted with masters of workshops, as is mentioned in the entry devoted to the *zajjāj* (glass maker). See al-Qāsimī, al-Qāsimī and al-ʿAẓm, *Qāmūs*, 163–65.

attentive to the social and economic context of the crafts. For example, they are interested in the confessional or ethnic communities who controlled specific crafts and often give evaluations of the relative profitability of each activity. Second, for all the limitations of the descriptions in the individual chapters, the *Qāmūs al-ṣināʿāt al-Shāmiyya* (from now on, the *Qāmūs*) is unparalleled in the sheer number of crafts surveyed, providing summaries of more than 430 operating in and around the city of Damascus. This comprehensive coverage allows the researcher to construct networks across the city, with one craft sometimes relying on several others for the supply of fuel, raw materials, and processed items. The information from the *Qāmūs* can be correlated with other broadly contemporary sources, such as travel accounts, economic surveys, and early ethnographic research, along with primary documentation, including tax records and daily summaries from construction projects.

The *Qāmūs* provides descriptions of several crafts that were directly connected to the provision and storage of fresh water and the removal of spent water and other waste.[13] In the following section information about five groups of artisans is discussed: the *qasāṭilī* (maker of pipes), the *qanawātī* (layer of water pipes and drainage systems), *qunayāṭī* (the person who maintained sewers and associated infrastructure), the *nawāʿīrī* (maker of norias and *sāqiya*s), and the *saqqāʾ* (water carriers and sellers). Obviously, there were others who worked closely with some of these artisans, though space does not allow for a detailed discussion of each one. For example, the chapter describing the activities of the *qanawātī* mentions the use of bricks and lime (calcium oxide) for building the channels into which drainpipes were set under the ground and caulking the joints between the pipes. These materials would presumably have been supplied by the *ṭawwāb* (brick-maker) and the *kallās* (lime burner) respectively,[14] either on site in cases of larger projects or on the open market for smaller ones. The *qanawātī* also employed other materials, which would have been purchased from workshops in the city (see below). Potters played a crucial role in this sector across the Middle East, with some producing the pipes and others making water jars and the basket-shaped pots that were sometimes attached to norias.[15] The chapters vary in the extent of the information included in the descriptions. Even in cases where this is rather sparse, they remain valuable for their notes on such issues as the socio-economic context, the clientele, and the affiliations of the craftspeople.

13 The *Qāmūs* is arranged alphabetically, meaning that these entries are not grouped together. The authors do sometimes provide cross-references to alert the reader that there might be related crafts elsewhere in the text, though this practice was not followed consistently. The first published edition (1960) contains French and Arabic indices.

14 al-Qāsimī, al-Qāsimī and al-ʿAẓm, *Qāmūs*, 294–95, 392.

15 For example, see Pentz, *Hama*, 48, figs. 45, 46. For a similar *sāqiya* pot: 65, fig. 83.

In the following translations square brackets are used to supply words that clarify the meaning of the Arabic text. In cases where no obvious translation can be given for an Arabic term, it has been transliterated and placed in italics. Cross references and poetic quotations in the chapters have not been translated.

The maker of interlocking ceramic pipes was known as the *qasāṭilī*. The chapter on this artisan reads as follows:

> It is the craft of making pipes (sing. *qasṭal*). And [for the] pipe, make them from the red earth (*al-turrāb al-aḥmar*), mix it with sufficient water, and from it [i.e. the mixture] all kinds of pipe can be made by means of a special hollowed-out mould. When it is done, it is allowed to dry in the sun until it is ready. It is then baked in a special oven (*furn*). And there are many types [of pipe] both large and small. And of them, the one known as *al-zamr* has the smallest dimensions. Then *al-sharkas* is the largest of the first group. Then the biggest is known as *al-īrānī*. And from them are [also] *al-sabīlī*, *al-mujīr*, and *al-zinjārī*.[16] And [the last of] these is the largest in dimensions. They [the different types of pipe] are designed for drawing sweet water (*al-māʾ al-ḥalwa*) from cisterns (*ḥawḍ*, pl. *ḥiyāḍ*) to the houses and other [buildings].
> [...] It is a very important craft in Damascus, and it has a special neighbourhood devoted it. It is widely known as "al-Qasāṭiliyya". A living wage can be earned from it.[17]

Without descriptions of their precise forms and dimensions it is impossible to correlate the names given to the types of pipe mentioned above with extant examples from excavations or identified in situ.[18] However, the chapter does, at least, illustrate that pipes of different shapes and sizes were made within single workshops. These variations may have responded to practical considerations, such as the pressure of the water likely to be flowing through them, though they may also have resulted from learned practice within individual workshops (as seems to be the case with the different rim profiles often encountered in Islamic ceramics from archaeological contexts). On this second issue, there is considerable variability in, for example, the profiles of the joining sections of pipes found on excavations, which appear to have little impact on their functional characteristics (Fig. 8.3).

[16] The organization of the types is rather confusing and, in the absence of comparative textual sources, difficult to resolve. The "groups" indicated in the discussion of *al-sharkas* might refer to the shape of the junctions that would join pipes together. In which case, *al-īrānī* would refer to a distinctive type of junction, with the following designations (*al-sabīlī*, *al-mujīr*, and *al-zinjārī*) being pipes of different dimensions within this larger grouping. The wide variety of junctions can be seen in published archaeological examples. See Milwright, "Out of Sight," figs 7.2, 4, 6, 7.

[17] al-Qāsimī, al-Qāsimī and al-ʿAẓm, *Qāmūs*, 351. For further discussion of this text, see Milwright, "Out of Sight," 198–204.

[18] Milwright, "Out of Sight."

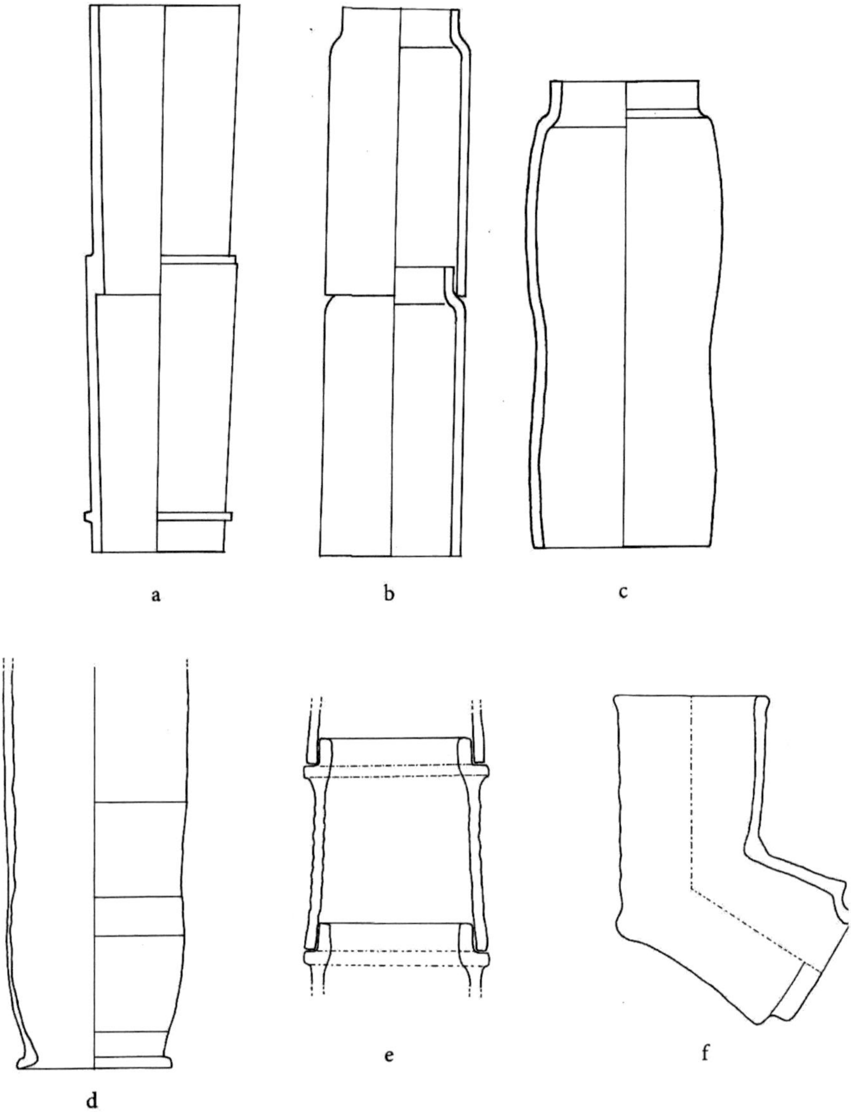

Figure 8.3: Ceramic pipes found during excavations. a–c: from the Amman Citadel, Umayyad and early Abbasid period (after Arce, "Umayyad Hydraulic System"); d–f: from the citadel at Hama (after Pentz, *Hama*). Drawings by Genevieve Neelin and Marcus Milwright.

No mention is made of the manufacture of ceramic vaulting tubes, even though these items have a similar profile to drainage pipes.[19] The authors of the *Qāmūs* state that the pipes themselves were cast in moulds, though the later observations of a potter in Antioch in the 1930s by Pierre Bazantay show that these same objects could be thrown on a wheel.[20] Whether this was, in fact, the normal practice of the Antiochene potter is not apparent. Evidence is currently lacking, but careful examination of pipes from excavations and within standing buildings could help to establish how the craft adapted over the centuries.[21]

The laying of pipes and the creation of drainage systems in Damascus was the speciality of another group of artisans. These men are identified in the *Qāmūs* as the *qanawātī*. The chapter devoted to them is translated below:

> He is the one who maintains the water distribution hubs (*al-ṭawāli'*) and the flow of *al-dumna* (pl. *al-duman*).[22] He is also known in Damascus as "*al-shāwī*." He inspects the water systems under his supervision and always cleans them of dirt and leaves so that they do not enter the pipes and clog them. The aforementioned *qanawātī* comes and separates the dirt to one side and churns it [the interior of the pipes] with five-cubit[23] long reed in order to drain the water. Much of [the existing] *al-duman* are old, causing the loss of water in them. He brings scraps of leather (*qaṣāṣ al-julūd*) and places these scraps, one by one, in the nozzle of *al-ṭāli'* (pl. *al-ṭawāli'*),[24] where the water descends, and so on, which the water drives it [the leather scraps] to the place from whence the water leaks. This place will [then] be clogged with the scraps, and allowing the water to drain away along its [correct] path. Each *ṭāli'* is shared [between] several co-owners (*muqāsim*) with the water going from it [to them]. As a result, the *qanawātī* always serves these *ṭawāli'*, so that the water is continuous to each place. The *qanawātī* has a fixed fee per house, which he takes monthly or yearly.
>
> In past times, they used animal dung instead of leather scraps. The aforementioned *al-shāwī* has the skill of drawing water from rivers and streams into cisterns (*ḥawḍ*, pl. *ḥiyāḍ*) of homes for anyone who desires it.

[19] Dayoub, Kasmo and Mollenhauer, *Culture of Building*, I, 60, fig. 3; 62, figs. 3, 4; II, 39, figs. 45, 46, 49, 50; 84, fig. 127; 85, fig. 128. Also Petersen, "Ottoman Terracotta."

[20] Bazantay, *Antioche*, 38–40, pl. 37.

[21] See the checklist in Milwright, "Out of Sight," 204–15.

[22] This appears to be a Syrian dialect word referring to drainage lines. On the system installed in the late 6th/late 12th century for the distribution of water in the city, see Abd al-Razzaq and Takieddine, "The *Tali'*."

[23] The Syrian cubit (*dhirā'*) is 0.68m. 5 cubits = 3.4m.

[24] The word literally translates as the ascent. This probably refers to the ascending section of pipe that can be accessed above ground.

> As the available supply of water belongs to the town, whoever has a right to the water, or has recently gained the right and wants to withdraw it to his house, or divert it in another direction, or [finds that] the path of this water has been blocked, calls upon the aforementioned *al-qanawātī*. He [i.e. the *qanawātī*] clears the path to the water and searches for the path of the old *dumna*. If the man is a new buyer or wants to bring water to a new place, he [the *qanawātī*] builds a new *dumna* that does not touch anyone else's *dumna*. He builds a new channel made of stones and bricks. He places pipes (*qasṭal*, pl. *qasāṭil*) on top of the channel and inserts one *qasṭal* into another *qasṭal*, where the head of the *qasṭal* is narrow, and on its other side is open, then the head of the second *qasṭal* is inserted into it, and so on. The pipes are extended and connected into each other with the structure underneath them. Pieces of *laqūna* are added onto these pipes, which is made from crushed cotton and lime with pure oil (*al-kils ma'a al-zayt al-khālis*).[25] Then they are pounded together well, and these connections are attached to them, thus controlling them, and preventing them from leaking at all. Then bricks and mortar are also built on both sides of the pipes, so that when the building rises on either side of the ceramic [pipes], they are covered with bricks. Thus, the pipes are preserved within that structure, and it [the construction method] is continued in this way until the cisterns [of the houses] are connected to that water [supply].
>
> His daily wage is fifteen piastres (*ghirsh*) or more.
>
> This craft is important in Damascus. Only Christians work in it, and they have perfect skill in it. It produces good profits.[26]

The thriving craft of the *qanawātī* involved both the construction of new lines for the supply of water and the maintenance of existing ones. These lines of pipes fed into domestic cisterns.[27] The build-up of residues in such systems would have necessitated regular cleaning and the replacement of broken sections. The costs of both construction and cleaning were evidently shared among those householders who accessed the given line of drainage pipes. The chapter also makes clear the care required during the construction of new lines, such that existing ones were not damaged. In areas where buildings and streets were constructed over centuries, this must have been a significant issue requiring considerable skill and training to

25 *Laqūna* appears to be the dialect word for this combination of cotton fibre, lime and oil used for caulking the joints between pipes. Traces of lime have been found caulking pipes on excavations. For example, see: Pentz, *Hama*, 89; Milwright, "Out of Sight," 204, 206, 212 (Acre, Amman citadel and Palmyra).

26 al-Qāsimī, al-Qāsimī and al-ʿAẓm, *Qāmūs*, 363–4. Translated by Hala Qasqas.

27 On domestic cisterns in Aleppo, see Dayoub, Kasmo and Mollenhauer, *Culture of Building*, II, section 2.2.9.

navigate.[28] The reference to scraps of leather (*qaṣāṣ al-julūd*) may indicate that the *qanawātī* was supplied by either the tanners (*dabbāgh*) or by other artisans who made items from the prepared animal skins.[29] This points to the active manner in which waste from one craft would be employed by another in Islamic urban centres.[30] The claim that the craft of the *qanawātī* was exclusively Christian is only partially supported by documentation of building projects (see below). The authors of the *Qāmūs* remark on other crafts that had a strong Christian component; for example, the Christian jewellers (sing. *ṣāʾigh*) in the district around the Umayyad Mosque had "no equal in the city."[31]

In many instances the *qanawātī* would need to work in coordination with other specialists. This would obviously occur in large-scale construction or renovation projects. The interactions with other artisans during the building process can sometimes be traced through surviving documents of the later Ottoman period. One such record was written in Damascus for the High Advisory Council (*majlis shūrā al-shām al-ʿālī*), covering the renovation of the Treasury and the office of the official overseeing the expenditure of *waqf* funds (*nāẓir muʿajjalāt al-awqāf*). The document deals with worksite personnel over the four days of 16–19 Jumādā al-Awwal 1260 (3–5 June 1844) and has been translated and analysed by Randi Deguilhem. The findings are significant in the present context in showing the religious affiliations of the master artisans and skilled workers (*faʿʿālā*)—a majority of Muslim workers, with about 20% Christians.[32] The *qanawātī* on this project comprised two masters and two assistants. One, at least, was Christian, according to his given name (Butrus). They were paid the same wage of 7 piastres per day as the master masons (*miʿmār*), the carpenters (*najjār*), and the lime burners (*kallās*). They were, however, only paid for work on the first two days of the project. Skilled labourers by contrast only received 4 piastres per day. These sums are all lower than those given by Sir John Bowring in his commercial survey of Syria, published in 1840, but using data from 1836. He reports that masons received 8–12 piastres per day and carpenters, 9–15 piastres per day.[33] The sources of his data are not revealed, however, nor is it apparent whether these were the typical or average wage levels across the region or only in urban areas.

[28] For example, see: Scanlon, "Housing and Sanitation"; Vallejo Triano, "Madinat al-Zahraʾ," 8–12; Arce, "Umayyad Hydraulic System."

[29] On these crafts, see Milwright, "Leather."

[30] Discussed in Milwright, "Rubbish, Recycling and Repair."

[31] al-Qāsimī, al-Qāsimī and al-ʿAẓm, *Qāmūs*, 264–5.

[32] Deguilhem, "Shared Space." The translation of the document giving the site personnel appears on 275–8.

[33] Bowring, *Report on Syria*, table on 51.

A related craft dealt with in the *Qāmūs* is the *qunayāṭī*. The chapter on this states:

> The term *qunayāṭī* in Damascus is the sweeper of the restrooms (*kāsiḥ al-akhliyya*). In the past, it was called "*al-sarābātī*." The public calls it *mu'azzil al-khawārij*. *Al-khawārij* are the restrooms (*akhliyya*) and places of filth (*aqdhār*). Most of the restrooms in the Damascene houses have sewers (*siyāqāt*), and each area contains a larger sewer (*siyāq*) into which these gather. Then all of those sewers flow into the watercourses known as the Barada river and the Qulayt river. Due to the abundance of water in Damascus, no house is devoid of water. When the water comes out of the cisterns (*al-aḥwāḍ*), it goes into the restrooms of the interior (*ḍimn*) of the house. It [then] goes down to the sewer and it is there that the filth is gathered. Due to the large accumulation of filth, these aforementioned sewers become clogged, and sometimes their channel is destroyed.
>
> As for the houses that do not have water and are far from the town, their restrooms have pits (*bi'r*, pl. *ābār*),[34] and so do the villages of Damascus, and those wells fill up every now and then, so people who practice this craft are brought in, and they repair the ruins of these sewers, clean them of filth and facilitate their [open] channels.
>
> Only Jews (*al-yahūd*) work in this craft in Damascus, and they have complete experience with it. Others may work in it too. Al-Tāj [al-Dīn] al-Subkī warned in *Mu'īn al-Ni'am*[35] that the owner of this craft has the responsibility to be diligent in cleaning the drains and canals and to report their fullness and emptiness and clean them truthfully because they are hidden from their owners, and they are often unable to discover this and deal with it themselves.
>
> This craft cannot be dispensed with.
>
> The word "*qunayāṭī*" could be a distortion of "*qulayāṭī*" with the letter *lām* having been replaced by the common people with a *nūn*. [The latter term] may refer to the Qulayṭ river, the dirty river in Damascus. Everything which is drawn into the sewers is given this name, because the master of the craft works in them.[36]

Like the *qanawātī*, the *qunayāṭī* offered a crucial service to the inhabitants of Damascus. Their importance to urban life in earlier centuries can also be inferred from references to them in manuals of *ḥisba* (market law) and associated legal texts. The authors of the *Qāmūs* refer to one such manual, produced by the prolific Mamluk-period jurist, Tāj al-Dīn Abū Naṣr ʿAbd al-Raḥmān al-Subkī (d. 1370).[37] It

34 The authors appear to be referring here to cesspits.

35 The title should read: *Mu'īd al-ni'am*.

36 al-Qāsimī, al-Qāsimī and al-ʿAẓm, *Qāmūs*, 365–6. Translated by Hala Qasqas.

37 On this work, see: al-Subkī, *Mu'īd al-ni'am*. The biographical information on this author and the list of his books appear in the editor, David Myhrman's introduction: 8–34 (English text).

is unclear, however, which sections of the *Muʿīd al-Niʿam* is being referred to.[38] The relative profitability of the craft is not mentioned by the authors of the *Qāmūs*, though one can assume that a job of this nature would have been reasonably remunerated given that it involved dealing with human waste and other refuse in sewers, cesspits, and the associated channels.

The chapter is also interesting for its reference to dominant confessional affiliation. The authors of the *Qāmūs* identify other crafts in Damascus that were largely operated by Jews, including the *samkarī* (tinner) and the *naqqāsh* (engraver of metalwork).[39] Notably, the former is described as not being a "noble" (*sharīf*) craft, despite being much in demand across Damascus.[40] This sense in which tinning was regarded as somewhat demeaning by the authors of the *Qāmūs* may point to a wider pattern of Muslims avoiding certain crafts on the grounds of their low status or that they required contact with impure materials.[41] There is, however, no specific mention of the craft of the *qunayāṭī* being ritually unclean (*istiqdhār*), as is claimed, for example, in the chapter devoted to the *jallād* (skinner). Here, the justification is that the practitioner comes into regular contact with blood and other bodily fluids.[42] Others, such as the *dabbāgh* (tanner) are found to have elements of impurity (*najāsa*), without being unclean.[43] In neither case, however, is there any indication that the crafts were shunned by Muslims (or Jews) on religious grounds. The *dabbāgh* was another activity that involved contact with faeces (required for the swelling of the skins prior to the addition of the tanning agents). In Damascus, the tanners employed dog faeces, collected for them on the streets of the city by "special people" (*unās makhṣūṣūn*), though nothing is revealed about their religious or ethnic identity.[44]

As I have discussed elsewhere, the Geniza archive contains several detailed records pertaining to the installation and renovation of drains of buildings in Cairo.[45] This sort of work was generally undertaken by a specialist contractor. Dating to the late 1030s, one document records the cleaning of a pipe (for 8 dirhams),

38 The chapters that relate broadly to the topic are: *qayyim al-ḥammām* (keeper of the bath), *tāṭūr* (watchman of the bath), and *kāsiḥ* (sweeper). See: al-Subkī, *Muʿīd al-niʿam*, 191–92, 195, 209 (Arabic text).

39 al-Qāsimī, al-Qāsimī and al-ʿAẓm, *Qāmūs*, 239, 476–7.

40 al-Qāsimī, al-Qāsimī and al-ʿAẓm, *Qāmūs*, 239. The classification of crafts as "noble" is culturally defined, though it can relate to the commercial value of the materials involved in a given manufacturing process or the social worth accorded the activity, such as midwifery. For example, see the listing of noble crafts in Ibn Khaldun, *Muqaddima*, 2:347–48.

41 It is possible that tinning was regarded as impure as it involved the use of salammoniac (ammonium chloride)—which was commonly manufactured from animal urine—as a flux.

42 al-Qāsimī, al-Qāsimī and al-ʿAẓm, *Qāmūs*, 82. Milwright, "Leather," 250–1.

43 al-Qāsimī, al-Qāsimī and al-ʿAẓm, *Qāmūs*, 140–3. Milwright, "Leather," 253.

44 al-Qāsimī, al-Qāsimī and al-ʿAẓm, *Qāmūs*, 142–3. Milwright, "Leather," 253–4.

45 Gil, "Maintenance, building operations, and repairs," 176–9; Gil, *Jewish Pious Foundations*, *passim*; Milwright, "Out of Sight," 201–203.

and the construction, at a price of 18 dirhams of "the drainpipe of Dār al-Quṭayṭ ("House of the little Tomcat") and its outlet into the Nile." Another from this decade refers to the completion of a drainpipe and lateral channels (*mizrāb*, pl. *mazārīb*)." The *mizrāb* was designed catch the rainwater from the roof in order that it could be directed down to the drains.[46] The demolition of a pipe for 1.5 dirhams is recorded in another document, dating to c. 1040.[47] Giving only the total cost, these references do not allow for a calculation of the daily wages of the contractor and his assistants, though some comparisons can be made with the pay of other construction workers that appear in the Geniza archive.[48]

Some records of construction work provide comparative data for the prices of drainage construction and maintenance. The "cleaning of a pipe (*tanẓīf al-qanāt*) in the court (*qāʿa*) behind the *sukkā* in the Synagogue of the Palestinians is recorded in one document. This job was priced at 31.5 dirhams, which may be compared, for example, to the construction of a bathhouse (154 dirhams) or the repair of an apartment (31 dirhams). The installation of two pipes, one in a house of a shaykh (23.5 dirhams) and the other in the little house (*duwayra*) "adjacent to the Synagogue of the Babylonians" (14 dirhams) is recorded in a second document, dated 1183–4. These again look to be substantial sums when seen in relation to the wages paid by the synagogue and to everyday expenses such as the oil used for lighting.[49] The information in the Geniza provides, therefore, a general insight into the costs of installing and maintaining drainage systems relative to other aspects of construction, as well as to wages and other building expenses. Although it is perilous to compare economic data from different periods, we can, at least, suggest that the comparatively high daily wage quoted for the Damascene *qanawātī* in the *Qāmūs* finds support in these earlier records of construction work in Cairo.

Costs associated with restoration and cleaning appear also in later documents. For example, the *sijill* from Jerusalem detailing the restorations associated with Khān al-Faḥm (the charcoal khan or warehouse), dated 948/1541 is valuable for data on prices. The repairs to the latrines of the khan are listed as 300 *ʿuthmānī*s (gold coins), and the assessment of the repairs as a further 300. This is a considerable sum when seen in relation to other tasks such as the construction of a wall (80) and the lowering of arches in an arcade (160), but is small in comparison to the

46 It was one of the duties of the market inspector (*muḥtasib*) to ensure that the *mazārīb* jutting out from houses did not send their water directly into the street. See: Ibn al-Ukhuwwa, *Maʿālim al-qurba*, 179 (Arabic text).

47 Gil, *Jewish Pious Foundations*, 156.

48 On pay and the provision of food to skilled workers, see Goitein, *Economic Foundations*, 96, 113.

49 Gil, *Jewish Pious Foundations*, 302–303, 353–5. For example, the rabbis were paid 60, 9, and 7 dirhams over a three-month period, while the costs of oil for two synagogues over a month was 16 dirhams.

costs of construction materials and skilled labour. The total bill for all the work on the site amounted to 20,250 *ʿuthmānī*s.[50] Seen in the context of the evidence in the *Qāmūs*, it is clear that the specialists involved in installation and maintenance of drainage systems could command relatively high prices, reflecting the skill and difficulty of their work (including navigating around existing underground hydraulic infrastructure). Aside from the budgetary dimension, contractors also had to pay close attention to logistics, bringing in the *qanawātī* at the correct phase during the construction process.

The authors of the *Qāmūs* give an account of the *nawāʿīrī*, encompassing both the noria and *sāqiya*:

> It is the name of someone who makes norias (*nāʿūra*, pl. *nawāʿīr*). Some of them rotate over running water (*māʾ jārī*,) and some of them revolve around a large well (*biʾr*, pl. *ābār*) by using an animal (*dābba*), in order to extract water from the bottom of the well to the top. As for the waterwheels (*nawāʿīr*) that rotate on water, they are very few in Damascus, because the source of the water in Damascus is high above the level of the town, so there is no need for waterwheels to bring water to a high place. We saw a few waterwheels in Damascus, including two large waterwheels in an orchard known as Bustān al-Qaṣr, which draw water from the river passing through that orchard, in order to irrigate that orchard, due to the high level of the orchard's land above the river.[51] There is another waterwheel in the town known as Nāʿūra Sīdī Khalīl. This waterwheel extracts water from the river passing next to it in order to irrigate the homes and mosques surrounding it. There are other small waterwheels in various places. But there are many waterwheels that bring water out of wells and they are found in most of the villages in Damascus that do not have running water. They bring out water with this waterwheel for public baths. This waterwheel is small in size and is mounted on the mouth of the large well. It has a bucket of wood tied with fibre ropes and is mounted on top of that wheel. It is driven by an animal on a path attached to that wheel, so the water comes out, pours into a mill (*maṣnaʿ*), and then goes to the reservoir (*kazzān*).
>
> It is a craft from which many people make a living.[52]

Aside from the final comment that the craft was the source of income for many people, this chapter contains little about the activities of the *nawāʿīrī*, focusing attention instead on the locations and functions of the extant norias themselves. As the authors emphasise, these features were relatively rare in and around the

50 Salameh, "Aspects of the *Sijills*," 142 (appendix 10.1).

51 On the traditional practices of measuring the hours of access to canals, see Hehmeyer, "Diurnal Time."

52 al-Qāsimī, al-Qāsimī and al-ʿAẓm, *Qāmūs*, 489–90. Translated by Hala Qasqas.

city because the natural water was generally above the level of the city itself. The final part provides a useful description of the form and operation of the *sāqiyas* in the rural areas surrounding Damascus. Reference is made in this last part to the operation of a water-powered mill,[53] though no other details are given.

Bazantay is an important source on the makers of norias in the first half of the 20th century. He records that Antioch contained two workshops, employing eight workers, four of whom were masters, catalogued by Bazantay as Turkophone Sunnis.[54] Up to twenty people could be employed at any one time, indicating the commissioning of less skilled labour when it was required. Norias were the collective property of the owners of the land they irrigated. Nine of these mechanisms were installed in the vicinity of Antioch. Most of the work undertaken by the workshops comprised the maintenance and replacement of broken or rotten elements. Bazantay remarks that the waterwheels deteriorated very quickly, usually requiring the attention of the artisans every year, or more rarely every five or six. The replacement components would be fashioned in the workshops and installed on site. He records the cost of the construction of a noria of 10m diameter as 6,000 francs, while the salary of skilled worker in this field was about 4 francs per day.[55] By comparison, the daily wages of other broadly comparable artisans are stated as: 6–8 francs per day for tanners, 4.50 francs per day for tinsmiths, 5.50 francs per day for potters, and 9 francs per day for brickmakers (during the part of the year they were employed).[56]

Other regions provide evidence concerning the high cost of such projects; for example, Bowring offers some broad-brush insights into the costs of the *sāqiyas* (Fig. 8.4) that were a prevalent feature of the hydraulic infrastructure of pre-modern Egypt. He cites a report concerning calculations for *sāqiyas* in Lower Egypt.[57]

This estimates the existence of 50,000 *sāqiyas*.[58] Each requires three oxen and two men per mechanism (i.e. 150,000 animals and 100,000 people, working for 180 days per year). Oxen are reported as being worth 900 piastres and that they cost 1.5 piastres per day to hire (total = 40,500,000 piastres).[59] The men who tended the

53 On traditional penstock mills in rural Jordan, see McQuitty, "Watermills."

54 The book shows a consistent concern for the identification of artisans according to ethnicity, confessional allegiance, and language preference.

55 Bazantay, *Antioche*, 23–4. The individual salaries mentioned are 5.50, 3.80, 3.30, 3, 2, and 1.50 francs per day. Presumably, the differences related to seniority within given workshops.

56 Bazantay, *Antioche*, 34, 38, 40.

57 The author of the report is given by Bowring as one M. Linant, though no other information about this person is provided.

58 Bowring notes that this does not account for the use of *shādūfs* (simple water-lifting devices, using a pivoted pole carrying a bucket).

59 Bowring, *Report on Egypt*, 13–14. He converts these sums to British pounds at a rate of 100 piastres to the pound.

Figure 8.4: Stereographic photograph showing a *sāqiya* in Egypt, created 1896. Underwood & Underwood. (Library of Congress Prints and Photographs Division: LC-DIG-stereo-1s21678.)

animals and operated the *sāqiya* were paid 1 piastre per day (total = 18,000,000 piastres). *Sāqiyas* themselves cost 1,200 piastres to construct, creating a total cost of construction and animals amounting to 105,000,000 piastres. An annual interest rate of 12%, Linant estimates, costs of 65,520,000 piastres per year. While such numbers should be viewed with considerable caution, some aspects seem relatively plausible, such as the cost of the *sāqiya* (about one fifth of the sum reported later by Bazantay for the construction of a noria). The initial costs of construction in such cases suggest that there was a strong financial incentive for maintaining existing norias and other water-lifting mechanisms. Thus, the revenues generated by Damascene *nawāʿīrī* would likely have included more work on maintenance than on creation of new norias.

The last chapter from the *Qāmūs* reviewed here concerns the activities of the *saqqāʾ* (water carrier; Fig. 8.5). This person is described as follows:

> *Saqqāʾ* is the name of someone who transports water to houses during the days when the [regular supply of] water is cut off. He brings the water to the people [according to what] they need daily and charges a fixed amount for each waterskin (*qirba*). He goes to the river and fills water for some rich and wealthy people from some endless, special springs (*al-ʿuyūn al-makhṣūṣā*) in summer and winter, and he charges them a fixed fee for each waterskin. Some water carriers go around the markets, giving people water with a bowl that he fills from a jar he holds in his hand. This craft is very rare, except during the days when the rivers are cut off when it becomes widespread due to the need of the people for water.
>
> In the year 1327 AH (1909 CE), this craft became virtually abolished due to the fact that the water of ʿAyn al-Fīja, famous for its goodness and purity, was brought to Damascus using iron pipes (*qasāṭil min ḥadīd*) and the establishment of large cisterns (*aḥwāḍ*

> *kabīra*) in the district of al-Salihiyya into which it was directed. From these cisterns, water is distributed via iron pipes to all parts of the town, its markets, and alleys. With this delicious water, the water of rivers and other springs was dispensed with. Thus, diseases were reduced by seventy percent, as we are informed by the most famous doctors. The renowned minister (*wazīr*), Naẓīm Pāshā,[60] served Damascus and thanks to his tireless effort he brought water. This was among his services, and certainly [one of] his good deeds, for which God rewarded him and forgave his evil ones.
>
> In any case, the person who lives by this craft is very content and satisfied with what his Lord has apportioned for him, so we ask Him, the Almighty, for contentment and wealth from people.[61]

While the authors of the *Qāmūs* claim that this craft gives contentment to all who engage in it, at the time the chapter was completed (in 1909 or soon after) the *saqqā'* faced an uncertain future. In common with several other crafts described in the book, their livelihood was threatened by the onset of modernity. This latter phenomenon could take many forms, ranging from the adoption of practices originating in Europe and North America (such as cigarette smoking) to the large-scale importation of mass-produced goods via steamships. Examples of the latter phenomenon include Bohemian glass, Meissen porcelain, and Staffordshire ironstone wares. In each case, these imports had a deleterious impact on manual crafts in operating in Damascus and elsewhere in the Middle East.[62] Local crafts also started to create goods to cater for the growing market of tourists, while the *antikjī* (antiques merchant) also sought out ancient and medieval items for wealthier buyers.[63] For the *saqqā'* the crucial challenge came with the use of industrially-manufactured iron pipes to carry supplies of fresh water.

The chapters of the *Qāmūs* were written over a long period—nearly two decades—and it is probable that some were revised and updated years after they were initially drafted. This may well be the case for the *saqqā'*. Although the new water supply is not mentioned in the other chapters surveyed above, it seems likely that some of these crafts, most particularly the *qanawātī*, would have experienced changes as the result of Naẓīm Pāshā's intervention. The maker of drainpipes would also disappear during the early 20th century, his skills unable to compete with mass-produced cement pipes made in local factories.[64]

60 Referring to the author and adminstrator Ḥüseyin Naẓīm Pāshā (d. 1927), who became governor of the *vilayets* of Beirut and Syria in 1896.

61 al-Qāsimī, al-Qāsimī and al-ʿAẓm, *Qāmūs*, 185–6. Translated by Hala Qasqas.

62 Milwright, "Imported pottery," 135–9.

63 al-Qāsimī, al-Qāsimī and al-ʿAẓm, *Qāmūs*, 40–41.

64 A factory of this type existed in Beirut from the 1940s. See Milwright, "Out of Sight," 201.

Figure 8.5: Water carrier, probably in Cairo. Photograph taken between 1860 and 1890. (Library of Congress Prints and Photographs Division: LC-USZ62-104877).

Conclusion

The *Qāmūs* was written at a time of profound change in the social, cultural and economic life of Damascus. As noted above, many chapters contain signs of the encroachment of modernity into the practices of manufacturing, the changing expectations of the urban clientele, and the types of imported goods that were appearing in the markets of the Syrian capital. As such, the descriptions in the *Qāmūs* often represent a written record of crafts that were soon to disappear from the urban landscape. However, the text clearly needs to be employed critically because of the backgrounds and biases of the three authors and the extended time period over which they were written and revised. Additional caution is required in trying to assess the importance of this evidence as a source for earlier practices relating to the provision of water and the transportation of waste in the urban centres of Syria and the wider Middle East. There is certainly evidence for some continuity in craft practice, both in the archaeological and in primary sources (though this observation does not preclude the periodic introduction of new manufacturing practices). For example, there are records of the making of pipes in Damascus in urban topographies dating to as early as the $7^{th}/12^{th}$ century.[65]

The selected chapters from the *Qāmūs* present different evidence concerning the economic viability of the crafts associated with the supply of water and the removal of waste. These data are again specific to their time, in that the *saqqāʾ* was already become a rare craft in Damascus. As noted above, other changes would occur in the decades after the completion of the *Qāmūs*, leading to the eradication, for example, of the *qasāṭilī*. The documents relating to construction projects—ranging from the $5^{th}/11^{th}$ to the $13^{th}/19^{th}$ centuries—suggest a degree of long-term continuity in terms of the logistical planning for substantial construction projects. Not surprisingly, the *qanawātī* features early in these projects, preparing the infrastructure below ground level and establishing the connections with latrines, baths, and the pipes running down the walls from the guttering. Not only did these artisans seem to be in high demand across the urban centres of the Middle East, their work also represented a significant part of the building cost, in terms of materials and labour. The costs of installing norias were also considerable, with the burden shared among those whose lands would benefit from the water supply. Lastly, the *Qāmūs* provides intriguing evidence concerning the role of confessional minorities in the construction and maintenance of this crucial infrastructure. These observations find some support in other textual sources, though it remains to be established whether the religious minorities dominated these crafts in other regions of the Islamic world.

65 Elisséeff, "Corporations," 69n18.

Bibliography

Primary Sources

Bazantay, Pierre. *Enquête sur l'artisanat à Antioche*. Les états du Levant sous Mandat Français. Imprimerie Catholique, 1936.

Bowring, Sir John. *Report on the Commercial Statistics of Syria*. Her Majesty's Stationery Office, 1840.

Bowring, Sir John. *Report on Egypt and Candia, addressed to the Right Honourable Lord Viscount Palmerston*. Her Majesty's Stationery Office, 1840.

Gil, Moshe. *Documents of Jewish Pious Foundations from the Cairo Geniza*. Brill, 1976.

Ibn Khaldun, Abu Zayd ʿAbd al-Rahman ibn Muhammad. *The Muqaddimah: An Introduction to History*, trans. Franz Rosenthal (Bollingen History Series 43, second edition with corrections). Princeton University Press, 1967.

Ibn al-Ukhuwwa. *The Maʿālim al-qurba fī aḥkām al-ḥisba of Ḍiyā' al-Dīn Muḥammad ibn Muḥammad al-Qurashī al-Shāfiʿī, known as Ibn al-Ukhuwwa*. Edited by Reuben Levy. Luzac & Co., 1938.

al-Qāsimī, Muḥammad Saʿīd, Jamāl al-Dīn al-Qāsimī and Khalīl al-ʿAẓm. *Dictionnaire des métiers damascains (Qāmūs al-ṣināʿāt al-Shāmiyya)*, edited by Ẓāfir al-Qāsimī. 2 vols. Mouton and Co., 1960.

al-Subkī, *Tāj al-Dīn Abū Naṣr ʿAbd al-Wahhāb. Kitāb muʿīd al-niʿam wa-mubīd al-niqam*, edited by David W. Myhrman. Luzac & Co., 1908.

Secondary Sources

Abd al-Razzaq, Moaz and Zena Takieddine. "The *Tali'* (Water System) of Damascus," in *Discover Islamic Art*, Museum With No Frontiers, 2026. https://islamicart.museumwnf.org/database_item.php?id=monuments;ISL;sy;Mon01;12;en. Accessed: 11 January 2026.

Arce, Ignacio. "The Umayyad Hydraulic System at Amman Citadel—Collection, Storage, Distribution, Use and Sewage." In *Men of Dikes and Canals: The Archaeology of Water in the Middle East*, edited by Hans-Dieter Bienert and Jutta Häser, 41–58. Marie Leidorf, 2004.

Bienert, Hans-Dieter and Jutta Häser, eds. *Men of Dikes and Canals: The Archaeology of Water in the Middle East*. Marie Leidorf, 2004.

Dayoub, Dima, Ruba Kasmo and Anne Mollenhauer, eds. *A Culture of Building: Courtyard Houses in the Old City of Aleppo*. 2 parts in 2 volumes. Al Ayn Editions, 2023.

Deguilhem, Randi. "Shared Space or Contested Space: Religious Mixity, Infrastructural Hierarchy and the Builders' Guild in Mid-Nineteenth Century Damascus." In *Craftsmen of the Middle East: Fashioning the Individual in the Muslim Mediterranean*, edited by Suraiya Faroqhi and Randi Deguilhem, 261–82. I.B. Tauris, 2005.

Elisséeff, Nikita. "Corporations de Damas sous Nūr al-Dīn. Matériaux pour une topographie économique de Damas au XIIe siècle." *Arabica* 3 (1956): 61–79.

Gayraud, Roland-Pierre. "Fostat: Évolution d'une capitale arabe du VIIe au XIIe siècle d'aprés les fouilles d'Istabl ʿAntar." In *Colloque international d'archéologie islamique, IFAO, Le Caire, 3-7 février 1993*, edited by Roland-Pierre Gayraud, 435–60. Institut Français d'Archéologie Orientale, 1998.

Gil, Moshe. "Maintenance, Building Operations, and Repairs in the Houses of the Qodesh in Fustat: A Geniza Study." *Journal of the Economic and Social History of the Orient* 14 (1971): 136–95.

Goitein, Shlomo, *A Mediterranean Society. The Jewish Communities of the Arab World as portrayed in the Documents of the Cairo Geniza. Volume 1: Economic Foundations*. University of California Press, 1967.

Hehmeyer, Ingrid. "Diurnal Time Measurement for Water Allocation in Southern Yemen." *Proceedings of the Seminar for Arabian Studies* 35 (2005): 87–96.

Kubiak, Wladislaw and George Scanlon, eds. *Fustat Expedition Final report. Volume 2; Fustat-C*. ARCE, 1989.

Levanoni, Amalia. "Water Supply in Medieval Middle Eastern Cities: The Case of Cairo." *Al-Masāq* 20, no. 2 (2008): 197–205.

McQuitty, Alison. "Watermills in Jordan: Technology, Typology, Dating and Development." *Studies on the History and Archaeology of Jordan* 5 (1995): 745–51.

Milwright, Marcus. "The Manual Crafts and the Challenge of Modernity in Late Nineteenth-century Damascus." In *Making Modernity in the Islamic Mediterranean*, edited by Margaret Graves and Alex Dika Seggerman, 139–54. Indiana University Press, 2022.

Milwright, Marcus. "Out of Sight, Out of Mind? Preliminary Observations on the Production and Use of Ceramic Drainage Pipes in the Islamic Middle East." In *Made for the Eye of One who Sees: Canadian Contributions to the Study of Islamic Art and Archaeology*, edited by Marcus Milwright and Evanthia Baboula, 196–220. McGill-Queen's University Press and Royal Ontario Museum, 2022.

Milwright, Marcus. "Rubbish, Recycling, and Repair: Perspectives on the Urban Crafts of the Islamic Middle East." In *The Making of Islamic Art: Studies in Honour of Sheila Blair and Jonathan Bloom*, edited by Robert Hillenbrand, 138–55. Edinburgh University Press, 2021.

Milwright, Marcus. "The *Qāmūs al-Ṣināʿāt al-Shāmiyya* as a Record of Leather-working Crafts in Late Ottoman Damascus." In *The Arts and Crafts of Syria and Egypt from the Ayyubids to World War I: Collected Essays*, edited by Marcus Milwright, 247–69. Gorgias Press, 2018.

Milwright, Marcus. *An Introduction to Islamic Archaeology*. Edinburgh University Press, 2010.

Milwright, Marcus. "Imported Pottery in Ottoman Bilad al-Sham." *Turcica* 40 (2008): 121–52.

Milwright, Marcus and Evanthia Baboula. "Water on the Ground: Water Systems in Two Ottoman Greek Port Cities." In *Rivers of Paradise: Water in Islamic Art and Culture*, edited by Sheila Blair and Jonathan Bloom, 213–38. Yale University Press, 2009.

Pentz, Peter. *Hama IV. The Medieval Citadel and its Architecture*. Nationalmuseet, 1997.

Petersen, Andrew. "Ottoman Terracotta Vaulting Tubes." *Orient Express* 3 (1994): 89–91.

Quibell, James. *Excavations at Saqqara (1908–09, 1909–10), IV, The Monastery of Apa Jeremias*. Imprimerie de l'Institut Français d'Archéologie Orientale, 1912.

Salameh, Khadr. "Aspects of the *Sijill*s of the Shariʿa Court in Jerusalem." In *Ottoman Jerusalem, the Living City: 1517–1917*, edited by Robert Hillenbrand and Sylvia Auld, 103–44. Altajir World of Islam Trust, 2000.

Scanlon, George. "Housing and Sanitation: Some Aspects of Medieval Islamic Public Service." In *The Islamic City: A Colloquium*, edited by Albert Hourani and Samuel Stern, 179–94. Bruno Cassirer, 1970.

Vágner, Zsolt and Zsófia Csóka."New Research on the Medieval Water-Management System at Crac des Chevaliers." In *Bridge of Civilizations: The Near East and Europe, c. 1100–1300*, edited by Peter Edbury, Denys Pringle and Balász Major, 54–66. Archaeopress, 2019.

Vallejo Triano, Antonio. "Madinat al-Zahraʾ: Transformation of a Caliphal City." In *Revisiting* Al-Andalus*: Perspectives on the Material Culture of Islamic Iberia and Beyond*, edited by Glaire Anderson and Mariam Rosser-Owen, 3–26. Brill, 2007.

Wilkinson, John. "Muslim Land and Water Law." *Journal of Islamic Studies* 1 (1990): 54–72.

CHAPTER 9

Traditional Water Harvesting in Mamluk and Early Ottoman Bilād al-Shām: The Family Cistern

Bethany J. Walker

Abstract

This essay examines the smallest scale of water harvesting in rural and urban households of pre-modern Syria: the humble family cistern. Constructed, maintained, and privately owned by household units, they were the linchpin in hydrological systems for many communities lacking reliable, year-round sources of water. The state took little interest in such small units of water storage until the Mamluk and early Ottoman periods, when their financial benefits were fully exploited. This study combines textual and archaeological approaches to explore the reasons for the commodification of the household cistern, and to better understand the roles it played in daily life. Two archaeological sites stand as case studies in this regard: Tall Ḥisbān in central Jordan and Khirbet Beit Mazmīl in Jerusalem.

Keywords: cistern; water harvesting; *waqf*; sustainable water use; Jordan; Palestine

Introduction

The environment of Bilād al-Shām is characterized by marked variations in rainfall from year-to-year and from one region to the next, with many regions receiving too little for rainfed agriculture. In those areas that are located far from perennial rivers, water systems were developed from Antiquity that were comprised of springs, cisterns, check dams, and rock-cut channels that collected water from various sources, channelled them to fields and homes, and saved the remainder for future use, mostly in cisterns. These systems remain largely outside the coverage of the textual sources available to us for the Mamluk and Ottoman periods: they were not (directly) taxed, were of such a small scale that they escaped the attention of most chroniclers, and were rarely the focus of formal (documented) legal action. In comparison to the extensive irrigation networks of Egypt or those of Syrian urban centres, such systems were long invisible to government officials. They are ever-present, however, in archaeological studies, which document their role in sustainable water use, as cisterns captured winter rains and diverted water seasonally from wadis and springs. Household cistern construction and maintenance, moreover, was carried out independently of the state; it was an indigenous response to meeting water needs in semi-arid and arid environments. Three kinds of cisterns served the needs of local communities in this way: rain-filled communal and family cisterns and spring-fed run-off cisterns. Large multi-chamber cisterns, cut into the bedrock and with entrance corridors, were constructed in the Roman and Byzantine periods to service entire communities (large villages and towns). Such installations were built by the hands of skilled masons and engineers, would have required the financial investment of a wealthy patron or a public institution, and were filled through the capture of rainwater over a long period of time. Family cisterns in this region tend to be small, bell-shaped installations, generally covered with a stone or metal sheet. Their history of construction is as old as that of historical settlement in the region. They were built by local hands (usually by the family itself) and filled by rainwater captured and transported by channels cut into bedrock or gutters on house rooftops. Only spring-fed cisterns were filled by running water. In construction, they are indistinguishable from the small household cisterns found in villages; these, however, were used most often for irrigated agriculture. In addition to water capture and storage, cisterns—as rock-cut installations—were also built for or repurposed as grain silos (in Palestine, s. *maṭmūra*[1]) (Fig. 9.1).

In regions of normally adequate rainfall (300 mm annually or above), both rural and urban households in southern Syria were dependent on these small installations to survive the water shortages of the summer months, and to guarantee cereal supplies for family and flocks during periods of drought, through irrigation of these normally rainfed crops. The household cistern was the key to household resilience in places of irregular rainfall and political uncertainty.

The following essay explores the many roles of the household and agricultural cisterns in Transjordanian and Palestinian society in the late Mamluk and early Ottoman periods (roughly the 7th–11th/14th–17th centuries), focusing on their economic value. By the 11th/17th century such cisterns had become fully commodified, largely as a result of their storage potential, and they remain so today. We will trace this development in small communities and individual households located in areas that receive a minimum average 300 mm of rainfall annually.[2] In order to obtain as broad a picture as possible of the value placed on cisterns by local society, the essay combines ethnographic, archaeological, and textual data to reveal ways in which the family cistern came to dominate rural landscapes, raise property

[1] Dalman, *Work and Customs*, 545.

[2] At the 300 mm isohyet, rainfed agriculture is possible, but risky, necessitating the use of cisterns.

Figure 9.1: Drawing wheat from a cistern, Ashdod, c. 1908.
(LOT 13739-1, no. 78, courtesy Library of Congress.)

values, and became a commodity in its own right. Development projects in modern Jordan provide the background for interpreting patterns in cistern use that emerge from the archaeological and historical records. We begin, then, with the household cistern in modern use, as documented by NGOs and ethnographers, and then follow its history back to the Early Modern and medieval eras, drawing on traces of cistern use from texts and archaeological remains.

The Family Cistern as Sustainable Water Harvesting

Project Rainkeep and Today's Jordan

Resilience as a social concept is the ability of a social group to survive as a unit, adapting to threats from within and without. Outside of political conflict, it was primarily water shortages that led to the dissolution of small communities, as described by the chroniclers of Mamluk and Ottoman-era Syria.[3] Repeated drought was one of the main causes of village abandonment and peasant flight in this

[3] Walker, "Mobility and Migration."

period. Water scarcity drives much human migration today, too, pushing people to search for new homelands in other places.

There are many indices used by NGOs today to describe the water needs of modern Jordan. It is widely recognized as one of the most water-scarce countries in the world, as its renewable water resources provide less than 100 cubic metres per person annually.[4] According to USAID, the country's renewable water supply meets the needs of only two-thirds of the population, and groundwater is consumed twice as quickly as it can be replenished. Jordan is also a water-stressed country, a term in the development world used to describe water needs that exceed available volume.[5] Every summer, when many water reserves from the winter have dried up, Jordan needs to import water from abroad. During the rest of the year, in the capital city of Amman, which has the greatest water demands, there is frequently no running water for 12–24 hours a week. It is not only drinking water that is affected, but also water for agriculture, rain-fed and irrigated. Generally low and unpredictable rainfall causes the grain harvest to fail an average of once every five years.[6] Repeated years of inadequate rainfall then leads to drought and famine. Jordanians are well aware of the fact that there is not enough groundwater to meet their needs. Children are taught about and enculturated into water-conservation in school.[7] Nonetheless, pressure on groundwater supplies is too high, and alternatives need to be found.

One solution has been sought in the commonplace family cistern, which captures and stores winter rains, relieving pressure on groundwater reserves. Project Rainkeep was a community development initiative, launched in 1986 through a grant from the Canadian International Development Agency (CIDA).[8] Inspired by archaeological research on past water management practices at the site of Tall Ḥisbān, where ancient cisterns—widespread through the landscape—were regularly revitalized in the past, the goal was to reintroduce rainwater harvesting

[4] On this basis, UNICEF rated the country in 2019 as the world's second most water-scarce, second only to Qatar In 2024 the World Population Review ranked Jordan the fifth most water-scarce country in the world (https://worldpopulationreview.com/country-rankings/water-stress-by-country), accessed 19.08.2024. This trend is expected to continue.

[5] According to the World Resources Institute, water stress is caused when the annual supply of water for an individual drops below 1,700 cubic metres, or the equivalent of eight to nine glasses of water a day https://www.activesustainability.com/water/ranking-countries-highest-water-stress/?_adin=11734293023, accessed 19 August 2024.

[6] Palmer, "Following the Plough," 132.

[7] Jenssen, "Traditional Rainwater Harvesting," 165. Today the public schools hold workshops and distribute booklets to raise awareness about this issue (personal communication Elena Ronza, Wadi Musa, Jordan, email of 25 August 2024).

[8] Personal communication, Øystein LaBianca (email of 26 September 2024).

in cisterns as one way to address the country's increasing water shortages.[9] The work was carried out by ADRA (the Adventist Development and Relief Agency) in Amman, beginning with the restoration of cisterns in the village of Ḥisbān, and then extended to other villages in the nearby Madaba District in 1994-5. Thirty-two of the over three hundred defunct cisterns in the district were repaired and secure cement covers constructed for them.[10] In an evaluation of the long-term impact of this project, written ten years later, Jenssen points out that while project leaders considered the results a failure, because too few families chose rainwater harvesting over buying water from the municipality, the concept itself was a sound one, as restoration of household cisterns was embraced by those villages whose cisterns were repaired and put back into use with ADRA funding.[11] While some villagers did not use the cistern for water storage, disliked the salty taste of the water, or filled them with city-piped water, instead of rainwater, others continued to use them as they were originally designed. This provided a particular benefit to poorer households, who may not have been able to purchase trucked-in water during the summer months. That attempts to revive household cistern use were not successful country-wide is demonstrated by a recent study published by the World Bank that documents that by 2020, only 3.4% of Jordanian households were harvesting water in cisterns, and in the district of Amman only 0.4%.[12]

In the 1990s the Amman Municipality began to require all new building projects in the city to include household cisterns to harvest rainwater, or to revive ancient ones, if any were found on the building site. This was already practiced in Palestine as a way of acquiring water at no cost and preventing flooding.[13] More recently, there have been similar initiatives in the Hawran, led by local communities, to address mounting water shortages by reviving ancient water runoff systems that include reservoirs and supply channels.[14] Such projects are essentially a return to what was a long-standing, indigenous system of water harvesting and storage in water-scarce regions. Household and agricultural cisterns had been the mainstay of both rural and urban communities in southern Syria since Antiquity. A detailed development of cistern use and value can be traced, in particular, for the text-rich Mandate, Ottoman, and late Mamluk periods.

9 LaBianca, "On-Site Water Retention Strategies," 771–76.

10 One repeated concern of villagers about the old cisterns was the fear that children could fall into them.

11 Jenssen, "Traditional Rainwater Harvesting".

12 Ababsa and Abu Hussein, "Metropolitan Amman," 44.

13 Ababsa and Abu Hussein, "Metropolitan Amman," 44.

14 Braemer *et al.*, "Long-term Management" 49.

Traditional Water Harvesting in the Late Ottoman and British Mandate Periods

Travellers' accounts and the land registrations of the Ottoman Tanzimat and British Mandate periods document what cultural and financial value Palestinian and Transjordan villagers attributed to their household cisterns. They also provide us with important spatial data that provide analogies for interpreting the legal and fiscal records of earlier centuries. In his ethnographic account of the 1920s, Gustaf Dalman makes several important observations about household and agricultural cistern use in Palestine. The first is that no family could survive a year of low rainfall without its own cistern. In that regard, he argued that in 1924-5, a period of low rainfall, the government created a water shortage by not requiring the construction of cisterns in new houses.[15] He further describes the location of family cisterns: under and next to houses, in gardens, in agricultural fields, next to roads, in the desert. Finally, Dalman, echoing the concerns of Jordanian villagers documented by Project Rainkeep, makes frequent reference to the danger of empty, abandoned, and uncovered cisterns, into which people often fell and died.[16] The abandoned cistern had become a trope for abandoned land and financial loss. A functioning cistern, on the other hand, stood for property ownership and self-reliance.

The local value attributed to cisterns, however, went well beyond the symbolic. They raised the value of property, both houses and farmland. During the Tanzimat-inspired registration of farmland in northern Jordan in the late 19th century, peasants were eager to register everything they considered market-worthy, including cisterns on private land. For example, of the fifty-five entries in the list of registered private (*mulk*) properties in the village of Beit Ra's in 1880, nineteen were of houses and thirty-six of cisterns and wells.[17] Martha Mundy has suggested that it was the villagers that gave value to these cisterns, and not the state. Other interesting patterns emerge from these registers. Cisterns, as deep and plastered rock-cut installations, were multi-functional: only four of the cisterns were used for storing water; the rest were put to use for storing wheat, chaff, and barley. Additionally, not every household had a cistern. Some were shared by two or multiple households; others had only a share in a cistern. Some households owned multiple cisterns. One of the highest-valued houses in the village of Hawwara in 1883 had five wells and two cisterns.[18] In terms of function, wells gave access to ground water. Cisterns, on the other hand, could hold either water or cereals and cereal by-products. Their potential to store two critical resources—water and cereals—is what made them most valuable on a local level.

[15] Dalman, *Work and Customs*, 183.

[16] Dalman, *Work and Customs*, 544.

[17] Mundy and Smith, *Governing Property*, 111.

[18] Mundy and Smith, *Governing Property*, 129.

By the 1940s, when titles to land had been legally settled, household cisterns had acquired a fiscal value of their own, separate from the land, much like houses and caves.[19] In his study of land inheritance from the court records of Irbid District, Wåhlin notes that cisterns were valued from five to twenty Palestine pounds. Half of the properties in the study area had cisterns, some more than one cistern; all were cut into limestone (and not the local sandstone) outcrops; and it was the households of sedentary tribes (the Salṭīyya), as opposed to traditionally nomadic ones (the ʿAbbādī), that owned cisterns here.[20]

Ethnographic studies clearly document the economic value that cisterns had acquired as a result of their storage capability and multi-functionality. Whether a similar value was attributed to the family cistern in the deeper past, in the Mamluk and Early Ottoman centuries, requires recourse to different source material and different methods of analysis.

The Family Cistern in the Mamluk and Early Ottoman Periods

In the Mamluk and Ottoman periods, cisterns do not appear in the kinds of sources that are otherwise rich in information about water systems. The Ḥaram al-Sharīf documents of 8th/14th-century Jerusalem make occasional reference to water: water allotment could be used as collateral against a cash loan; shares in water, and the land irrigated by it, could be purchased; endowed villages, and the waters allotted to them, could be leased; and competing legal claims to "wells" (*abyār*, s. *biʾr*) did arise.[21] As of yet, I have found no references in these texts to "cisterns" (*ṣahārīj*, s. *ṣahrīj*), as such.[22] As the terminology in these documents is unclear, one cannot be certain if wells or cisterns were the focus of conflicts. Water treatises and *fatwā* manuals, containing jurists' legal advice, are another rich source on water issues. However, they deal primarily with the canal-based irrigation system of Damascus and the Ghūta, describing conflicts over diversion of water from the canals, blockage of canals (and who should clear it), right to use, right to sell water, and so on.[23] It was the canals that were the legal focus of this hydrological system, and not

[19] In Mamluk-era *waqfiyyāt*, the value of houses increased with the presence of caves (used for storage) and cisterns, but those amenities do not appear to have had a separate fiscal value (for sale, inheritance, endowment, etc.).

[20] Wåhlin, "Inheritance of Land," 57–84; Wåhlin, "The Family Cistern,".

[21] Little, *A Catalogue*. See especially pp. 379–80 (doc. #861, 712 H/1312 CE), 383 (doc. #859, 742 H/1342 CE), 384 (doc. #873, 721 H/1321 CE), and 242 (doc. #38, 765 H/1364 CE).

[22] These documents use the term *biʾr*. The term is used inconsistently in documentary sources of the Mamluk period. When opposed to the term "*sahrīj*", it generally refers to a well.

[23] Walker, "The Struggle Over Water," 292–93.

cisterns. Likewise, one can find no direct reference to water or water systems, of any form, in Ottoman tax registers (*defterleri*). The registers provide only indirect evidence of water harvesting, diversion, and storage: the tax value of fruit trees (which required irrigation), industrial facilities that were driven by water (such as mills), etc. The so-called "*filāḥa* treatises" (something akin to farmers' manuals) are more promising, and describe such practical tasks as how to dig a "*bi'r*", how to increase water capacity, best crops for certain watering (and soil) conditions, different categories of watering (quality), how to predict rainfall, how to raise water, and so on.[24] In his chapter on irrigation, the 12th/18th-century Damascene scholar al-Nābulūsī describes in some detail how and where to dig a *bi'r*, and how to remove the saltiness from its water.[25] In his treatise, the term *bi'r* refers to a cistern, rather than a well, as water flows into it (*al-bi'r idhā waṣala ilayhi al-mā'*). From the context of the chapter, we know that he is describing agricultural cisterns found in mountain orchards and used to hand-water trees. In this regard he specifically notes the watering of apple trees, grapes, quince, oranges, peaches, bananas, plums, and hazelnuts. His treatise represents the most mature form of this genre in Syria, which included much contemporary and indigenous, rather than curated and encyclopaedic, knowledge. By the 12th/18th century, texts such as these were in high demand among other scholars and educated landowners. This period also coincides with the apex of market gardening in the Jerusalem region, as elsewhere in the Ottoman Empire.[26] Land was developed, largely through local initiative, through the construction, and repair, of agricultural terraces and cisterns.

The richest textual sources on cisterns in the Mamluk and Ottoman periods, however, are *waqf* documents, narrative sources, and court registers. *Waqfiyyāt*, documents describing the foundation of a religious endowment (*waqf*), and *sijillāt*, court registers, document the fiscal value of cisterns as amenities of private land, connected to houses, gardens, and orchards. In narrative sources, such as chronicles and travellers' accounts, cisterns appear as one of most prominent markers in the rural landscape. The *waqfiyyāt* are particularly detailed in their descriptions of different kinds of cisterns, which of them are in use and which have become derelict, how many of them are in a village and where they are located, how they were used, and how they were filled with water (through rainfall collection, diversion from springs and wadis, etc.).

In the *waqfiyya* of the Mamluk Sultan al-Ashraf Shaʿbān (r. 764–78/1363–77), dated 777/1375, the Transjordanian village of Adar, located outside of Shobak, was

24 Walker, "The Struggle over Water," 295–98.

25 al-Nābulūsī, *Kitāb ʿilm al-milāḥa*, Ch. 2, 13–14.

26 Shopov, "When Istanbul Was a City of *Bostāns*," 279–307; Walker, "Peasants,"; Walker, "The Changing Face".

listed among the properties endowed for the financial support of his *madrasa* complex in Cairo.[27] The village was endowed in its entirety, its 83 houses (73 of these were in ruins, *kharāb*), listed as potential rental properties in the future; gardens, with roses and poplars; caves; a mill; and a bathhouse (*ḥammām*). It was the 157 village cisterns, however, that were described in the greatest detail in the document. 127 of them were still in use; all but two were house cisterns, with one cistern located in the courtyard of the village mosque, and another large cistern located in a private garden (fed by a spring). The notary distinguishes between two types of cisterns: the freshly-built family cistern (*ṣahrīj*), and "ancient" ones rehabilitated for various uses (*sahārīj rumānīyya*). All of the functioning cisterns in the document were *ṣahārīj rumānīyya*, filled by "flowing water" (*māʾ al-aṣīla*).[28] According to the document, they were filled either as stand-alone cisterns (rain-fed or hand-filled with jars and waterskins)[29] or were part of a larger network connected to one another through spring-fed canals. The thirty cisterns listed as being in ruins appear to have been included for their economic potential: if cleaned out and replastered, they could be put back into use. These cisterns, like the ruined houses, were "fixer-uppers": even derelict ones had economic potential and could be developed by new managers.

Cisterns also appear with great frequency in Syrian-authored chronicles, geographies, and travellers' accounts of the late Mamluk and Ottoman periods, where they are woven into detailed descriptions of the rural environment and village life.[30] Villagers often gave their cisterns nicknames that find their way into ethnographic accounts, as well as into proverbs.[31] These appellations could refer to the locally known names of plots of land or ruins (on which the cistern belonged), the name

[27] Dar al-Wathāʾiq, Cairo – Waqfiyya 8/49, *Hujaj umarāʾ wa-ṣalāṭīn*. Parts of this manuscript have been previously published in Ghawanmeh, "Al-Qarya" and *Al-Tārīkh al-Ḥaḍārī*, 290–92, as well as Walker, *Jordan in the Late Middle Ages*, 154–61; and Walker, "The Struggle Over Water," 299.

[28] It is not clear from the document why only the older, rehabilitated cisterns were functioning at the time of this survey of property. In modern Jordan, villagers regularly refer to such installations on their land as "Roman", emphasizing their antiquity as a matter of prestige or, arguably, to raise the value of the land.

[29] Ghawanmeh, "Al-Qarya," 366.

[30] Walker, "Struggle Over Water", 298.

[31] The proverbial use of cistern analogies is rich, to cite only a few from early 20th-century Palestine, in local dialect: "He has not placed it in a leaky cistern" (*mā ḥaṭṭāsh fī bīr kharīb*—essentially, he puts his trust in something solid); "Until the cistern is filled by dew" (*ḥatta yimteli el-bīr min en-neda*—meaning it is a hopeless situation); "Who has drunk from a cistern does not throw a stone in it" (*man shirib min el-bīr lā yirmi ḥajar fīh*—the equivalent of "people who live in glass houses should not throw stones"); (Dalman, *Work and Customs*, 543, 539, 541, respectively). In Dalman's narratives, the term *bīr* (*biʾr*—formally a well) is often used to denote a cistern.

of the owner(s), or the cistern's size, shape or other characteristics.[32] The practice of naming one's cisterns in this fashion continues today in Palestinian villages.[33]

The economic value of cisterns, however, is most visible in the Ottoman-era Sharīʿa court records (*sijillāt sharīʿa*) of Jerusalem. Four hundred registers in this series are preserved, spanning a period of 388 years (936–1335/1529–1917). I have been surveying these sources for the last few years, primarily as part of a larger archaeological study on garden agriculture and agricultural terracing.[34] In addition to the rich documentation on orchards and orchard harvests, household and garden cisterns appear with such frequency in the same contexts that localized water use and evolving fiscal practices emerge from these documents, and it is possible to trace the development of these practices over time.

One two-year period (1081–83/1670–72) stands apart, during which time there was an active buying, selling, and leasing of cisterns, both household and field-based, as amenities of urban houses and their gardens and as valuable installations inside rural orchards.[35] Ninety-three entries in these registers include references to cisterns.[36] These are all financial transactions: primarily sales and leases, but also inheritances and mortgages of endowed properties (both urban and rural). Most of the cisterns listed were located in Jerusalem or in villages in its immediate outskirts. The term *ṣahrīj*, of Persian origin, is consistently used to designate cisterns, whether for household, garden, or field use. The majority (71) of the cisterns appearing in the register entries of these years are called only *ṣahārīj* (pl.). The remaining 22 references, however, specify that they are cisterns "for collecting rainwater" (sg. *ṣahrīj li-jamʿ māʾ al-ishtaʾiyya*). The question emerges, then, whether there was a functional difference between a cistern called a *ṣahrīj* and one referred to as a *ṣahrīj li-jamʿ māʾ al-ishtaʾiyya*. The latter term is used most frequently in the entries when spatial contexts are specified, and are related to household, garden, or field use. Whether the generic term *ṣahrīj* means that the exact function was unknown to the notary, that it was used for some other purpose than water storage,

33 For example, the names of two of the ancient cisterns found at Khirbet Beit Loya, an archaeological site in central Israel excavated under the co-direction of this author, include Bīr al-Ṭawāzāt and Bīr Abū al-Amad, both local names for plots of land (courtesy of Dr. Roy Marom, project ethnographer and historical geographer, February 2026).

33 Personal communication, Dr. Roy Marom, "Palestine Rural History Project" (June 2024).

34 This terrace study, with textual analysis, will appear in the forthcoming Walker, *Terrace Societies*.

35 The *sijillāt* record an intensive exchange of many kinds of rural properties, and harvests, in the late 11th/17th century. The economy of the rural hinterland of Jerusalem prospered as a result of the intensive, market-oriented agriculture that was its foundation.

36 The following uses ʿAlāwinah's edition of registers 172 and 173 (1081–83 H/1670–72 CE), published in 2014 and 2017, respectively. They will be cited below according to both the volumes and pages of this edition, and the document and page numbers from the original manuscript.

or that it was filled some other way than through capturing rainwater deserves further study. In only four entries was the term *bi'r li-jamʿ māʾ al-ishtaʾiyya* (a "well" for collecting rainwater) used to designate a cistern.[37] This latter term was used for residential complexes, not fields; two of these entries specifically state that the cisterns were attached to *iwāns* built of stone and lime (*ḥajar wa-jīr*).[38] This relatively rare term may refer to how the structure was built—a construction rather than a rock-cut installation—or to its physical relationship to a built structure. A wider range of registers would have to be examined to determine such a pattern.

The majority (59) of the entries deal specifically with household cisterns. They would have functioned for household use (drinking water, water for cleaning, and so on). Ten entries mention cisterns located in gardens (s. *ḥakūra*), separate from the house.[39] The gardens generally took the form of walled courtyards and were used for aromatic plants, flowers, decorative trees, and the cultivation of vineyards and fruit trees (such as oranges). Two properties had gardens with cisterns that supported the cultivation of a commercial crop: *suwaydā* (glasswort), a succulent that grows in saline conditions, the ashes of which were used as a flux in the manufacture of glass.[40]

Many of the larger residential complexes, which accommodated multiple households, were equipped with stables. Eleven of these properties had cisterns, and six of the documents mention that the cisterns were located inside the stables.[41] Cisterns in stables could have had two functions: the first was for the watering of animals, and the second was as urinals for the animals stabled there.[42] The urine (like animal dung) was later collected from the stables and transported to the agricultural fields, to be used there as a common form of fertilizer.

37 *Sijillāt* 173.1, p. 71–91 (doc. 3, pp. 35–32), pp. 159–60 (doc. 2, p. 62), and pp. 244–45 (doc. 1, p. 100), and 173.2, pp. 59–60 (doc. 2, p. 222).

38 The term *jīr*, as used here, may refer to nari (a heavily eroded and chalky limestone) or mortar; in the latter case, the document describes architecture of stone and mortar, rather than dry-stone construction. As for the entries, one mentions property outside the village of Lifta (*sijill* 173.2, pp. 302-303; doc. 3, p. 326) and another in the village of al-Baqāʾ, to the east of Hebron (*sijill* 173.2, pp. 313–14; doc. 3, p. 331).

39 *Sijillāt* 172, p. 269 (doc. 5, p. 134); 173.1, pp. 132–34 (doc. 2, p. 51), pp. 146-147 (doc. 1, p. 57), pp. 266–67 (doc. 2, p. 108), and 283–84 (doc. 2, p. 114); 173.2, pp. 220–21 (doc. 1, p. 291), pp. 221–22 (doc. 2, p. 291), pp. 301–302 (doc. 1, p. 326), pp. 323–24 (doc. 1, p. 335), and pp. 332–35 (doc. 1, p. 338).

40 *Sijill* 173.2, pp. 59–60 (doc. 2, p. 222) and pp. 295–96 (doc. 1, p. 323)

41 *Sijillāt* 172, pp. 350 (doc. 10, p. 176) and 173.1, pp. 247–49 (doc. 1, p. 101), pp. 329–31 (doc. 3, p. 133), pp. 331–34 (doc. 2, p. 134), pp. 361–62 (doc. 1, p. 148), and pp. 409–10 (doc. 4, p. 167)

42 Archaeological excavations frequently uncover interconnected, rock-cut cavities in the bedrock floors of ancient stables (built for that purpose) and natural caves (used for that purpose). These cavities are often identified as urinals or *shunas* (for cereal storage). Some may have been cisterns repurposed for those functions.

The registers also record the exchange of agricultural land on which there were cisterns. Six involve the sale of orchards cropped in a wide range of fruit trees (olives, figs, quince, pomegranates, apricots, pears, and peaches), as well as berries and lilies, which were, apparently, irrigated from the cisterns on the land.[43] Five other registers document the endowment, sales, and leases of orchard harvest, along with the use of the *quṣūr* (hilltop residences, s. *qaṣr*) and cisterns on that land.[44] Here, the cisterns were part of the amenities of the estate (*qaṣr*).

Clearly, by the 11th/17th century, cisterns had been fully commodified. While not listed individually in these registers as properties for financial exchange, they were an indispensable part of landed property and raised its value for commercial purposes, much like terraces did for agricultural land. Cisterns could be jointly owned, be parts of endowments, and their shares inherited, endowed, and leased. The question arises as to what exactly constituted a cistern in these documents. Water capture and storage seem to have been only one, though the most important, function. It appears as if, from a bureaucratic perspective, the cistern was simply a bedrock-cut installation (regardless of function), or any feature that collected rainwater. It was its storage potential that ultimately gave it its value, much like the caves listed in the same documents.

An Archaeology of the Family Cistern

Cisterns honeycomb the rural landscape of southern Bilād al-Shām. Geology has much to do with this. Extensive outcrops of exposed bedrock are found throughout the highlands of Transjordan and Palestine, where erosion is pronounced and slopes steep. Ancient cisterns are generally located on visible outcrops of soft limestone and chalk (nari) deposits. The bell-shaped cistern, which is regularly described and illustrated in archaeological reports, reflects the way the installations were carved out of these limestone and chalk layers. Because limestone is porous, cisterns carved into the rock had to be lined with hydraulic plaster.

Archaeology is another way of analyzing the value and use of these cisterns. Household cisterns are ubiquitous at archaeological sites that had been settled on a long-term basis. Here the cisterns were reused over and over again, as these places were abandoned and then resettled. It is increasingly apparent that the majority of long-abandoned sites resettled in the 7th/13th century, for example, were chosen

[43] *Sijill* 173.2, pp. 59–60 (doc. 2, p. 222), pp. 200–201 (doc. 2, p. 283), pp. 234–35 (doc. 2, p. 296), pp. 295–96 (doc. 1, p. 323), pp. 302–303 (doc. 2, p. 326), and pp. 313–14 (doc. 2, p. 331).

[44] *Sijillāt* 172, p. 172 (doc. 6, p. 83); 173.1, pp. 363–64 (doc. 1, p. 149); 173.2, p. 50 (doc. 3, p. 318), pp. 99–103 (doc. 1, pp. 241–42), pp. 313–14 (doc. 2, p. 331).

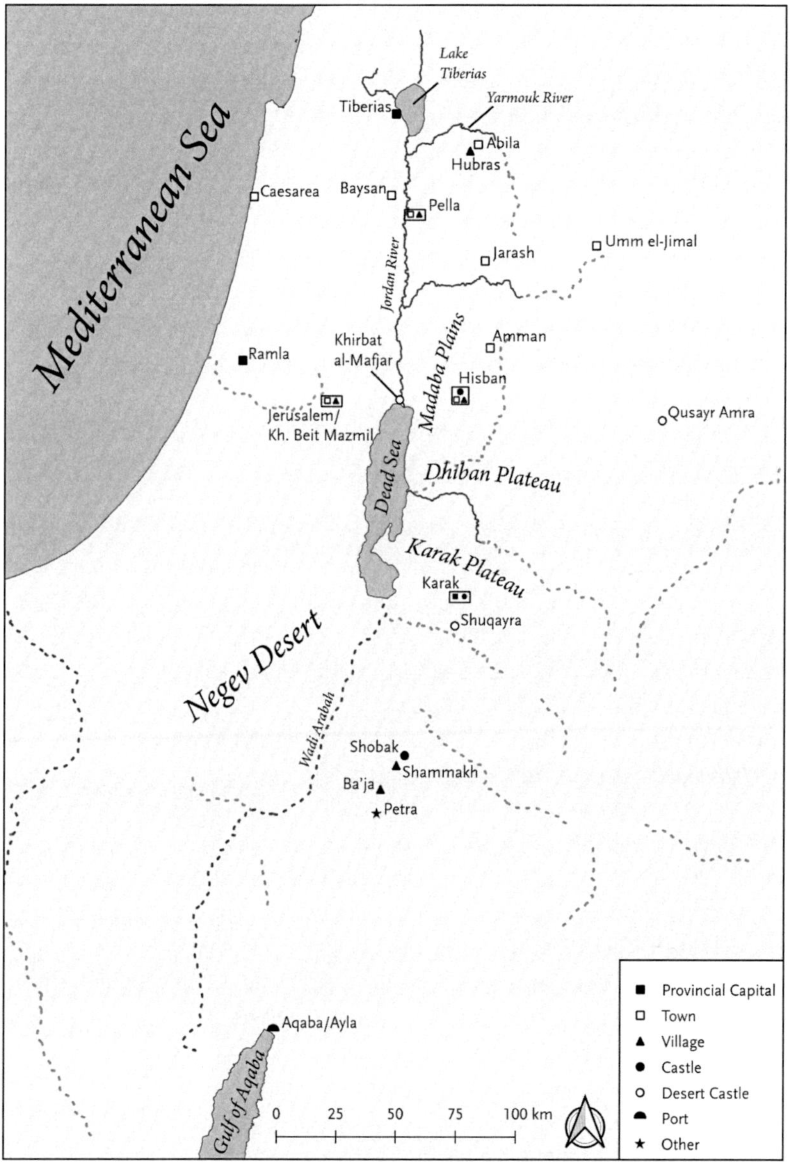

Figure 9.2: Map of selected Late Byzantine and Early Islamic sites reoccupied in the Middle Islamic period. Courtesy of Luigi Pinchetti, PhD Universität zu Köln.

specifically because of the presence of many cisterns.[45] The 19th-century resettlement of the Madaba Plains in central Jordan, and the foundations of modern agriculture in the country, are largely attributable to the revitalization of ancient cisterns in archaeological ruins there.[46]

While textual sources of the Mamluk and early Ottoman eras document the commodification of household cisterns, the archaeological record provides the physical evidence of water harvesting practices, suggesting ways in which cisterns created neighbourhoods, claimed territory, and grew wealth. Two case studies—highland villages in southern Bilād al-Shām—are here compared for their revival and development of ancient cistern systems: Tall Ḥisbān in the Madaba Plains of central Jordan, and Khirbet Beit Mazmīl in Jerusalem (Fig. 9.2).

The two sites were excavated by the author, are contemporary with one another, and occupy comparable environmental and hydrological zones. They represent, however, different site typologies and field systems, which have conditioned the use and valuation of cisterns over time. The transformation of large urban and small, rain-fed household cisterns at Ḥisbān into *shunas* (cereal storage vats) was intimately tied to the commercialization of cereal cultivation on the Madaba Plain by the 18th and 19th centuries. In contrast, the communal water structure at Beit Mazmīl and the numerous small, spring-fed cisterns distributed throughout the terraced fields cultivated by the peasants of this village supported commercial gardening from the 15th and 16th centuries.

The Household Cisterns of Tall Ḥisbān

Tall Ḥisbān is arguably the most important archaeological site of the Mamluk period in rural Bilād al-Shām. Having been the focus of scientific excavations on and off since 1968, it is the longest-running, foreign-led archaeological excavation in Jordan; the archaeological site has been well-documented and is well-known. It is an exceptionally well-preserved medieval site, with a small Mamluk citadel on the summit of the hill, and clusters of vaulted domestic units on the slopes and beyond. The place, called at times a village (*qarya*) and at others a town (*madīna*), appears with great frequency in Arabic narrative sources of the 8th/14th and 9th/15th centuries. It was, for a time in the 8th/14th century, the capital of the Balqā district of central Transjordan, housed a Mamluk garrison of modest size, hosted its own regional market (that served, we are told, 300 villages in the region), and had its own judge and a madrasa (school of law), built in the late Ayyubid period.[47] It had

[45] Walker, "Looking for Home".

[46] Abujaber, *Pioneers Over Jordan*, 8.

[47] For a review of the textual sources on the site, see Walker, *Jordan in the Late Middle Ages*.

Figure 9.3: Aerial photo of Tall Ḥisbān, taken in October 2021. Courtesy of Robert Bewley, APAAME (APAAME_20211022_FB-0097).

close ties to both Jerusalem and Damascus, and was well-networked with villages across Palestine and southern Syria. Two Mamluk sultans (al-Nāṣir Muḥammad—third reign 709–41/1310–41—and Barqūq, r. 784–91/1382–89 and 792–801/1390–99) spent time there and cultivated close relations with the local Bani Mājid. It was, significantly, an agricultural centre, renowned for its diversified cereal production and nuts. Ḥisbān was a place of relative affluence in the Mamluk period, its standard of living achieved, in part, through state incentive, but largely by local investment in water harvesting.

The eight-hectare, multi-period archaeological site is in the village of the same name, located twenty kilometres south of Amman and nine north of Madaba.[48] At 800 meters above sea level, it offers views, on clear days, of the Dead Sea, Jericho, and the Old City of Jerusalem. The site occupies a strategic position on the Madaba Plains, a highland plateau in the centre of modern Jordan (Fig. 9.3).

Since Antiquity this region had been a breadbasket of the ancient world, and it remains today a centre of cereal production. The tell sits at the intersection of the King's Highway and a well-travelled road connecting central Transjordan with

[48] The Biblical site was called "Heshbon", and in the Roman and Byzantine periods, it was "Esbus". Ḥisbān was the medieval Islamic name and is what the village continues to be called today. While this essay focuses on the Mamluk-era site, the town was an important one in the Abbasid and Ayyubid eras as well.

Jerusalem. These routes continued to serve the Madaba Plains commercially as late as the 19th century, when enterprising landowners invested in large-scale cultivation of cereals for markets in Palestine and abroad.[49]

Its environmental conditions are a mixed blessing for agriculture. Falling mostly in the winter months (in January), average annual rainfall is 392 mm a year, which is generally sufficient for dry farming of cereals. However, droughts are common, impacting the ability to retain seed for the following year. In addition to unpredictable and highly seasonal rainfall, it is the particular geology of this part of the Madaba Plains that has made cistern-construction a necessity for settled life and agriculture of any form. Tall Ḥisbān sits on the edge of the slopes that descend from the Madaba Plains to the Ghūr (Jordan Valley). The tell itself is comprised of chalky limestones, marls, and cherts, while the plateau is formed by resistant limestones, sandy dolomites. This geology has impacted the hydrology of the area. As a result of the overwhelming presence of permeable carbonate bedrocks, soils that absorb water, and the semi-arid climate, there are no perennial surface water resources near the site.[50] The wadis al-Majjar and al-Marbat, which encircle the tell, rarely carry water today, and when they do it is seasonally, fed by seven local springs. The most reliable freshwater sources for the peoples living at Ḥisbān in the past have been the perennial springs of the Wadi Ḥisbān (the head of which is 3.5 kilometres northwest of the tell)[51] and winter rainwater captured in bedrock-hewn cisterns or diverted to reservoirs.[52]

The cisterns are largely constructed into outcrops of chalky ("soft") limestone and nari (degraded bedrock), which dominate the landscape of the region. Such cisterns are found not only on the tell, but across the landscape of this part of the Madaba Plains, where they are associated with domestic ruins, field and agricultural terrace walls, and in natural caves.[53] The tell itself is a quarry, making use of the interwoven beds of chalky limestone, marls, and cherts.[54] The surface bedrock,

49 Abujaber, *Pioneers Over Jordan*.

50 Lacelle, "Bedrock," 61.

51 Ethnographic accounts attest to the use of this wadi, and its springs, as a source of drinking water for people living in Ḥisbān in the early 20th century (Abujaber, *Pioneers Over Jordan*). Archaeological reconstruction of the deeper chronology of the use of these springs, and their associated agricultural terraces, is ongoing (Walker and Hamarneh, "The Agricultural Terraces," in print).

52 In addition to the Iron Age reservoir still visible at the southern base of the tell, and two others in the Wadis Ma'jar and Muḥtariqa, nineteen reservoirs were mapped in a ten-kilometre radius from the tell during the hinterland surveys of the 1980s. The estimated holding capacity of the Iron Age reservoir at the tell has been calculated at two million litres (LaBianca, "On-Site Water Retention Strategies," 773). While this reservoir had fallen out of use by the Hellenistic period, many of the cisterns were still in use in the Mamluk period, as described below.

53 Ibach, *Archaeological Survey*.

54 Lacelle, "Bedrock", 26–31.

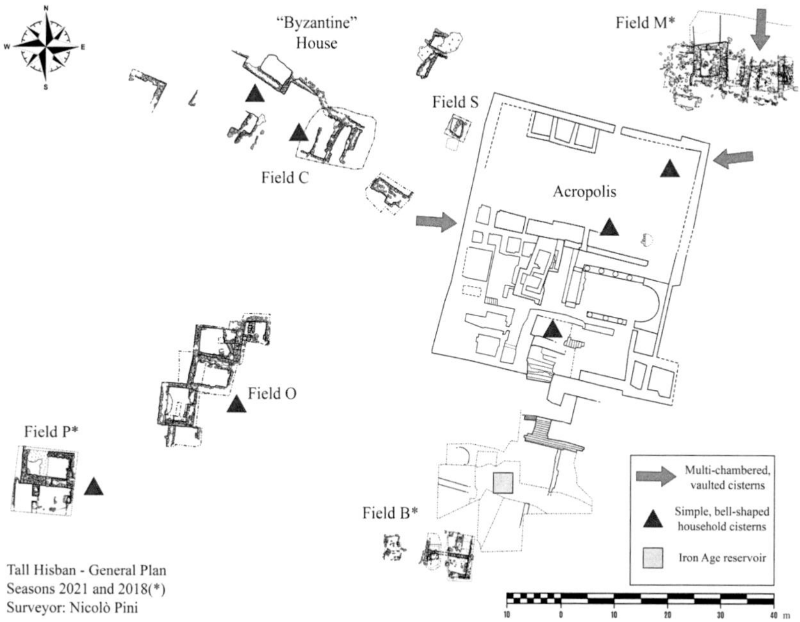

Figure 9.4: Site plan of Tall Ḥisbān with locations of cisterns and reservoirs, 2021 season. Courtesy of Nicolò Pini, Université libre de Bruxelles.

where exposed, is primarily composed of nari, a material burned in large kilns throughout Antiquity to produce the lime plaster used to waterproof these cisterns, as well as to plaster walls and floors of houses, and pave roads.[55] Because of its unique geology, the tell is also a vast bedrock-cut, subterranean water tank, with cisterns of various sizes, many of which appear to be interconnected.[56]

Over fifty years of fieldwork at the site has documented scores of water storage facilities of various forms, including two reservoirs (Iron Age-Hellenistic and Early Byzantine in date); four communal, urban-scale, covered cistern complexes (tentatively dated to the Byzantine era); and numerous smaller bell-shaped cisterns that served households (spanning the Roman through Mamluk periods) (Fig. 9.4).[57]

[55] The remnants of an industrial-size lime kiln, used in the Byzantine period, can still be seen on the site on the southern slope, above the main reservoir.

[56] Walker, "The Struggle Over Water," 303.

[57] Andrews University in Michigan (USA) launched the Heshbon Expedition in 1968, with an original focus on what they believed was the Biblical site of Heshbon. The Expedition continued for four more seasons, ending in 1976. Excavations began again in 1996, with a consortium of American, and now European, universities. This fieldwork continues today, and has a specific research focus on the Islamic periods, particularly the Mamluk era. There have been fourteen field seasons since 1998, when the author joined the project.

The Heshbon Expedition began to map the large, urban cisterns on the summit of the tell in the 1970s. Since the focus of excavations shifted from the medieval citadel (on the summit of the tell) to the village below in 2010, there has been a concerted effort to map, measure, and better understand the household cisterns associated with the so-called farmhouse clusters on the tell slopes and beyond.[58] Many of the cisterns on the summit of the tell appear to be interconnected, having been modified from natural caves and subterranean passageways formed over millennia by the action of rainwater on soft limestone. It has been difficult to fully map them and estimate their holding capacities, as they are filled with deep sediment, and the ceilings of many have collapsed or are on the verge of collapsing. They have been dated through a variety of methods. For initial construction, we have relied on architectural style and stratigraphic relationship to independently dated contexts. To reconstruct a history of use, reuse, and abandonment, we have made use of a combination of methods: the pottery used in the sealant plaster, ceramics in post-depositional fill, C14 analysis of the charcoal added to the sealant plaster, and OSL (optically simulated luminescence)-dating of alluvial deposits. On this basis, it appears that at least three of the urban cisterns (two on the summit of the tell and one on the west slope) were built anew in Mamluk period; at the same time, most of the ancient ones were cleaned out, replastered, and put back into use.[59] This revitalization of ancient cisterns was a common practice in the region, and examples can be found for most time periods. When these cisterns stopped being used for water storage, they were generally repurposed as middens (for communal garbage disposal).

After decades of excavations, and several field seasons devoted to the study of the local water systems, we can make several observations related to a typology of cisterns and their functions.[60] Two kinds of cisterns were used at the site: large, urban-scale cisterns of multiple chambers, the bedrock ceilings supported by arches and pillars; and much smaller, bell-shaped, single-chambered cisterns, closed by a large stone. The communal, urban cisterns have been tentatively dated to the Early Byzantine period and are all modifications of natural caves. The largest identified thus far is a four-chamber complex at the base of the tell on its southern face. Dubbed the "Abu Nur Cave" by project participants, and published under this name, this urban-scale cistern developed out of centuries of modification of a natural cave. The results of a geomorphological study of the complex in 2013,

58 Study of the spring-fed cisterns in the nearby Wadi Ḥisbān is part of an ongoing landscape project led by the University of Bonn (Walker and Hamarneh, "The Agricultural Terraces").

59 Walker, "The Struggle Over Water," 303–304.

60 The Heshbon Expedition in the 1970s cleared one of the large, covered cisterns on the summit and mapped it. The preliminary reports can be found in different issues of the *Andrews University Seminary Studies*.

Figure 9.5: 3D model still of the Abu Nur cistern interior, Field G, Tall Ḥisbān. Courtesy of Thomas Mewes and Henning Nitschke, Berliner Hochschule für Technik, Berlin.

followed by laser and 3-D mapping the following year (Fig. 9.5), suggest that at some point the cistern may have functioned as a *qanāt* (here, a subterranean conduit), and that its channels connect to other subterranean passages to the south of the plaster-lined, Iron Age reservoir.[61]

At some point a smaller, bell-shaped cistern was constructed into the front chamber (immediately inside the entrance), and the southernmost chamber went out of use for water collection sometime in the Mamluk period. It has been modified and redesigned numerous times over the centuries, revealing complex changes in water harvesting needs and strategies.

The bell-shaped installations are the household cisterns, located in the shared, open-air courtyards of the "farmhouse clusters", and cut into exposed bedrock. Every housing cluster had at least one such cistern, which would have likely been shared by an extended family. On the basis of the stratigraphic relationship with the houses, the cisterns appear to have been built at various times in the past (Roman, Byzantine, Abbasid) and simply taken back into use in the Mamluk period. At least three new household cisterns were constructed in the Mamluk era.

Like everything else at this heavily spoliated site, the cisterns were frequently reused, recycled, and repurposed, finding a second or third life for cereal storage, garbage collection, and burials (of animals and humans). In the modern village of Ḥisbān, cisterns located under houses were repurposed to hide goods from tax collectors during the latter years of Ottoman rule.[62]

[61] Walker *et al.*, "Tall Hisban 2013 and 2014 Excavation Seasons," 482–523.

[62] This is known through ethnographic interviews with local villagers by excavation project staff, who have also recorded the use of subterranean caves repurposed in the same manner (Carroll, Fenner, and LaBianca, "The Ottoman Qasr at Hisban," 141).

Cisterns and Terracing at Khirbet Beit Mazmīl

The lights of Jerusalem are visible at sunrise, on a clear day, from the summit of Tall Ḥisbān. Mamluk-era Ḥisbān and Jerusalem maintained close commercial and social ties. Even though both highland towns had their own *sūq*s (markets) and courts, peasants from the villages outside Jerusalem brought their produce to the market of Ḥisbān[63], and Ḥisbānīs bought property in the Jerusalem region and visited the Holy City.[64] Among the largest of the villages in the immediate vicinity to the west of the Old City was the village of Beit Mazmīl. In the 8th/14th century it was surrounded by terraced orchards, many of which were part of religious endowments. In the early Ottoman era, its own commercial agriculture prospered, with local peasants actively buying and selling orchard crops in the village market, as well as those in Jerusalem and beyond.[65]

The 0.3-hectare archaeological site of Khirbet Beit Mazmīl is what remains of this village.[66] It is a walled farmstead perched on the top of the second highest hill in metropolitan Jerusalem, at an elevation of 819 meters above sea level. While the average rainfall for Jerusalem is 590 mm per year (falling mostly in the winter months), this particular site benefits from its location on the western watershed that provides runoff water to Jerusalem and villages to the east. Several spring-fed wadis pass nearby, but they run with water only seasonally. If winter rains are captured and stored, there is sufficient water even for irrigated agriculture.

As throughout the highlands of central Palestine, the geology of the Jerusalem area consists of sedimentary limestone, dolomite, and dolomitic limestone. Like the central highlands of Transjordan, where the layers of softer limestone are accessible, cisterns are numerous; hard limestone (dolomite), on the other hand, provided the main building material. Agricultural cisterns dot the rural landscape outside the Holy City; household cisterns, long fallen out of use, riddle the ruins of domestic spaces inside and outside the walled city. While the site of Khirbet Beit Mazmīl is comprised of hard dolomite, and so are the nearby hills and ridges, the wadis between them have outcrops of mixed dolomite and soft limestone, as

[63] To cite one example, in the autumn of 795/1393, people from the village of al-Quṣūr, near Jerusalem, were attacked in Jericho while on route to the market at Ḥisbān to sell their raisins, pomegranate seeds, and other produce (Müller, "Crimes without Criminals?").

[64] Walker, *Jordan in the Late Middle Ages*, 70.

[65] Walker ed., *Life on the Farm*.

[66] The excavations, co-directed by the author and Benjamin Dolinka, took place during five seasons between 2015 and 2019. OSL-dating of the nearby relic terraces began as an extension of the project in 2017. The final report of all seasons of fieldwork, and historical studies, appears in Walker ed., *Life on the Farm*.

Figure 9.6: View of entrance to Khirbet Beit Mazmīl, in 2019. Photograph by the author.

well as marl and chalk.[67] Those hills and valleys are covered with ancient cisterns, scattered among old field walls.

What is visible at the archaeological site (the *khirbeh*) today is the 19th-century *qaṣr*—a walled farmstead of drystone construction with agricultural towers (*manāṭīr*) and a central courtyard (Fig. 9.6).[68] That complex sits directly on top of, and spoliates, much construction material from a Mamluk-era complex of vaulted rooms, which we have identified as an amiral estate (the residence and farm of a *muqṭaʿ*).[69] Roman, Byzantine, and Abbasid wall spurs and cisterns were repurposed in the 7th/13th century with the resettlement of this long-abandoned place.

The walled, hilltop residences known locally as *quṣūr*, which overlooked terraced hillsides, were one of most visible features of the landscapes of rural Palestine and southern Syria in the Mamluk and Early Ottoman periods.[70] In Palestine, the terraces largely supported orchard cultivation, where trees often had

67 Sneh and Avni, "Jerusalem: Sheet 11-II."

68 Walker and Dolinka, "Ḫirbet Bēt Mazmīl." This kind of "hybrid manor" is also known in Palestinian Arabic as a *eʿzbeh* or *seerah*.

69 Walker, "Peasants", 103; Walker, "The Changing Face"; Walker, *Life on the Farm*. The identification of the medieval complex as a Mamluk officer's estate is based on its location (on a hilltop outside Jerusalem), the architectural style (monumental with complex vaulting systems, suggesting an official function), and the copious equestrian equipment and accessories among the metal finds.

70 al-ʿUlaymī, *al-Uns al-jalīl*, 81–83.

Figure 9.7: Ancient cistern reused in Mamluk industrial complex, Field C, Khirbet Beit Mazmīl, 2019. Photograph by the author.

Figure 9.8: Field D cistern complex, Khirbet Beit Mazmīl, 2019. Photograph by the author.

Figure 9.9: Ein Kerem terrace system with cisterns, 2019. Courtesy of Omer Ze'evi-Berger, University of Bonn.

to be watered by hand, from in-field cisterns, filled by run-off from springs, and by rain. By comparison, the walled orchards of the Ghūta of Damascus were watered by channels running from the main irrigation canals fed by the Barada River. At the centre of these orchards, in the Mamluk and Ottoman periods, were the same walled residences, called *quṣūr*.[71]

Khirbet Beit Mazmīl was likely selected for resettlement in the 7th/13th century because of its water infrastructure (Fig. 9.7 and 9.8). There are six extant household cisterns at Khirbet Beit Mazmīl: one Late Roman in date, two interconnected Abbasid cisterns (A11 in the Southern Sector, dismantled in the Mamluk period), one covered cistern (under the vault-supporting walls of the Mamluk-era North Building of B5), and two cisterns in a roofed complex in the Field D of the Mamluk period (one was a repurposing of a Late Roman ritual bath). During the last season of excavation (in 2019), when surprised by late season rains, these ruined cisterns quickly filled with water. The largest in the "cistern complex" of Field D held water year-round; neighbourhood children swam in it on hot days while we were in the field. Only the Field D system was still in use by the 19th century.

The coordinated study of the *khirbeh* and its terraced agricultural fields gave us the opportunity to study the use of two different kinds of cisterns, used at the same

[71] Eychenne, Meier, and Vigouroux eds., *Le waqf de la mosquée des Omeyyades de Damas*, 90 and n°60.

time, by the same community. Cisterns in agricultural fields are part of the larger field systems, which are comprised of agricultural terraces, plot fences, agricultural watch towers (*manāṭīr*), threshing floors, and roads (Fig. 9.9).

OSL-dating of the terraces helped to date, as well, at least one stage of use of the cisterns, if not their original construction. While many of the field cisterns may have been dug in Antiquity, they were certainly used in the Mamluk and Ottoman periods, providing ready access to water for the irrigation of tree crops.

Contemporary *filāḥa* treatises may shed some light on how these cisterns functioned in the cultivation of fruit trees, which required irrigation. Tree crops could have been watered by hand or through the capture of rainfall run-off from specially prepared terraces. Terraced fields with cisterns could serve both functions. In discussing "best practices" in planting olive trees, al-Nābulusī, citing a much earlier Byzantine source, advises digging a hole, placing a layer of small stones at the bottom of it, placing the small olive tree on top of these stones, and filling the rest of the space with soil. He claims this practice was to be used to keep the soil in place and to regulate temperature.[72] This has been interpreted at other terrace systems in the region as a method of water management, to control the flow of run-off irrigation and to prevent root decay.[73] It is also a method for diverting rainwater to fields farther off, and, eventually, cisterns downslope.

The agricultural cisterns near Beit Mazmīl were an important condition for the village's commercial successes in the 11th and 12th/17th and 18th centuries. It is at this time, as well, that the cisterns acquired a fiscal value, as part of domestic and agricultural real estate, according to the Jerusalem *sijillāt*. The commercialization of garden agriculture may have been one factor driving this development.

Conclusions

In those localities of Bilād al-Shām that are far from perennial water courses, local communities have had to rely on micro-catchment of rainwater runoff to meet their daily needs. The most widespread, durable, and resilient localized water harvesting has made use of household and agricultural cisterns to survive the summer seasons and years of unpredictable rainfall. These small-scale installations have always been locally constructed and designed, and locally used and revived, independently of the state.

The social organization of these systems was the village. There were occasional conflicts over access to shares in cisterns, but these were localized and usually

72 al-Nābulūsī, *Kitāb ʿilm al-milāḥa*, 38.

73 Hamarneh, "Investigating Ancient Man-Made Terraces," 115.

settled within the family or with the manager of an endowment. They were not a point of conflict with the state; we rarely read about disputes over them in chronicles, court records, or *fatwā* manuals, for example, while, on the other hand, conflicts over access to irrigation canals were common.

It was village communities that first acknowledged the economic value of household cisterns, long before the state recognized their potential to raise real estate costs and taxable revenues. On the local level, they were an asset because of their multifunctional potential for storage. Cut into bedrock and easily accessible, they could be used for storage of water, cereals, or anything else of value. They were the most practical and common storage facility for households.

The expansion of endowments in the late Mamluk period ushered in the early commercialization of agricultural cisterns. They gained even more economic importance with the growth of market agriculture in the Palestinian highlands, where orchard cultivation relied on irrigation. The state came to evaluate cisterns accordingly. They raised the value of the property on which they were located (as a house's "amenity", and a component of the developed agricultural land), and eventually they became commodities of their own, separate from the land. As the textual sources that document these developments—*waqfiyyāt*, *tapu defterleri*, and *sijillāt sharīʿa*—have generally not been preserved for Transjordan in the same periods, we need to develop other methods to identify the catalysts for these developments there. Ethnographic studies of later periods, combined with carefully designed archaeological fieldwork at settlements and agricultural fields alike, have the potential to reconstruct the economic background of cistern use that is otherwise not visible in written sources for Mamluk and Ottoman Jordan.

Bibliography

Primary Sources

Archival Sources (manuscripts)

Waqfiyya 8/49, *Ḥujaj ʿumarāʾ wa-salaṭīn*—endowments of Sultan Shaʿbān for his madrasa in Cairo, 777/1375, Dār al-Wathāʾiq, Cairo.

Primary Sources (edited, published manuscripts)

ʿAlāwinah, S.Z., ed. *Sijillāt Maḥkamat al-Quds al-Sharʿiyya al-ʿUthmāniyya. Sijill (172) – (1081-1083 H/1670-1672 M.)*. Al-Quds Open University, 2014.
ʿAlāwinah, S.Z., ed. *Sijillāt Maḥkamat al-Quds al-Sharʿiyya al-ʿUthmāniyya. Sijill (173) – (1081-1083 H/1670-1672 M.), Part One*. Al-Quds Open University, 2017.

ʿAlāwinah, S.Z., ed. *Sijillāt Maḥkamat al-Quds al-Sharʿiyya al-ʿUthmāniyya. Sijill (173) – (1081-1083 H/1670-1672 M.), Part Two*. Al-Quds Open University, 2017.

Eychenne, M., A. Meier, and E. Vigouroux, eds. *Le waqf de la mosquée des Omeyyades de Damas: Le manuscrit ottoman d'un inventaire mamelouk établi en 816/1413*. Institut Français du Proche-Orient, 2018.

al-Nābulūsī, ʿAbd al-Ghanī ibn Ismāʿīl. *Kitāb ʿilm al-milāḥa fī ʿilm al-filāḥa*. Nahaj al-Suwāb, 1882.

al-ʿUlaymī, Mujīr al-Dīn. *al-Uns al-jalīl bi-tārīkh al-Quds wa-Khalīl*. Edited by M.ʿA. al-Kaʿābinat. Maktabat Dandīs, 1999.

Secondary Sources

Ababsa, M. and A. Z. Abu Hussein. "Metropolitan Amman: Comprehensive Climate Plans." In *Greater Than Parts: A Metropolitan Opportunity*, vol. II, edited by S. Mehrotra, L.L. Lewis, M. Orloff, and B. Olberding, 1–50. The World Bank, 2020.

Abujaber, R.S. *Pioneers Over Jordan: The Frontier of Settlement in Transjordan, 1850-1914*. I.B. Tauris & Co, Ltd., 1989.

Braemer, F., D. Genequand, C. Dumond Maridat, P.-M. Blanc, J.-M. Dentzer, D. Gazagne, and P. Welch. "Long-term Management of Water in the Central Levant: The Hawran Case (Syria)." *World Archaeology* 41, no. 1 (2009): 36–57. DOI: 10.1080/00438240802666424.

Carroll, L., A. Fenner, and Ø.S. LaBianca. "The Ottoman Qasr at Hisban: Architecture, Reform, and New Social Relations." *Near Eastern Archaeology* 61, no. 3–4 (2006): 138–45.

Dalman, G. *Work and Customs in Palestine,* Vols. 1&2. Translated from German into English by N. Abdulhadi-Sukhtian. Dar Al Nasher, 2013.

Ghawanmeh, Y.D. "Al-Qarya fī Janūb al-Shām (al-Urdun wa-Filasṭīn) fī al-ʿAṣr al-Mamlūkī fī Ḍawʾ Waqfiyyat Qaryat Adar." *Studies in the History and Archaeology of Jordan* 1 (1982): 363–71.

Ghawanmeh, Y.D. *Al-Tārīkh al-Ḥaḍārī li-Sharqī al-Urdun fī al-ʿAṣr al-Mamlūkī*. Dār al-Fikr, 1982.

Hamarneh, C. "Investigating Ancient Man-Made Terraces of Petra–Jordan." PhD diss., Humboldt-Universität zu Berlin, 2022. https://edoc.hu-berlin.de/handle/18452/25400. Accessed March, 2026.

Ibach, R.D., Jr. *Archaeological Survey of the Hesban Region*. Hesban 5. Andrews University Press, 1987.

Jenssen, S.M. "Traditional Rainwater Harvesting in Jordan: A Qualitative Study of Project Rainkeep (1994-1995)." Unpublished MA thesis, University of Bergen, 2006.

LaBianca, Ø.S. "On-Site Water Retention Strategies: Solutions from the Past for Dealing with Jordan's Present Water Crisis." *Studies in the History and Archaeology of Jordan* 5 (1992): 771–76.

Lacelle, L. "Bedrock, Superficial Geology, and Soils." In *Environmental Foundations: Studies of Climatical, Geological, Hydrological, and Phytological Conditions in Hesban and Vicinity*, edited by Ø.S. Labianca and L. Lacelle, 25–58. Andrews University Press, 1986.

Little, D.P. *A Catalogue of the Islamic Documents from al-Ḥaram aš-Šarīf in Jerusalem*. Beiruter Texte und Studien, Band 29. Franz Steiner, 1984.

Müller, C. "Crimes without Criminals? Legal Documents on 14th c Injury and Homicide from the Haram." In *Legal Documents as Sources for the History of Muslim Societies: Studies in Honour of Rudolph Peters*, edited by M. van Berkel, L. Buskens, and P. Sijpesteijn, 129–79. Brill, 2017.

Mundy, M. and R.S. Smith. *Governing Property, Making the Modern State. Law, Administration and Production in Ottoman Syria*. I.B. Tauris, 2007.

Palmer, C. "Following the Plough: The Agricultural environment of Northern Jordan." *Levant* 30 (1998): 129–65.

Sneh, A. and Y. Avni. "Jerusalem: Sheet 11-II." In *Geological Map of Israel 1:50,000*. Geological Survey of Israel, 2016.

Shopov, A. "When Istanbul Was a City of *Bostāns*: Urban Agriculture and Agriculturists." In *A Companion to Early Modern Istanbul*, edited by S. Hamadeh and Ç. Kafescioğlu, 279–307. Brill, 2021.

Wåhlin, L. "Inheritance of Land in the Jordanian Hill Country." *British Journal of Middle Eastern Studies* 21, no. 1 (1994): 57–84.

Wåhlin, L. "The Family Cistern: 3000 Years of Household Water Collection in Jordan." In *Ethnic Encounter and Culture Change: Papers from the Third Nordic Conference on Middle Eastern Studies, Joensuu, June 1995*, Nordic Research on the Middle East No. 3, edited by M. Sabour and K. Vikør, 233–49. Center for Middle East and Islamic Studies, University of Bergen, 1997.

Walker, B.J. *Jordan in the Late Middle Ages: Transformation of the Mamluk Frontier*. Chicago Studies on the Middle East Monograph Series 8. Middle East Documentation Center, University of Chicago, 2011.

Walker, B.J. "Mobility and Migration in Mamluk Syria: The Dynamism of Villagers 'on the Move'." In *Proceedings of the Conference "Everything is on the Move: The 'Mamluk Empire' as a Node in (Trans-) Regional Networks"*, edited by S. Conermann, 325–48. Bonn University Press, 2014.

Walker, B.J. "The Struggle Over Water: Evaluating the 'Water Culture' of Syrian Peasants under Mamluk Rule." In *Developing Perspectives in Mamluk History*, edited by Y. Ben-Bassat, 287–310. Brill, 2017.

Walker, B.J. "Looking for Home in Long-abandoned Places: The Resettlement of Late Medieval Syria." In *Humanistische Anthropologie. Ethnologische Begegnungen in einer globalisierten Welt. Festschrift für Christoph Antweiler zu seiner fünfundsechzigsten Geburtstag von seinem Freund*innen und Kolleg*innen*, edited by T-D. Vu, O. Pye, H.D. Olschleger, and G. Distelrath, 451–72. V&R Press, 2021.

Walker, B.J. "Peasants, Rural Economy and Cash Crops in Medieval Islam." In *Markets, Money and Exchanges: their Economic Logics in Pre-modern Societies*, edited by J.C.M. García, 91–113. Oxbow Books, 2021.

Walker, B.J. "The Changing Face of Agricultural 'Estates' in 15th and 16th-c Palestine: the Commercialization of Khirbet Beit Mazmīl." In *The Mamluk-Ottoman Transition: Continuity and Change in Egypt and Bilad al-Sham in the Sixteenth Century, Vol. II* edited by S. Conermann and G. Şen, 325–57. University of Bonn Press, 2022.

Walker, B.J. ed. *Life on the Farm in Late Medieval Jerusalem: The Village of Beit Mazmil, its occupants, and their industry over five centuries*. Monographs in Islamic Archaeology 2. Equinox, 2025.

Walker, B.J. ed. *Terrace Societies: Modelling Peasant Behavior in Terraced Agriculture in the Ancient and Medieval Levant*. Monographs in Mediterranean Archaeology. Equinox, 2027, forthcoming.

Walker, B.J., R. Bates, J. Hudon, and Ø.S. LaBianca. "Tall Hisban 2013 and 2014 Excavation Seasons: Exploration of the Medieval Village and Long-Term Water Systems." *Annual of the Department of Antiquities of Jordan* 28 (2017): 482–523.

Walker, B.J. and B. Dolinka. "Ḫirbet Bēt Mazmīl, Investigations of Medieval Jerusalem's Hinterland: Interim Report on the 2015 and 2017 Field Seasons." *Zeitschrift der Deutschen Palaestina-Verein* 136, no. 2 (2020): 191–234.

Walker, B.J. and C. Hamarneh. "The Agricultural Terraces of Wadi Ḥisbān, 2024 and 2025 Field Seasons," *Annual of the Department of Antiquities of Jordan* 63 (2026): in print.

CHAPTER 10

"From a Spring Where Allah's Servants Will Drink": Reassessing the Social Significance of *Sabīls* in Mamluk-Era Cairo*

Angela Isoldi

Abstract

This article reassesses the socio-economic role of Mamluk-era *sabīls* (charitable water dispensers) in Cairo through the analysis of three 15th-century *waqf* documents. Challenging the traditional definition of *sabīls* as "public fountains", the analysis reveals these structures were components of larger foundations designed as strategic investments for the ruling network (*ahl al-dawla*) to manage private assets. *Sabīls* functioned through a donor-driven model of piety, remaining entirely dependent on existing water-carrier networks and restrictive operating schedules rather than providing a reliable water supply. Crucially, their proliferation decentralized urban water management by creating significant employment for unskilled labourers employed as *sabīls* attendants, granting this vulnerable social group unique agency in the city's water management.

Keywords: Mamluk Cairo; *sabīl*; charity; *waqf* (perpetual endowments); socio-economic history

Introduction

From the early Mamluk period (648–923/1250–1517), monumental water dispensers known as *sabīls* (Ar. *sabīl*, pl. *asbila*) were established through perpetual endowments (*waqf*, pl. *awqāf*) as charitable facilities. An early example was the *sabīl* added to the Ẓāhiriyya madrasa, founded by Sultan al-Ẓāhir Baybars in 662/1263.[1] These structures featured a cistern for storing fresh water transported from the Nile or Cairo's canals—the city's only sources of potable water, as local well-water was typically saline. Within the *sabīl*, a specialized attendant called a *muzammalātī* distributed this water free of charge to passers-by through iron-grated windows

* This publication is part of the project "Source of Life: Water Management in the Premodern Middle East (c. 7th–15th centuries)" with file number VI.C.181.016 of the research programme Vici SGW, which is financed by the Dutch Research Council (NWO).

[1] al-Maqrīzī, *Sulūk*, 2:4; *Khiṭaṭ*, 4:224; Ibn Taghrī Birdī, *Nujūm*, 7:121.

facing the street. The charitable purpose of these buildings was deeply rooted in Islamic tradition, which regards the provision of water as a highly meritorious act; indeed, a *hadith* attributes to the Prophet Muhammad the statement that "giving water" is the best form of charity.[2] *Sabīls* became increasingly prevalent in Cairo from the $8^{th}/14^{th}$ century onward, and a significant number of these Mamluk-era structures remain standing today.[3]

Earlier scholarly works have tackled the significance and functions of *sabīls* within the Mamluk city. In his seminal work on the role of *waqf* foundations in Mamluk-era Cairene social life, for instance, Muḥammad Amīn highlighted the role of *sabīls* as structures commonly financed by perpetual endowments to contribute to the water supply of the city, which due to its climate and geographical position struggled with the provision of potable water.[4] Saleh Lamei Mostafa brought to the foreground the symbolic decorative elements that characterise these water structures and the connection of such elements to the Qur'anic descriptions of water.[5] While acknowledging this important symbolic aspect of the Cairene *sabīl*s, Lamei Mostafa defined these structures as "public drinking fountains", a definition that continued to be used in later publications as well. For instance, Doris Behrens-Abouseif's entry in *The Encyclopœdia of Islam* also defines *sabīl*s as providers of drinking water "for free public use",[6] and Sylvie Denoix's study on conceptions of "public" and "private" in Islam describes *sabīl*s as "public (drinking) fountains".[7] Various studies have opted for such a definition of *sabīl*s, whereas some scholars have otherwise articulated the argument that these water structures were essentially meant for the urban population as contributions to the water supply of the city.[8]

However, the terminology we use to describe *sabīls* is not neutral; it fundamentally shapes—and potentially distorts—our understanding of water management in medieval Cairo. The validity of this "public service' argument remains poorly documented. The perceived "public" function of the *sabīl* often rests on the misleading assumption that *waqf* foundations were primarily tools for the "common good".[9] In fact, recent scholarship has demonstrated that *awqāf* were frequently

[2] Ibn al-Malik, *Sharḥ al-maṣābīḥ*, 2:475.

[3] On the conservation of Mamluk and Ottoman-era *sabīls* in modern Cairo, see the contribution of Bakhoum to this volume.

[4] Amīn, *al-Awqāf*, 148-150.

[5] Lamei Mostafa, "The Cairene Sabil," 37–38.

[6] Bosworth and Behrens-Abouseif, "Sabīl."

[7] Denoix, "Les notions de 'privé' et de 'public'," 131–49.

[8] See, for instance: Blair and Bloom, "Introduction," 15; Stillman, "*Waqf* and the Ideology of Charity," 369; al-Harithy, "Charity and Piety,", 162.

[9] Frenkel, "The Waqf System," 221.

exploited as lucrative investments and effective assets management vehicles for ruling networks.[10]

Despite the relatively abundant research dealing with *sabīls*, the nature, function, and impact of these water structures must be reassessed in light of recent findings regarding the political and economic role of *waqf* foundations. As early as 2010, Valentine Denizeau's doctoral thesis on water management in Mamluk Cairo argued that *sabīls* were not established to satisfy the water demands of the urban population, nor did they represent a tool for the "common good". Instead, these structures held a distinct religious value and served as vehicles for socio-political messaging on behalf of their founders.[11] This significant argument challenges previous assumptions and compels us to adopt a more nuanced perspective when evaluating the true significance of *sabīls* within Mamluk Cairo.

This chapter proposes a context-specific analysis that illuminates elements previously overlooked in the scholarship on *sabīls*. By juxtaposing common patterns emerging from various Mamluk-era *waqf* documents, this study seeks to formulate more precise conclusions regarding the socio-economic roles and ideological significance of these water dispensers. Rather than beginning with overarching assumptions, the argument is constructed through three case studies drawn from 9th/15th-century *waqf* deeds held at the National Archives in Cairo (Dār al-Wathāʾiq, hereafter DW). These cases were selected from a broader corpus analysed in my doctoral dissertation, which revealed consistent patterns among such foundations—specifically regarding the founders' affiliations with contemporary political networks (the *ahl al-dawla*), the use of *waqf* endowments as strategic investments, and the establishment of charitable water facilities.[12] The three case studies selected for this article serve as representative models for the establishment, financing, and *modus operandi* of *sabīls*, finding significant parallels in other contemporary documentation.

Through analysis of these case studies, I propose a more balanced perspective on *sabīls*—one that considers these structures as components of larger foundations designed as lucrative asset management vehicles for their founders, while simultaneously acknowledging them as water facilities that functioned not to alleviate urban water scarcity, but as expressions of a donor-driven charitable outlook. Despite their private motivations, these endowments nonetheless exerted a significant impact on water management within the Sultanate's capital. Before analysing these case studies in detail, however, a few remarks regarding the nature of the primary documentary sources—the *waqf* deeds—are in order.

[10] Petry, "A Geniza for Mamluk Studies?," 51–60; Igarashi, *Land Tenure and Mamluk Waqfs*, 5–41; *Land Tenure, Fiscal Policy*, 177–205; "*Waqf* as a Means," 277–88.

[11] Denizeau, "Conduire l'eau," 445–50.

[12] Isoldi, "Confluent Waters," chapters 2–4.

Navigating the *Waqf* Deeds

As previously noted, *sabīls* were established as charitable components within *awqāf*. In Islamic law, these perpetual foundations are created when a founder (*wāqif*, pl. *wāqifūn*) endows a set of properties (*mawqūf*), thereby legally prohibiting any further transactions—such as sales, inheritance, or gifts—involving these assets. A portion of these properties must be revenue-generating to provide the necessary funds for the foundation's maintenance, ensuring it remains financially self-sustaining. These revenues also finance the charitable facilities or initiatives designated by the founder for the benefit of specific or general groups (*mawqūf ʿalayhi*), ranging from the founder's own family to broader social categories, such as "the poor".[13] For example, a *waqf* foundation might include a communal *ḥammām* (bathhouse) which, when leased to a manager (*ḥammāmī*), generates the income required to maintain the bathhouse itself and to fund a charitable facility like a *sabīl*, which does not generate its own revenue.

A *waqf* document—generally designated in modern scholarship as a *waqfiyya* (pl. *waqfiyyāt*), though Mamluk-era sources refer to it as a *kitāb waqf* or *maktūb waqf*[14]—contains essential data regarding the founder, including their name, *nisba*, and various titles. It also provides detailed descriptions of the endowed properties (*mawqūf*), their precise locations, and the designated beneficiaries of the foundation. Furthermore, the deed establishes the criteria for appointing a manager (*nāẓir* or *mutawallī*) responsible for ensuring that the foundation's operation aligns with the founder's stipulations. Because these documents held binding legal authority once validated by religious officials, they contain minute regulations governing the charitable structures and initiatives they funded. Consequently, the *kitāb waqf* does more than merely enumerate the state of properties at the moment of endowment; it prescribes the intended operational framework of the entire foundation. For this reason, *waqf* deeds serve as the most exhaustive sources for reconstructing the location, architectural features, and *modus operandi* of *waqf*-endowed structures such as *sabīls*.

The wealth of detail regarding *sabīls* found in *waqf* documentation is unparalleled in narrative sources, such as chronicles, which mention these dispensers only sporadically.[15] However, because *waqf* deeds are normative sources, they

13 For a more detailed general introduction to *waqf* law see: Abbasi, "The Classical Islamic Law of Waqf," 124–30; Peters, Behrens-Abouseif, "Waḳf."

14 Livingston, *Managing Paperwork*, 46.

15 For example Ibn Iyās and al-Malaṭī mention the establishment of Sultan Qāytbāy's *sabīl* in the Qashshāshīn neighbourhood, stating that the structure was particularly beneficial for that quarter. Ibn Iyās, *Badāʾiʿ*, 3:43; al-Malaṭī, *Nayl*, 6:409. Ibn Iyās reports that in 881/1476, the quality of Nile water deteriorated so significantly in both taste and colour that a group of wealthy individuals opted to

describe the *modus operandi* of charitable structures as intended by their founders at the time of recording; they do not necessarily reflect subsequent realities on the ground or the evolution of the foundations after their inception. Fortunately, many of these documents—typically written on long scrolls—contain marginalia recording specific developments, such as the addition of new properties or the exchange of assets (*istibdāl*). These marginal records are fundamental for reconstructing the economic dynamics of a foundation and tracing its development over time.

Furthermore, to fully grasp the social and economic significance of these foundations, it is essential to integrate documentary records with the information provided by narrative sources, where available. Chronicles, topographies, and biographical dictionaries offer invaluable insights into the social and political backgrounds of the founders; such context is vital for reconstructing the motivations behind the establishment of larger foundations, of which *sabīls* were a constituent part. As other scholars have rightly noted, the integration of narrative sources is crucial for reconstructing the "living background" of documentary evidence.[16] The analysis of the case studies presented here, and the resulting perspective on *sabīls*, stems from this integrative approach, which utilizes narrative sources to validate, contextualize, or contrast the evidence found within the *waqf* deeds.

The exhaustiveness of the available data varies across the three case studies selected for this article. Narrative sources do not always provide detailed information regarding the backgrounds and political careers of every founder. In the first case study—the foundation established in 810/1408 by Fatḥ Allāh ibn Muʿtaṣim ibn Nafīs—a wealth of information exists due to the high status and political prominence of the founder. As discussed below, Fatḥ Allāh was one of the most powerful figures in the sultanic administration during the first decade of the 9th/15th century.[17]

Significantly less is known regarding the *wāqif* of the second case study, Sharaf al-Dīn Mūsā ibn al-Burhān, whose foundation was established in 830/1427. Consequently, the *waqf* deed and the successive records annotated on its margins represent the primary sources of information for this case.[18] Finally, narrative sources provide no data to help reconstruct the specific context of the third case study: the foundation of the *amīr* Sayfī Qarājā al-Jamālī al-Khāṣṣakī al-Malikī al-Ashrafī. For this founder and his endowment, one must rely solely on data

source their drinking water from cisterns (likely those of the *sabīls*). Ibn Iyās, *Badāʾiʿ*, 3:119. This incident suggests that stored water was regarded as a secondary, rather than a preferred, option for those with the means to choose.

16 Reinfandt, "Mamlūk Documentary Studies," 300.

17 DW355. The wealth of details on Fatḥ Allāh recorded in the narrative sources is also due to the fact that he was the most important patron of the historian al-Maqrīzī. See: Rabbat, *Writing Egypt*, 80-99.

18 DW91.

gleaned from documentary records—namely, two *waqf* deeds dated 880/1475 and 882/1477, respectively.[19]

Despite the disparity in the volume of data available for each case study, a combined analysis of these three samples illuminates critical aspects of Mamluk-era *sabīls* that must be prioritized to better understand their nature, function, and urban impact. After detailing the backgrounds of the founders and the circumstances surrounding the establishment of their respective foundations—factors that reveal both the founders' motivations and their strategic exploitation of these endowments—the subsequent section of this paper will analyse the divergent and common patterns emerging from the documentation. Ultimately, this leads to the formulation of a new perspective on the social significance of *sabīls*.

Two *Sabīls* in the *Waqf* of Fatḥ Allāh

The earliest foundation examined in this paper is the one Fatḥ Allāh ibn Muʿtaṣim ibn Nafīs established on the 10th of Ramadan 810 (February 1408).[20] By that date, Fatḥ Allāh held the high-ranking post of *kātib al-sirr* (chief secretary) at the court of Sultan Faraj ibn Barqūq (r. 801–15/1399–1412). Fatḥ Allāh was the scion of an influential family of Jewish physicians from Tabriz that established itself in Cairo in the mid-8th/14th century.[21] Fatḥ Allāh gained influence at court after Sultan Barqūq (r. 784–91/1382–89 and 792–801/1390–99) appointed him as *raʾīs al-aṭibbāʾ* (chief physician) in 797/1395 and as *kātib al-sirr* (chief secretary) of the sultanic administration in 801/1399. Fatḥ Allāh kept this position through the end of Barqūq's and the beginning of Faraj ibn Barqūq's first reign. In 808/1405, however, Fatḥ Allāh's political rivals managed to get him arrested: only after paying the substantial sum of 500,000 silver *dirhams* to the sultan was Fatḥ Allāh able to regain his freedom and influential position at court.[22]

At this point, Fatḥ Allāh's political influence was at its zenith: reinstated to the post of *kātib al-sirr*, he was able to obtain a series of properties in Ḥārat Zuwayla, a prestigious quarter in the north-western section of Fatimid Cairo (al-Qāhira),

[19] DW177 and DW184.

[20] This *waqf* document was studied and partially edited in: Behrens-Abouseif, *Fatḥ Allāh and Abū Zakariyya*, 20–21, 27–30, 39–44. In this article, as in my PhD thesis, I have analysed parts of this document that were not included in Behrens-Abouseif's book.

[21] Fatḥ Allāh's paternal grandfather converted to Islam and established ties with Sultan al-Nāṣir Ḥasan (r. 748–52/1347–51 and 755–62/1354–61), while Fatḥ Allāh's uncle held the position of *raʾīs al-aṭibbāʾ*, the chief physician at the service of the sultan and nominated by sultanic decree. al-Maqrīzī, *Durar*, 3:12–13.

[22] al-Maqrīzī, *Durar*, 3:12; al-Sakhāwī, *Ḍawʾ*, 6:166; Ibn Taghrī Birdī, *Nujūm*, 12:326 and 13:43.

including the palace he had formerly dwelled in as a tenant and which he expanded after acquiring it through exchange (*istibdāl*) from an earlier *waqf*—a manoeuvre he was able to execute thanks to his political influence.[23] This palace, expanded through an extensive spatial reconfiguration of its surrounding neighbourhood,[24] was endowed alongside other profitable assets—including a *ḥammām* and a horse stable—in the *waqf* foundation established by Fatḥ Allāh in 810/1408. Because these properties could be leased to tenants, they functioned as revenue-generating properties for the endowment.

Crucially, the *waqf* deed stipulates that Fatḥ Allāh appointed himself as the foundation's manager (*nāẓir*) for life, with the bulk of the income allocated to his sister and descendants. Such an arrangement unequivocally demonstrates that the primary objective of this endowment was to secure a substantial income for his family and possibly to safeguard his estate from further extortion and confiscation. However, this goal was only partially achieved: in 815/1413, Sultan al-Mu'ayyad Shaykh (r. 815–24/1412–21) imprisoned Fatḥ Allāh, extorted 50,000 *dīnārs*, and forcibly diverted the foundation's revenues by naming himself the new beneficiary. The endowment only returned to the family after the Sultan's death, by which time Fatḥ Allāh had already been assassinated.[25]

Along with the lucrative properties, Fatḥ Allāh's foundation also included two *sabīls* for charitable water distribution. One of these structures was located within the same Ḥārat Zuwayla; while the other was attached to the funerary complex (*turba*) Fatḥ Allāh endowed in the cemetery area outside Bāb al-Barqiyya, one of the eastern gates of Fatimid Cairo. In this extramural area, Fatḥ Allāh had invested in the restoration of the *khānqāh* (Sufi lodge) established in the first half of the 8th/14th century as part of the *waqf* of *amīr* Ṭughāy Tamur al-Najmī (d. 748/1347).[26]

Although this paper focuses primarily on the socio-economic significance of *sabīls* rather than their architectural organization, several relevant details emerge from Fatḥ Allāh's *waqf* document. First, the deed reveals that the *sabīl* endowed

[23] Fatḥ Allāh acquired the palace in Ḥārat Zuwayla through *istibdāl* from an 8th/10th-century *waqf*. *Istibdāl* was a widespread yet legally controversial practice which involved exchanging damaged revenue-generating *waqf* properties for other properties or for a sum in cash, theoretically with the purpose of restoring the revenue flow of the *waqf* from which the asset was taken, but in practice allowing the acquisition of endowed assets by the people requesting the exchange. Al-Maqrīzī suggests that the *istibdāl* through which Fatḥ Allāh obtained the palace was facilitated by the damages the palace had suffered as a consequence of the environmental and economic crisis that befell Cairo after 806/1403. al-Maqrīzī, *Khiṭaṭ*, 3:115–16.

[24] al-Maqrīzī, *Khiṭaṭ*, 3:115. Al-Maqrīzī states that, after obtaining the palace through *istibdāl*, Fatḥ Allāh purchased adjacent properties, evicted the residents, and demolished the buildings to expand his residence.

[25] al-Maqrīzī, *Durar* 3:14–15; Behrens-Abouseif, *Fatḥ Allāh and Abū Zakariyya*, 27.

[26] al-Maqrīzī, *Khiṭaṭ*, 4:299.

at Ḥārat Zuwayla was a freestanding structure; water was supplied to it from a nearby cistern (*ṣahrīj*) that Fatḥ Allāh had established in the same neighbourhood.[27] Today, Sultan Qāytbāy's *sabīl* in the Khalifa quarter is the only extant example of such a freestanding dispenser, as the majority of surviving Mamluk-era *sabīls* have been integrated into the corners of larger religious complexes, such as mosques or Sufi lodges (*khānqāhs*). Secondly, none of the *sabīls* established by Fatḥ Allāh are mentioned in the document as being paired with a *maktab* (a Qur'anic school for orphans), an architectural feature commonly found atop most extant Mamluk *sabīls*. While the combination of a *maktab* and *sabīl* appeared early in the Mamluk period,[28] surviving evidence suggests that from the late 8th/14th century onward, these two elements were typically synthesized into a single vertical unit including an underground cistern at the base, the *sabīl* at street level, and the *maktab* above.[29]

Finding this different spatial arrangement in the *waqf* deed of Fatḥ Allāh might indicate that, in fact, extant structures do not account for all the possible layouts with which Mamluk-era *sabīls* were built. Perhaps these were more adaptable architectural elements than the material evidence would suggest, and what remains today is mainly what was attached to mosques and religious buildings because these places of worship were spared from the many spatial changes the city has undergone since the later Middle Ages. These different spatial organizations may have been more common than the material evidence suggests; *sabīls* likely represented relatively versatile monumental buildings that founders could adapt to the space at their disposal. These details, beyond their architectural value, clearly show how *waqf* documents can reveal aspects of *sabīls* that cannot be grasped by looking at the material remains alone. The other two *waqf* documents discussed in this paper similarly reveal architectural and spatial arrangements of endowed *sabīls* that differ from the "standard" layout displayed by extant structures (Fig. 10.1).

[27] DW355. Fatḥ Allāh's document describes the cistern (*ṣahrīj*) storing drinking water for this *sabīl* as being located near the water dispenser (*mujāwir lahu*) rather than beneath the distribution chamber as usual in extant *sabīls*. For the *sabīl* at Fatḥ Allāh's funerary complex, water was to be drawn from two cisterns, one of which was restored as part of the *khānqāh* of Ṭughāy Tamur al-Najmī while the other was a newly constructed cistern at Fatḥ Allāh's *turba*.

[28] A *maktab* was built together with the *sabīl* of the Ẓāhiriyya madrasa already in 662/1263. al-Maqrīzī, *Sulūk*, 2:4; al-Maqrīzī, *Khiṭaṭ*, 4:224; Ibn Taghrī Birdī, *Nujūm*, 7:121.

[29] Most of the extant Cairene *sabīls* are organized as three-floors units, with an underground cistern (*ṣahrīj*), a room for water distribution at street level (*ḥānūt al-sabīl*), and a *maktab* on the upper floor. The earliest extant example of this layout is the *sabīl* at the mosque *amīr* Asanbughā endowed in Darb al-Saʿāda (within Fatimid Cairo) in 772/1370. Monument no. 185. al-Harithy, "Charity and Piety," 165; Bosworth and Behrens-Abouseif, "Sabīl"; Behrens-Abouseif, *Cairo of the Mamluks*, 221; Lamei Mostafa, "The Cairene Sabil," 38. Earlier extant *sabīls* such as the one attached to the Manṣūrī complex in Bayn al-Qaṣrayn or the one of amīr Shaykhū al-ʿUmarī al-Nāṣirī (built in 755/1354) do not display this "standard" layout.

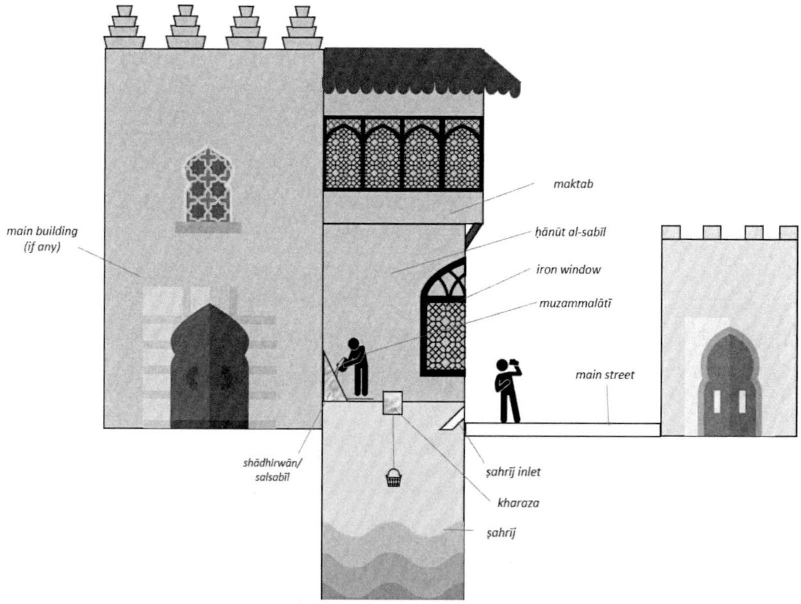

Figure 10.1: Schematic representation of a *sabīl*'s "standard" layout. It features the cistern (*ṣahrīj*), distribution chamber (*ḥānūt al-sabīl*), and *maktab* organized as a vertical unit. Inside the distribution chamber, the *muzammalātī* draws water from the cistern through a floor opening (*kharaza*), and pours it over a slanted marble slab (*shādhirwān*) before offering it to passers-by through the iron-grilled window. Image created by the author.

More importantly, the document analysed here shows that the locations of these two *sabīls* corresponded to the urban areas where Fatḥ Allāh had intervened with other construction projects. The intramural area where Fatḥ Allāh endowed one of the two *sabīls* was the same quarter he had heavily reshaped after becoming the owner of the residential palace he had previously rented from another *waqf* foundation. Meanwhile, the other *sabīl* was part of Fatḥ Allāh's funerary complex in the cemetery. Thus, the choice of the areas in which to build these structures was not driven by a need for water (in fact, Ḥārat Zuwayla had easier water access than other urban areas, as it was located near one of Cairo's canals); rather, the *sabīls* were strategically placed in spaces already shaped by the patronage of their founders.

In addition, the document contains important indications regarding the resources allotted to the *sabīls* and their *modus operandi*. In his deed, Fatḥ Allāh allocated an indefinite budget for the annual filling of the *sabīls*' cisterns; the deed stipulates that the foundation must disburse the funds required to fill these three cisterns with water every year "at any price" (*bālighan mā balagha*), as was customary in such cases.[30] Regarding the staff of the water dispensers, Fatḥ Allāh's foundation allotted a yearly salary of 1,560 silver *dirhams* to the attendant (*bawwāb*/ *muzammalātī*) working at the *turba*, and a yearly salary of 600 silver *dirhams* to the *muzammalātī* in Ḥārat Zuwayla.[31]

The deed also lists the tasks of these employees: these responsibilities included opening and closing the *sabīls*, cleaning them, and transporting water from the cisterns to the dispensing chamber. The *muzammalātī* at the *sabīl* in the cemetery was required to distribute water every day from forenoon until the time of the afternoon prayer (*min waqt al-ḍuḥā ilā waqt al-ʿaṣr*). During the month of Ramadan, the schedule shifted to the period between the sunset prayer and the end of the *tarāwīḥ*—the prayer performed on Ramadan nights following the evening prayer. The attendant in charge of the *sabīl* in Cairo, in turn, was to distribute water every day from noon to afternoon (*min waqt al-ẓuhr ilā waqt al-ʿaṣr*), while in Ramadan he had to follow the same schedule as the former.[32] Both Fatḥ Allāh's *sabīl* at the cemetery and the one in Cairo were, therefore, operative for only a limited number of hours per day.

In sum, analysis of Fatḥ Allāh's *waqf* document reveals that his foundation was primarily intended to secure his properties against confiscation and provide his family with a stable source of income. The foundation also served as a vehicle to manage the properties Fatḥ Allāh had accumulated in Cairo and its surroundings, some of which had been acquired from earlier *waqf* foundations through *istibdāl*. Within this context, Fatḥ Allāh's two *sabīls*, strategically located in areas already shaped by his patronage, were funded by a portion of the foundation's revenues. Their locations and limited operating schedules, however, suggest they were not intended to contribute significantly to the city's general water provision. As the next two case studies will demonstrate, such structures possessed a different type of impact and significance.

Two *Sabīls* in the *Waqf* of Mūsā al-Burhānī

Most of the themes that emerged from the analysis of Fatḥ Allāh's *waqf* deed can also be observed in the second document examined in this article. With this deed, issued on 25 Shaʿbān 830 (June 1427), Sharaf al-Dīn Mūsā ibn al-Burhān (or al-Burhānī) established a *waqf* foundation that he later expanded in 839/1436. This expansion was formalized through a second deed recorded on the same scroll, which added

[30] DW355; Isoldi, "Confluent Waters," 290.

[31] DW355. Of the 1,560 silver *dirhams*, 720 silver *dirhams* are allotted to compensate the janitor's services, while 840 silver *dirhams* are allotted for the *muzammalātī*. The deed leaves it up to the *nāẓir*'s discretion whether to employ one or two persons to fill these roles. Isoldi, "Confluent Waters," 291.

[32] DW355; Isoldi, "Confluent Waters," 291–92.

more properties to the foundation and stipulated new operating conditions.[33] The addition of properties, together with a 831/1427 marginal log recording the purchase of an extensive land lot from the *bayt al-māl* (the Treasury), suggest that Mūsā had been expanding his capital in the years between the issuing of the first and the second *waqf* deed. The brief accounts recorded in the narrative sources endorse the evidence from the document—they attest that Mūsā, who had migrated to Cairo from Kazerun, served as the primary agent of Zayn al-Dīn ʿAbd al-Bāsiṭ ibn Khalīl (d. 854/1451), the influential qadi who held the post of *nāẓir al-jaysh* (secretary of the army) between 824/1421 and 842/1439.[34]

With the first deed of 830/1427, Mūsā endowed several revenue-generating assets within Cairo and its suburbs, including commercial facilities within Fatimid Cairo and a lavish palace on the shores of Būlāq. The inclusion of this palace is significant, as it is described in the document as being built on a *ḥikr* lot—land suitable for urban development rented via emphyteutic leases orchestrated by the *bayt al-māl*. This shows that Mūsā, in his capacity as secretary to the *nāẓir al-jaysh*, had direct access to the resources of the Sultanate's Treasury.[35] According to the conditions stipulated in this deed, Mūsā appointed himself as the manager of the foundation and also as the sole beneficiary of its income for life. The appointment of the founder as the beneficiary of a foundation was not considered legitimate by any Sunni school of law other than the Ḥanafī school. Yet, during the Mamluk period, this form of self-benefitting *waqf* (*waqf ʿalā al-nafs*) became particularly common.[36] This meant that the founder (in this case, Mūsā) remained in full control of his foundation and profited from it, utilizing it as a form of investment that was legally "safe" from taxation and confiscation. It seems, however, that this foundation did not fully succeed in maintaining Mūsā's capital safe for long. The chronicles report that when Mūsā's powerful patron, ʿAbd al-Bāsiṭ ibn Khalīl, was arrested by Sultan al-Ẓāhir Jaqmaq (r. 842–57/1438–53) on the 28th of Dhū al-Ḥijja 842 (June 1439), Mūsā was also arrested and mulcted of the substantial sum of 10,000 *dīnārs*.[37] Following this date, no information on Mūsā is available in the narrative sources. A log in the same scroll of the *waqf* deed, dated 848/1444, records that the lavish palace Mūsā had endowed in Būlāq was transferred to the *wakīl bayt al-māl* (the secretary of

[33] Isoldi, "Confluent Waters," 87–90.

[34] al-Sakhāwī, *Ḍaw'*, 10:175; Ibn Ḥajar, *Inbā'*, 4:120.

[35] Isoldi, "Confluent Waters," 161.

[36] Igarashi, "*Waqf* as a Means," 280–81. The spread of the self-benefitting *waqf* was allowed by the quadruple system of justice introduced by Sultan al-Ẓāhir Baybars (r. 658–76/1260–77) in 663/1265. Such system implied the possibility of solving legal matters in accordance with law schools other than the Shāfiʿī, which had been the only *madhhab* formally valid in Egypt before that. Rapoport, "Royal Justice," 76–79.

[37] al-Maqrīzī, *Sulūk*, 7:422 and 429; Ibn Ḥajar, *Inbā'*, 4:120; al-Sakhāwī, *Ḍaw'*, 10:175.

the Treasury) via *istibdāl*.[38] The log mentions that the transaction was authorised by Mūsā's son ʿAbd al-Raḥman, in charge of the foundation's administration at the time, which reveals that Mūsā had died by then and the foundation's management had passed to his children, as stipulated in his *waqf* deeds.

While the bulk of the foundation's revenues remained allocated to Mūsā and his family—who also retained control of the *waqf* administration—the second deed issued in 839/1436 saw Mūsā allot a portion of the foundation's increased resources to charitable works. Specifically, these funds were dedicated to the maintenance and operation of two *sabīls* that had been added to the endowment. One of these *sabīls* was located in the same area of Ḥārat Zuwayla where Mūsā had already endowed residential palaces and commercial facilities in 830/1427. The document describes this *sabīl* as being attached to a mixed-use residential-commercial building, which featured five shops and a horse stable on the ground floor, and two rental apartments on the upper floor.[39] This could indicate that *sabīls* attached to secular buildings might have been more common than what is suggested by the material remains of these structures. Nowadays, Sultan Qāytbāy's *sabīl* behind al-Azhar—which is attached to a caravanserai (*wikāla*) (Fig. 10.2)—stands as a rare surviving example of a *sabīl* integrated into a commercial building.[40] The second *sabīl* Mūsā established and financed with the deed of 839/1436 was instead embedded in the funerary complex he endowed in the cemetery outside Bāb al-Naṣr (one of the northern gates of the Fatimid walls). As in the previous case study, neither of these *sabīls* was attached to a *maktab* for orphans. Furthermore, as with Fatḥ Allāh, both of Mūsā's *sabīls* were built in areas of the city already shaped by the founder's patronage and existing investments.

Part of the revenues Mūsā assigned to these charitable structures was allotted to the yearly filling of the underground cisterns (*mabnī fī tukhūm al-arḍ*) built below each *sabīl*.[41] While also in this case no specific amount was indicated for this purpose, Mūsā's deed specifies that 400 large waterskins (*rāwiyas*) had to be purchased yearly for filling the cistern of the *sabīl* at the *turba*.[42] Furthermore, the

[38] DW91. It is worth mentioning that this exchange took place despite the fact that Mūsā had stipulated a condition forbidding the *istibdāl* of the endowed properties after his death.

[39] DW91; Isoldi, "Confluent Waters," 299–300.

[40] Monuments no. 75 and 76.

[41] DW91; Isoldi, "Confluent Waters," 299–301.

[42] DW91; Isoldi, "Confluent Waters," 301. The *rāwiya* was a large waterskin transported on camelback by the water carriers, who regularly carried water from the Nile to the city and sold it in the streets or delivered it to the dwellings of their customers. A single *rāwiya* likely had a capacity of roughly 200 litres. Water carriers charged their customers per *rāwiya* or per *qirba* (smaller water skins). Their service was crucial for the supply of drinking water, but their fees must have been unaffordable for the poorest social groups. Isoldi, "Confluent Waters," 115–30.

Figure 10.2. Sultan Qāytbāy's *sabīl* behind al-Azhar is attached to a caravanserai (*wikāla*). Photograph by the author.

deed allots a monthly salary of 20 silver *dirhams* for the *bawwāb*/*muzammalātī* at Mūsā's *turba*, in charge of cleaning the mausoleum, opening and closing it when necessary, and offering water from the *sabīls* without a pre-determined schedule. The *muzammalātī* in Cairo was instead assigned a monthly salary of 15 silver *dirhams* for opening and closing the *sabīl*, cleaning the cups belonging to the *sabīl* (which were chained to its window), and serving water according to a schedule not specified in the document.[43] In both cases, the salaries of these employees included fees to be paid for the items belonging to the endowed *sabīls*, such as the cups and the cleaning tools, as well as the oil for the *turba*'s lamps.[44]

When compared with the previous case study, Mūsā's document further demonstrates that *sabīls* were designed with a greater degree of spatial freedom than the extant "standard" layout would suggest. Moreover, this case reinforces the pattern of placing charitable structures within specific urban clusters where the endowers had already invested in other properties. Ultimately, Mūsā's foundation served as a vital asset-management tool, allowing him and his family to control and profit from the wealth he accumulated during his administrative career. Within this legal framework, only limited resources were allotted to the water dispensers, the operation of which was dictated entirely by the specific stipulations of the founder rather than actual urban demand.

The *Sabīl* in the *Waqf* of Qarājā al-Jamālī

The context of the final *sabīl* selected for this article—endowed in the *waqf* of *amīr* al-Sayfī Qarājā al-Jamālī al-Khāṣṣakī al-Malikī al-Ashrafī—can only be reconstructed on the basis of the two deeds kept at Cairo's Dār al-Wathā'iq, as the narrative sources do not seem to report any data related to this individual.[45] From the endower's name and titles, it is only possible to assume that he had been a *mamlūk*—presumably of Sultan al-Ashraf Īnāl (r. 857–65/1453–61), considering the date of the *waqfiyya*—and that he was a member of the Sultanic bodyguards (the *khāṣṣakiyya*). The *waqf* deeds of this *amīr*, however, offer a valuable glimpse into his capital and the role of his foundation.

With the *waqf* deed issued in 880/1475, Qarājā endowed a land lot in Minyā and several profitable properties in Cairo, including three residential buildings and a

[43] The document states that Mūsā's *sabīls* in Cairo operated during "customary hours" (*al-awqāt al-muʿtāda*), but it remains unclear whether it refers to a schedule common among other *sabīls* or to a schedule established within this particular foundation. DW91; Isoldi, "Confluent Waters," 301.

[44] DW91; Isoldi, "Confluent Waters," 301.

[45] Isoldi, "Confluent Waters," 91.

horse stable between Bāb al-Khawkha (one of the western gates of Fatimid Cairo) and the nearby Suwayqat al-Ṣāḥib. In this deed, Qarājā appointed himself as the sole beneficiary and administrator of the foundation for life; after his death, the revenues accrued by the foundation's properties were to be distributed among his wife and children.

In the deed issued in 882/1477, Qarājā added a small piece of land located near Asyūṭ to the revenue-generating properties of the foundation, leaving the conditions stipulated two years earlier—including the revenues allotted to charitable initiatives—essentially unchanged. This indicates that the addition of revenue-generating properties served to increase the amount Qarājā was able to draw from the foundation's revenues, a pattern that may be observed in several other contemporary cases.[46] As with the previous case studies, therefore, this *waqf* represented a strategic arrangement that served, first and foremost, the financial interests of the founder and his descendants.

The foundation Qarājā established in 880/1475 included his *turba* in the cemetery area outside Bāb al-Naṣr. Attached to this funerary complex was a *sabīl* equipped with an underground cistern, for the filling of which an unspecified amount of money was allotted annually.[47] The 880/1475 deed also stipulates that a *maktab* was to be built in Cairo near the properties endowed at the Suwayqat al-Ṣāḥib—though not on top of his *sabīl*, which, once more, was not associated with a *maktab*. The deed further specifies the conditions for its operation and the allowances for ten orphans and their educator. In the 882/1477 deed, however, this *maktab* is still referred to as a planned structure, and it is impossible to determine whether it was ever actually constructed.

In the 880/1475 deed, Qarājā allotted a monthly salary of 150 copper *dirhams* (*dirham fulūs*) to the *bawwāb*/*muzammalātī* employed at his *turba*. As in the previous case studies, this employee was assigned the duties of opening and closing the founder's shrine, cleaning the premises, maintaining the *ṣahrīj*, and transporting water from the cistern to the distribution chamber to be distributed for free to passers-by according to an unspecified daily schedule. Additionally, Qarājā's deed indicates that this employee was permitted to reside at the *turba* rent-free.[48] With

46 For instance, Sultan al-Ashraf Barsbāy (r. 825–41/1422–37) added profitable properties to his foundation with successive deeds without increasing the charitable expenditures stipulated in his first deed, thereby increasing the proportion of the surplus which was assigned to himself and his descendants. Amīn, *Izdihār al-awqāf*, 250. A similar pattern may be observed in the *waqf* foundations of Sultan al-Ashraf Īnāl (r. 857–65/1453–61), his son al-Mu'ayyad Aḥmad, and the influential Jawhar al-Lālā (d. 842/1438). Igarashi, "*Waqf* as a Means," 279; Reinfandt, *Mamlukische Sultansstiftungen*, 87–91.

47 DW177; Isoldi, "Confluent Waters," 314.

48 DW177; Isoldi, "Confluent Waters," 314.

the deed of 882/1477, a daily bread allowance of two *raṭls* (approximately one kilogram) was added to this employee's salary.[49]

This final case study shares several critical elements with those analysed previously. It represents a foundation established primarily to manage assets accumulated by the founder and to channel profits toward him and his family. Within this framework, a small portion of resources was allotted to a charitable water dispenser located at the founder's shrine. This *sabīl*, like those mentioned above, was maintained and operated by an attendant whose duties included distributing water according to a schedule likely established by the foundation, even if not explicitly specified in the document. These recurring patterns across the case studies provide a basis for broader observations regarding the function and significance of the *sabīl*, which will be detailed in the following section.

Beyond Charity: The Socio-Economic Function of the *Sabīl*

Although based on three specific case studies, this analysis has identified key aspects of these foundations that help redefine the function of the *sabīl*s within the endowments and their broader impact on the city. This new formulation must account for several elements revealed by the documents and narrative sources: namely, the social status of the founders and their shared endowment strategies; the significance of the *sabīl*s within these foundations based on their locations and operating conditions; and the recurring patterns governing the *modus operandi* of the *sabīl* and the responsibilities of its attendants.

The first striking feature of these case studies is a shared characteristic among the founders: despite their diverse backgrounds, all three were closely linked to the contemporary configuration of power (*dawla*) at the time they established or expanded their foundations. Consequently, they possessed significant social and financial capital. In each case, the *waqf* served as a strategic mechanism to ensure that the founders and their families retained access to revenues generated by various profitable assets—such as rental housing and commercial facilities—accumulated throughout their political careers. As the self-benefiting foundations of Mūsā and Qarājā particularly illustrate, these endowments functioned as a form of financial investment. However, as the chronicles demonstrate in the cases of Fatḥ Allāh and Mūsā, once the connection of the endowers to the *dawla* was severed, their assets were seized, and their *waqf* foundations could lose significant portions of their revenue streams.

49 DW184; Isoldi, "Confluent Waters," 318.

Within this context, *sabīls* appear as charitable endeavours financed by these private investments. A closer examination reveals, however, that despite their charitable character, these structures were also "investments"—albeit of a spiritual nature. One of the defining features of these case studies is the placement of water dispensers in urban and suburban areas already characterized by the patronage and investments of their endowers. Specifically, the association between the founder's funerary complex (*turba*) and the *sabīl* is a recurring feature in all three cases. This shows that *sabīl*s maintained a profound connection with their founders, marking both the urban neighbourhoods where they resided and the sites of their final resting places.

The *Sūrat al-Insān*, quoted in the title of this article, offers an essential interpretative key for understanding the charitable significance of these structures. This *sūra* draws a clear line between those people who, created to be tested by Allah, saw the right path and followed it, and those who ignored it. While shackles and hellfire await the latter, the righteous people, who for the love of Allah perform acts of almsgiving, are promised the fresh waters of the spring in Paradise called Salsabīl. It is no coincidence that verses from the Surat al-Insān are reported from the *sabīl* at the *khānqāh* of Sultan al-Ashraf Īnāl, at the *sabīl* of the *khānqāh* of Sultan al-Ashraf Qāytbāy (r. 872–901/1468–96), and at that of the funerary complex of *amīr atābak* (commander of the army) Qurqumās (d. 916/1510).[50] The evocative symbolism of this *sūra* illuminates the primary objective of these *sabīls*: securing spiritual merit for the founders through the act of giving water—traditionally regarded in Islamic thought as the most virtuous form of charity.[51] The name of the structure itself, *sabīl*, derives from the phrase *fī sabīl Allāh* (in the path of God), indicating a charitable deed performed for the sake of the Almighty.[52]

The operating conditions of these water dispensers, as stipulated in the *waqf* deeds, further suggest that the charitable services of the *sabīl* stemmed from a donor-driven model of piety rather than a systematic effort to enhance the city's water infrastructure. Indeed, specific clauses reported in the *waqf* documents—such as the limited annual budget for cistern maintenance and the discretionary power given to the founder's appointed attendants—reveal that *sabīls* embodied a private religious enterprise with an outward-facing façade, rather than a public service tailored to urban demand. One of the elements supporting this view is the stipulation, in all these case studies, of a flexible budgetary allocation for filling

[50] Thesaurus d'Epigraphie Islamique, fiche 42797; fiche 42825; fiche 42915.

[51] That medieval Islamic charity was characterised by this donor-driven perspective has been emphasised also by: Stillman, "*Waqf* and the Ideology of Charity," 359; Sabra, "Public Policy or Private Charity?," 95–96.

[52] Bosworth and Behrens-Abouseif, "Sabīl."

the *sabīls*' cisterns, crucial for storing the drinking water that was to be distributed from the distribution chamber of the *sabīl*. As emphasized above, none of these case studies contains specific indications regarding the funds allotted to the yearly filling of the cisterns, and only Mūsā's *waqf* deed indicates that a specific amount of 400 *rāwiyas* was to be purchased yearly for filling the cistern at the cemetery. Qarājā and Mūsā's deeds also indicate that the cisterns were to be filled during the annual plenitude of the Nile, when the water from the river flowed in the canals (*fī zaman jariyān māʾ al-nīl fī al-khuljān*).[53] During the yearly plenitude (between mid-August and October), water was in fact more abundant and of better quality, while when the Nile's level was low, water was scarcer and stagnant. Filling the cisterns during the plenitude season offered significant economic advantages: the abundance of water lowered the fees charged by water carriers, reduced the distances required for transport, and decreased the cost of fodder for the carriers' draft animals.[54]

While it is impossible to calculate the exact scale of this annual expense or the total volume of water these cisterns could store, this flexible budgetary allocation is indicative of a critical, yet often neglected, aspect of the *sabīl*: its fundamental reliance on the services of urban water carriers. Rather than being self-sufficient utilities, these structures were deeply dependent upon the fluctuating fees and logistical capabilities of the water carriers, who supplied water to the city.

Indeed, the water carriers' fees varied constantly based on the Nile's flood level. There is not enough data to reconstruct how their fees fluctuated in normal times, but various Mamluk-era historians do record how the water carriers' fees soared in times of water scarcity.[55] Thus, for the *waqf* foundations financing *sabīls*, leaving an indefinite budget for filling the cisterns represented a solution to the problem of price fluctuations, to a certain extent at least.[56] When water carriers' fees soared excessively, however, even the foundations could not guarantee the *sabīls*' water supply. For instance, in 892/1487, as the price of a *rāwiya* reached 4 *dirhams niṣf*,

53 DW91; DW177; Isoldi, "Confluent Waters," 301 and 314.

54 Isoldi, "Confluent Waters," 223.

55 For instance, al-Maqrīzī records that during the reign of al-Kāmil Shaʿbān (r. 746–47/1345–46) the price of a *rāwiya* rose from 0.75 to 2 silver *dirhams*, and in 749/1348 it soared to 8 silver *dirhams*. al-Maqrīzī, *Sulūk*, 4:90. In 802/1399, the price of a *rāwiya* rose from 1.5 to 4 *dirhams* due to an unprecedentedly low Nile level. Ibn Ḥajar, *Inbāʾ*, 2:105; al-Maqrīzī, *Sulūk*, 6:30.

56 Some of the foundations that financed water dispensers actually allocate a fixed amount for replenishing the cistern every year. For instance, the *waqf* deed of *amīr* Azdamur min ʿAlī Bāy (d. 913/1507) allocates 12,000 copper *dirhams* yearly for filling the cistern of his *sabīl*. DW241; Isoldi, "Confluent Waters," 347. Comparably, Sultan al-Ashraf Barsbāy's *waqf* allocated 1,000 *dirhams* yearly for the purchase of water for the *sabīl* of the *khānqāh* at the cemetery. Thesaurus d'Epigraphie Islamique, fiche 9755.

some *sabīls* were forced to close, likely due to financial constraints.[57] Consequently, while the *sabīls* provided essential spaces for storing and distributing water, they did not offer the city an alternative supply system. Instead, they remained entirely reliant on the same network of traders that supplied the rest of the urban population. This dependency meant that *sabīls* were subject to the same market vulnerabilities as any other consumer; in periods of scarcity or exorbitant prices, these structures were often unable to provide water at all, effectively ceasing their charitable function when the city needed it most.

Further insights into the role and functioning of *sabīls* can be gained from the conditions regulating the distribution of drinking water from the iron windows of the distribution chamber. The documents analysed in this chapter indicate that the opening schedules of water dispensers varied depending on the stipulations of the founders. In the case of Fatḥ Allāh's *sabīls*, it seems that the time slots during which water was distributed were limited to a few hours each day only. Other *waqf* foundations, such as the one of Sultan al-Ghawrī, stipulated longer schedules for the charitable distribution of water from the *sabīl*—every day "from sunrise until people return to their homes in the evening" and, during Ramadan, from sunset until dawn.[58] Comparably, the *sabīl* attached to the mosque of *amīr* Qarāquja al-Ḥasanī at Darb al-Jamamīz, was to offer water daily from the beginning to the end of the day (*fī kulli yawm min awwal al-nahār wa-ilā ākhirihi*),[59] while Sultan Faraj ibn Barqūq's *sabīl* outside Bāb Zuwayla was to distribute water every day between forenoon (*ḍuḥā*) and sunset (*maghrib*).[60] Nevertheless, Fatḥ Allāh's limited schedule was not an isolated case; similar abbreviated hours appear in other deeds as well, suggesting that restricted service was a common practice in many endowments.[61]

These variations in distribution schedules reveal that *sabīls* did not operate under a unified policy; rather, their functioning was entirely contingent upon the specific stipulations of their founders. Furthermore, the restricted timeframes of certain *sabīls*—such as those of Fatḥ Allāh—clearly indicate that these structures

57 al-Sakhāwī, *Wajīz*, 3:1001; al-Malaṭī, *Nayl*, 8:60-3. The *dirham niṣf* (pl. *anṣāf*), originally a Syrian coin weighing half (*niṣf*) a standard silver *dirham*, circulated in Egypt from the reign of al-Mu'ayyad Shaykh onward. Initially high in silver content, the coin underwent progressive debasement under his successors, reaching its lowest silver percentage and value during the reign of Jaqmaq. Petry, *The Mamluk Sultanate*, 202.

58 al-Hazmeh, "Late Mamluk Patronage," 161.

59 Ibrahim, "Wathīqat al-amīr akhūr kabīr Qarāqujā al-Ḥasanī," 211.

60 Lamei Mostafa, "The Cairene Sabil," 40.

61 For instance, Azbak min Ṭuṭukh's *waqf* deed sets a similar water distribution schedule at the *sabīl* of Azbakiyya: each day between the noon prayer (*al-ẓuhr*) and the afternoon prayer (*al-ʿaṣr*), and during Ramadan between the sunset prayer (*al-maghrib*) and the end of the *tarāwīḥ* prayer. DW198; Isoldi, "Confluent Waters," 230 and 327.

were not intended to guarantee a steady or reliable water supply for the city. Indeed, it is difficult to imagine that they played such a role given that water was distributed only for a limited number of hours and from within a confined space accessible exclusively to the *sabīl* attendant—a design that placed water giving under the attendant's control, creating a barrier between the water supply and the people who needed it. Instead, *sabīls* served as localized acts of charity whose availability was dictated by the private priorities and financial constraints of the endowers. In sum, the operating conditions stipulated in the *waqf* documents reveal that *sabīl*s were expressions of a donor-driven model of piety. For the founders, the distribution of water was primarily a vehicle for attaining spiritual reward rather than an effort to systematically improve the city's water infrastructure. Ultimately, these structures remained bound by the specific—and often restrictive—stipulations of their endowers, functioning as private monuments of charity rather than reliable public utilities.

The proliferation of these structures nonetheless likely exerted an influence on Mamluk Cairo's water management extending far beyond the specific services they provided. To start with, each *sabīl* required the employment of a specific worker to manage its maintenance and daily operation. These attendants, described in the documents as *bawwāb* or *muzammalātī*, appear to have been drawn from the city's vast pool of unskilled labourers. The primary qualification for the role was not technical expertise, but rather moral standing—specifically that they be "pious" (*min ahl al-dīn*) or "righteous" (*min ahl al-khayr*).[62] As the case studies in this paper demonstrate, *waqf* deeds stipulated a regular salary for these attendants. In my doctoral thesis, I found that while these wages fluctuated between foundations, they generally aligned with the average monthly rates calculated by Adam Sabra for *waqf* servants across the Mamluk period.[63] These figures are summarized in Table 10.1.

Consequently, for these unskilled labourers, a position at a Cairene *sabīl* represented a valuable job opportunity and a reliable source of income, providing them with what Sabra identifies as the basic means of sustenance. This was particularly true for those employed by foundations like Qarājā's; by providing daily bread rations and onsite accommodation, such *awqāf* shielded their staff from the fluctuating market prices and the rising cost of living. The proliferation of *sabīls* throughout the city thus created a substantial number of employment posts for unskilled workers—a social group that was otherwise particularly vulnerable during times of economic crisis.

[62] DW355; DW91; Isoldi, "Confluent Waters," 291 and 301. Faraj ibn Barqūq's and al-Ghawrī's *waqf* deeds required higher aesthetic and hygienic standards from the *muzammalātīs* employed at their respective *sabīls*; these requirements, however, should not be interpreted as standard rules, but as an expression of the aesthetic character of monuments under royal patronage.

[63] Sabra, *Poverty and Charity*, 116-129; Isoldi, "Confluent Waters," 249-256.

Table 10.1. Salaries of *sabīl* attendants compared

Foundation	Salary for the *sabīl* attendants	Average of the period (from Sabra's survey)
Fatḥ Allāh (810/1408)	**1,560 silver *dirhams* yearly** (= approx. 945 dirham *fulūs* monthly) for the *bawwāb/muzammalātī* at the cemetery **600 silver *dirhams* yearly** (= approx. 675 dirham *fulūs* monthly) for the *muzammalātī* in Cairo	Between 270 to 832 *dirhams fulūs* monthly for the period 1401-1450
Mūsā (830/1427)	**20 silver *dirhams* monthly** (= approx. 270 *dirhams fulūs*) for the *bawwāb/ muzammalātī* at the cemetery **15 silver *dirhams* monthly** (= approx. 202.5 *dirhams fulūs*) for the *muzam-malātī* in Cairo	Between 270 to 832 *dirhams fulūs* monthly for the period 1401-1450
Qarājā (880-2/1475-7)	**150 *dirhams fulūs* monthly** for the *bawwāb/muzammalātī* at the cemetery +Free accommodation at the *turba* and supplement of daily bread ration of 2 *raṭls*	Between 150 to 300 *dirhams fulūs* monthly for the period 1451-1500

A final, but crucial, observation can be made on the basis of the tasks assigned to these employees as described in the *waqf* documents. As the case studies demonstrate, these workers were responsible for opening and closing the dispensers, transporting water from the underground cisterns to the distribution chamber, and overseeing its distribution according to the founder's schedule. Furthermore, they were tasked with the sanitation of both the structure and the vessels used for service. These tasks reveal that the unskilled labourers employed as *bawwāb/ muzammalātī* were given significant responsibilities over these charitable water structures, gaining therefore a degree of access to water control that did not have parallels in the city outside the professional trade of the water carriers. This is perhaps one of the most significant aspects of the present analysis: the widespread establishment of these structures further decentralized water control by opening up new avenues for the city's labouring groups to secure roles in the management and delivery of water.

Conclusions

The three case studies analysed in this article offer valuable insights that allow for a reassessment of the socio-economic role and impact of Mamluk-era *sabīls*. An examination of the founders' biographies reveals, first and foremost, that these structures were components of foundations conceived as strategic investments. Through these *awqāf*, members of the ruling networks (the *ahl al-dawla*) managed and profited from assets accumulated during their tenure in power, seeking to guarantee a secure and substantial revenue flow for themselves and their descendants. Within these strategically planned foundations, *sabīls*—situated in urban areas where the founder's physical legacy was most tangible—served as expressions of a donor-driven model of charity. This model was rooted in the Quranic principle that good deeds are done in order to attain Allah's mercy, as beautifully articulated in *Surat al-Insān*.

While we cannot calculate exactly how much water these *sabīls* distributed—since we lack data on cistern sizes or budgets—the *waqf* records exhibit a clear pattern. These structures did not create a new water-supply system; instead, they relied on the same water carriers who supplied the rest of the city. The water was kept out of reach in underground cisterns, accessible only to the attendants who handed it out through iron grilles. Because water distribution was restricted to specific hours and mediated by the *muzammalātī*, calling a *sabīl* a "public fountain" is misleading. Rather, these were private charitable monuments tied to personal investments and the specific spiritual goals of the donor.

Reassessing the impact of these water dispensers requires looking beyond their role as mere water providers and considering the job opportunities and livelihoods they offered to the unskilled workers employed as attendants. The multiplication of these roles across numerous *waqf* foundations did more than just alleviate the economic hardships of the city's less privileged; it integrated these workers into the foundations' self-sustaining economic cycles. By granting them significant responsibility over the operation and maintenance of these structures, the *sabīls*' founders gave the labouring group a unique form of agency in the urban water management. Ultimately, the impact of these structures was far more reaching than the simple provision of water—it fundamentally shaped urban water management by creating and boosting a decentralized, polycentric model of control throughout the medieval capital. The volume containing this article originated with a 2024 conference in Cairo that analysed water management in the pre-modern Middle East by examining "forces from above" and "forces from below". *Sabīls* represent a unique arena where these concurrent forces intersected: here, the charitable act of water-giving—sponsored by wealthy members of the ruling networks—was placed under the direct responsibility and daily management of much humbler employees.

Bibliography

Primary Sources

Archival Material

DW91.
DW177.
DW184.
DW355.

Inscriptions

Thesaurus d'Epigraphie Islamique, fiche 9755.
Thesaurus d'Epigraphie Islamique, fiche 42797.
Thesaurus d'Epigraphie Islamique, fiche 42825.
Thesaurus d'Epigraphie Islamique, fiche 42915.

Editions

Ibn al-Malik, Muḥammad ibn ʿAbd al-Laṭīf al-Kirmānī al-Rūmī al-Ḥanafī. *Sharḥ al-maṣābīḥ li-Ibn al-Malik.* Idārat al-Thaqāfa al-Islāmiyya, 2012.

Ibn Ḥajar al-ʿAsqalānī, Aḥmad ibn ʿAlī, *Inbāʾ al-ghumr bī abnāʾ al-ʿumr*. Edited by Ḥasan Ḥabashī. Wizārat al-Awqāf, 1998.

Ibn Iyās, Muḥammad ibn Aḥmad. *Badāʾiʿ al-zuhūr fī waqāʾiʿ al-duhūr*. Edited by Muḥammad Muṣṭafā. Franz Steiner Verlag, 1960–1975.

Ibn Taghrī Birdī, Abū al-Maḥāsin Yūsuf. *al-Nujūm al-zāhira fī mulūk Miṣr wa-l-Qāhira*. al-Muʾassasa al-Miṣriyya al-ʿĀmma, 1963–1972.

al-Malaṭī, ʿAbd al-Bāsiṭ ibn Khalīl ibn Shāhīn. *Nayl al-amal fī dhayl al-duwal*. Edited by ʿUmar ʿAbd al-Salām Tadmurī. al-Maktaba al-ʿAṣrīya, 2002.

al-Maqrīzī, Aḥmad ibn ʿAlī. *al-Mawāʿiẓ wa al-iʿtibār fī dhikr al-khiṭaṭ wa al-athār*. Dār al-Kutub al-ʿIlmiyya, 1998.

al-Maqrīzī, Aḥmad ibn ʿAlī. *al-Sulūk li-maʿrifat duwal al-mulūk*. Dār al-Kutub al-ʿIlmiyya, 1997.

al-Maqrīzī, Aḥmad ibn ʿAlī. *Durar al-ʿuqūd al-farīda fī tarājim al-aʿyān al-mufīda*. Edited by Muḥammad al-Jalīlī. Beirut: Dār al-Gharb al-Islāmī, 2002.

al-Sakhāwī, Muḥammad ibn ʿAbd al-Raḥmān. *al-Ḍawʾ al-lāmiʿ li-ahl al-qarn al-tāsiʿ*. Dār Maktabat al-Ḥayāh, 1966.

al-Sakhāwī, Muḥammad ibn ʿAbd al-Raḥmān. *Wajīz al-kalām fī al-dhayl ʿalā duwal al-Islām*. Edited by Bashshār ʿAwwād Maʿrūf, ʿIṣām Fāris al-Ḥarastānī, and Aḥmad al-Khuṭaymī. 1995.

Secondary Sources

Abbasi, Muhammad Zubair. "The Classical Islamic Law of Waqf: A Concise Introduction." *Arab Law Quarterly* 26 (2012): 121–53.

Amīn, Muḥammad. *al-Awqāf wa-l-ḥayyā al-ijtimā'iyya fī Miṣr.* Dār al-Kutub, 2014.

Amīn, Muḥammad. *Izdihār al-awqāf fī 'aṣr salāṭīn al-Mamālīk: dirāsa tārīkhiyya wathā'iqiyya.* Wizārat al-Awqāf, 2001.

Behrens-Abouseif, Doris. *Cairo of the Mamluks. A History of the Architecture and its Culture.* AUC Press, 2007.

Behrens-Abouseif, Doris. *Fatḥ Allāh and Abū Zakariyya: Physicians under the Mamlūks.* IFAO, 1987.

Blair, Sheila, and Jonathan Bloom. "Introduction." In *Rivers of Paradise. Water in Islamic Art and Culture*, edited by Sheila Blair and Jonathan Bloom, 1–26. Yale University Press, 2009.

Bosworth, Clifford E., and Doris Behrens-Abouseif. "Sabīl." *Encyclopaedia of Islam New Edition Online*, edited by P. Bearman et al. Brill, 2012. https://doi.org/10.1163/1573-3912_islam_COM_0954.

Denizeau, Valentine. "Conduire l'eau dans Le Caire mamlûk: installations hydrauliques et politiques d'aménagements dans la capitale égyptienne (1250-1517)." PhD diss., Université de Provence, 2010.

Denoix, Sylvie. "Les notions de 'privé' et de 'public' dans le monde Musulman sunnite medieval." In *Public et Privé en Islam. Espaces, autorités et libertés*, edited by Mohamed Kerrou, 133–51. Institut de recherche sur le Maghreb contemporain, 2002.

Frenkel, Yehoshua. "The Waqf System During the Last Decades of Mamluk Rule." In *The Mamluk-Ottoman Transition Continuity and Change in Egypt and Bilād al-Shām in the Sixteenth Century 2*, edited by Stephan Conermann and Gül Şen, 221–72. Bonn University Press, 2022.

al-Harithy, Howayda. "Charity and Piety. *Sabil-Kuttabs* and the conception of water during the Mamluk Period." In *Rivers of Paradise, Water in Islamic Art and Culture*, edited by Sheila Blair and Jonathan Bloom, 161–83. Yale University Press, 2009.

al-Hazmeh, Khaled Amad. "Late Mamluk Patronage: Qānṣūh al-Ghūrī's *waqf* and his foundation in Cairo." PhD diss., The Ohio State University, 1993.

Ibrahim, Abd al-Latif. "Wathīqat al-amīr akhūr kabīr Qarāqujā al-Ḥasanī." *Majallat Kulliyat al-Adāb* 18, no. 2 (1956): 183–251.

Igarashi, Daisuke. "*Waqf* as a Means of Securing Financial Assets: The 'Self-Benefiting *Waqf*' in Mamluk Egypt and Syria." In *Egypt and Syria under Mamluk Rule: Political, Social and Cultural Aspects*, edited by Amalia Levanoni, 277–91. Brill, 2022.

Igarashi, Daisuke. *Land Tenure and Mamluk Waqfs.* EB Verlag, 2014.

Igarashi, Daisuke. *Land Tenure, Fiscal Policy and Imperial Power in Medieval Syro-Egypt.* Middle East Documentation Center, 2015.

Isoldi, Angela. "Confluent Waters, Fragmented Powers. Social Geographies of Water in Late Medieval Cairo." PhD diss., Radboud University, 2026.

Lamei Mostafa, Saleh. "The Cairene Sabil: Form and Meaning." *Muqarnas* 6 (1989): 32–42.

Livingston, Daisy. *Managing Paperwork in Mamluk Cairo: Archives, Waqf and Society.* Cambridge University Press, 2025.

Peters, Ruud, Doris Behrens-Abouseif, D.S. Powers, A. Carmona, A. Layish, Ann K.S. Lambton, Randi Deguilhem, R.D. McChesney, G.C. Kozlowski, M.B. Hooker, and J.O. Hunwick. "Waḳf." In *Encyclopaedia of Islam New Edition Online (EI-2 English)*, edited by P. Bearman et al. Brill, 2012. https://doi.org/10.1163/1573-3912_islam_COM_1333.

Petry, Carl. "A Geniza for Mamluk Studies? Charitable Trust (*Waqf*) Documents as a Source for Economic and Social History." *Mamlūk Studies Review* 2 (1998): 51–60.

Petry, Carl. *The Mamluk Sultanate. A History*. Cambridge University Press, 2022.

Rabbat, Nasser. *Writing Egypt. Al-Maqrīzī and his Historical Project*. Edinburgh University Press, 2023.

Rapoport, Yossef. "Royal Justice and Religious Law: Siyasah and Shari'ah under the Mamluks." *Mamlūk Studies Review* 16 (2012): 71–102.

Reinfandt, Lucian. "Mamlūk Documentary Studies." In *Ubi sumus? Quo vademus? Mamluk Studies, State of the Art*, edited by Stephan Conermann, 285–310. Bonn University Press, 2013.

Reinfandt, Lucian. *Mamlukische Sultansstiftungen*. Klaus Schwarz Verlag, 2003.

Sabra, Adam. "Public Policy or Private Charity? The Ambivalent Character of Islamic Charitable Endowments." In *Stiftungen in Christentum, Judentum und Islam vor der Moderne*, edited by Michael Borgolte, 95–108. De Gruyter, 2005.

Sabra, Adam. *Poverty and Charity in Medieval Islam. Mamluk Egypt, 1250-1517*. Cambridge University Press, 2000.

Stillman, Norman A. "*Waqf* and the Ideology of Charity in Medieval Islam." In *Studies in Honour of Clifford Edmund Bosworth 1*, edited by Richard Netton, 357–72. Brill, 2000.

CHAPTER 11

Cairo's *Sabīls* during Egypt's Modern Period

Dina Ishak Bakhoum

Abstract

Cairo boasts numerous historic *sabīls* (water fountains) from the Mamluk and Ottoman eras as well as others from the 19th century. These were established as charitable endowments (*awqāf*) to provide free clean drinking water in perpetuity. Today, however, none of these *sabīls* maintain their original function. This chapter explores the possible reasons for this discontinuity in Egypt by tracing the condition of *sabīls* during the 19th and 20th centuries in light of the parallel transformations to the water supply infrastructure that were taking place at the time. By paying attention to diverse interest groups, including the Egyptian Comité de conservation des monuments de l'art arabe (1881–1950s), the chapter reveals the complex story of managing, restoring, conserving and sometimes relocating *sabīls* in modern Egypt.

Keywords: *sabīl*; Comité de conservation des monuments de l'art arabe; *waqf* (pl. *awqāf*); *saqqāʾ* (pl. *saqqāʾūn, saqqāyīn*); al-Khalīj al-Miṣrī; relocation of monuments; Cairo Water Company/La Société des eaux du Caire

The Nile and the Memory of Water in Cairo's Urban Landscape

The river Nile running through Egypt lies at the heart of the country's water story. It is thanks to the Nile, and the life and prosperity it brought to Egypt, that Herodotus described the country as being the "gift of the Nile".[1] Since antiquity, Egyptians have relied on the Nile's annual inundation, digging canals that connected the Nile to the settlements along its banks. Some of these canals remained in use until modern times, such as for example, the canal known as al-Khalīj al-Miṣrī (often pronounced as "al-Khalīg al-Maṣrī" and referred to simply as "al-Khalīg") in Cairo (Figs. 11.1a, 11.1b, and 11.2a).

Cairo (al-Qāhira), founded by the Fatimids in 358/969, was located on the east bank of the Nile and of al-Khalīj al-Miṣrī, and north of the earlier capitals of al-Fusṭāṭ, al-ʿAsqar and al-Qaṭāʾiʿ, which were also all built on the eastern bank of the Nile. The draining of al-Khalīj al-Miṣrī started in the late 19th century for hygienic

[1] For more on this description, see Griffiths, "Hecataeus and Herodotus," 57–61.

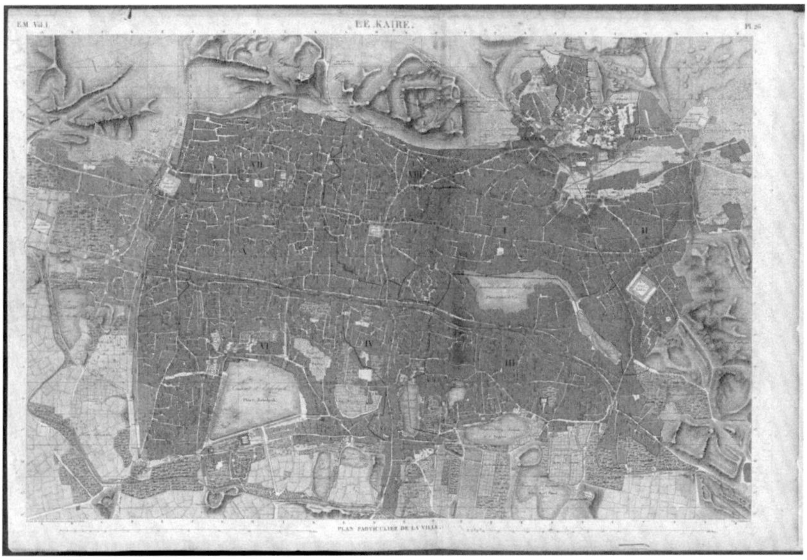

Figure 11.1a: Le Kaire. Plan particulier de la ville, Description de l'Égypte, 1809-1828.

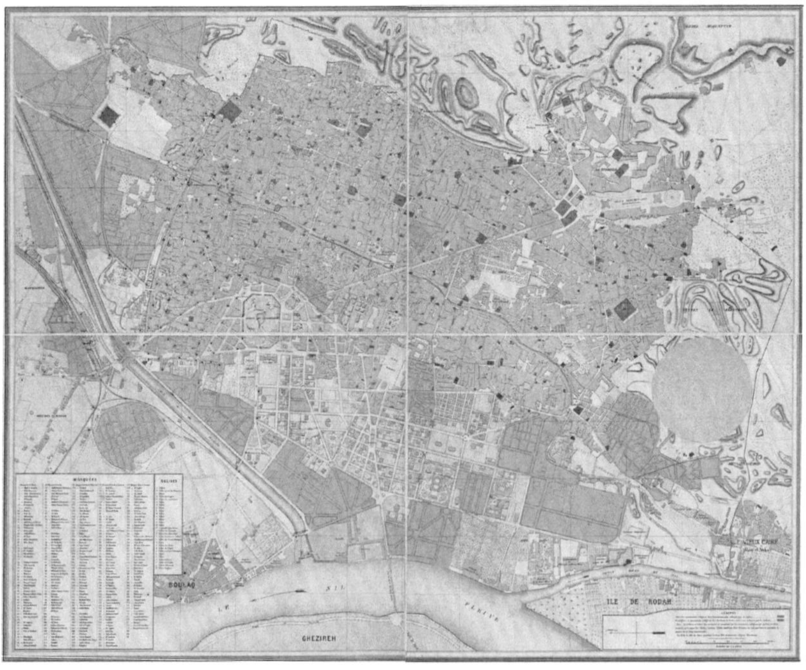

Figure 11.1b: Plan général de la Ville du Caire (also known as the Grand Bey map), 1874.

reasons and was completed shortly after.[2] Along the path of the Khalīj, a street with the same name was built. In the middle of the 20th century, it was renamed Port Said Street. Nonetheless, the presence of the Khalīj has remained in the urban memory of the city until today, not only through historic photographs, maps and written sources, but also through the area called Fumm al-Khalīg (Fig. 11.1b), where *fumm* (in classical Arabic *famm*) means the "mouth" of this canal where it joined the Nile. This name can still be found on contemporary maps and is widely used among Cairenes. In fact, many of the names of Cairo's streets and districts, along with what remains of its historic water infrastructure, preserve the memory of how water was managed in the city prior to the modern period.

For example, a small district and alley in Cairo called Ḥārat al-Saqqāyīn still keeps alive the memory of the water carriers (*saqqā'*; pl. *saqqā'ūn*; *saqqāyīn* more widely used in colloquial Arabic) who were in charge of bringing water from the Nile, canals, seasonal lakes and ponds to people's homes and *sabīl*s (water fountains/dispensers).[3] Maps of Cairo published in the *Description de l'Égypte* show that in this area there used to be a lake called "Birket el Saqqâyn" i.e. Birkat al-Saqqāyīn (also referred to in one of the maps as "Birket Sitty Nasrah" i.e. Birkat Sittī Naṣra), which no longer exists (Fig. 11.2a). André Raymond, in his article on the water carriers and their guilds, notes that the name Ḥārat al-Saqqāyīn was due to the large number of water carriers present in this part of Cairo.[4] And although during the modern period, the work of the *saqqāyīn* ceased to exist (more on this below) and the lake was filled in, the name of the alley remains to this day (Fig. 11.2b). Moreover, the narrative that the water carriers were living in this area found its way into popular culture in a 1966 Egyptian movie called *Ḥārat al-Saqqāyīn*, which also included a song of the same title.[5] The expression "don't go selling water in the water carriers' alley," coined in the lyrics of this song, is now part of everyday language in Egypt. It means you cannot sell something to the experts in that field.

Another part of Cairo that strongly highlights the importance of the Nile is the island of al-Maniyal (or Maniyal al-Rawḍa), where the Nilometer was constructed by the Abbasids in 247/861 on the southern tip of the island to measure the level of the Nile's inundation.[6] There are numerous other structures linked to water in the

[2] Alleaume, "Hygiène publique et travaux publics," 152–57; Ismail, "Engineering," 45–49; 158–59.

[3] For more on the water carriers, see Raymond, "Les porteurs d'eau" and Schultz, "Living and Dying in Water," 25–67.

[4] Raymond, "Les porteurs d'eau," 193.

[5] The film is available online. The song was written by Ḥusayn al-Sayyīd, composed by Munīr Murād and sung by Sharīfa Fāḍil. Earlier songs and novels dedicated to the water carriers will be discussed later in the text.

[6] For more on the Nilometer see H. Mostafa, "The Nile as Nexus". For the restoration of the Nilometer's cover structure by the Egyptian Comité de conservation des monuments de l'art arabe between the late 1920s and the early 1940s, see Bakhoum, "The Restorations of the *Comité*," 139–48.

Figure 11.2a: Al-Khalīj al-Maṣrī prior to its drainage during the late 19th century. Photo by Beniamino Facchinelli, “Canale El Khalig”. (Paris, Bibliothèque nationale de France, département des Estampes et de la photographie, BOITE FOL B-EO-1717.)

Figure 11.2b: A street sign with the same name as the area called Ḥārat al-Saqqāyīn (The alley of the water carriers). Photograph by the author.

city. There is the aqueduct connecting the Nile to the Citadel and its water intake tower next to the Fumm al-Khalīg area, as well as the waterwheels and other water structures at and around the Citadel itself.[7] Moreover, many *sabīls*, which are the focus of this chapter, still exist all over Cairo.

Cairo's *Sabīls*

Cairo's extant *sabīls* date predominantly to the Mamluk (648–923/1250–1517) and Ottoman (1517–1798 and 1801–1805) eras. Later ones were also constructed by Mehmed Ali (r. 1805–48) and subsequent khedives as well as other notables and individuals. Some *sabīls* are attached to mosques and/or madrasas (schools); others are attached to *kuttābs* (*kuttāb*—Qurʾānic school—also referred to as a *maktab*) and are referred to as *sabīl-kuttābs*; and some are attached to living units or other structures. They were mainly established as charitable endowments (*waqf*; pl. *awqāf*, Islamic religious and charitable endowments) to provide free clean drinking water. These historic *sabīls* have attracted the attention of many scholars who have studied their art and architecture, their establishment as charitable endowments (*awqāf*), their religious significance, as well as their social and economic importance.[8]

In my opinion, the history of each *sabīl* throughout the centuries—from the time of its endowment until the present day—depends on a number of variables, including but not limited to: the endower's position and status; the exact type and stipulations of the endowment; the available resources from other endowments for its maintenance and repair; and the way it was managed, historically as well as during the modern period. This means that despite general trends, each *sabīl* has

[7] A team from the Institut français d'archéologie orientale (IFAO) and ʿAin Shams University in collaboration with the Egyptian Ministry of Tourism and Antiquities is currently conducting archaeological excavations to reveal more information on water structures next to the Citadel. For more see "Citadelle du Caire", https://www.ifao.egnet.net/recherche/archeologie/citadelle-du-caire/

[8] *Sabīl*, pl. *asbila*. I will use *sabīls* instead of *asbila*. For some of the works written on Mamluk and/or Ottoman *sabīls* in Egypt, see Pauty, "L'architecture au Caire"; Ebeid, "Early *sabīls*"; Raymond, "Les fontaines publiques"; S.L. Mostafa, "The Cairene *Sabil*"; al-Ḥusaynī, *al-Asbila al-ʿuthmaniyya*; Bosworth and Behrens-Abouseif, "Sabīl"; Van den Bent and Isoldi, "Pleasing God" and see also the contribution of Isoldi in this volume. Archaeological excavations are constantly adding to our knowledge on how and by whom water was managed in the city and how *sabīls* functioned, where they were located, etc. For example, during works carried out by the Aga Khan Trust for Culture in the mosque of Umm al-Sultan Shaʿbān, the ablution area, the underground discharge water tunnels as well as the cistern of the *sabīl* were discovered; the cistern's location is not directly under the *sabīl* but under two rooms (rooms indicated as J and K on the AKTC plan; see Bakhoum, "The Madrasa of Umm al-Sultan," 108 (for the cistern) and 112–15 (for other archaeological discoveries linked to water), and the plan on page 100 showing the location of rooms J and K and the entrance to the cistern).

its own individual story. Today, however, what all of Cairo's still standing historic *sabīls* have in common is that their original function as water fountains has ceased to exist, despite stipulations in their endowment deeds that their function and benefits should continue in perpetuity. Today, many of these historic *sabīls* and *sabīl-kuttāb*s are registered as monuments and hence protected by law, although some are not. Some of the registered ones are open to visitors and tourists and/or have been refunctioned. Many, however, are closed. This in turn impacts their state of conservation, with some being well-maintained, others abandoned and in a state of decay, in need of solutions that will ensure their conservation for future generations.[9]

While it is generally accepted that with the modernization of Cairo and the introduction of modern water supply systems, the need for the water carriers and these *sabīls* began to gradually diminish (more on this below), it has also been argued that the Comité de conservation des monuments de l'art arabe (hereafter Comité), the body established in Egypt in the late 19th century to conserve Egypt's Islamic monuments, played a major role in the discontinuity in function—and hence endowment—of these *sabīls*. This chapter reconsiders this argument in an attempt to provide a more nuanced understanding of how the Comité dealt with these historic *sabīls* and the role it and other stakeholders played in safeguarding (or not) these structures and their function as water fountains. The case studies discussed in this chapter are therefore of *sabīls* that featured in the Comité's technical reports and where the Comité was in one way or another involved in decisions regarding actions to be taken. As noted, each historic *sabīl* has its own story and management system, which means that a properly in-depth understanding of why *sabīls* (individually and collectively) ceased to function would require a comprehensive survey. Nonetheless, the *sabīls* presented here were carefully selected to be representative of the Comité's involvement in these structures, demonstrating a variety of interventions the Comité carried out when it came to *sabīls*, whose condition and function during the late 19th century also varied. Moreover, tracing the history of *sabīls* during Egypt's modern period and the parallel transformation taking place regarding water supply, helps contextualize the fate of the *sabīls* and enables further exploration of the extent to which the Comité was a (major) player in this story.

9 El-Habashi and Nada, "Monuments or Heritage Buildings", focusing on *sabīls* in al-Darb al-Aḥmar area in Historic Cairo, offer some suggestions regarding methodology for evaluating the conditions of *sabīls* as well as ways to adapt, reuse and therefore preserve them. For the outcome of some *sabīl* restoration projects during the past two decades, see Dobrowolska and Fahmy, *Muhammad 'Ali Pasha and His Sabil*; Dobrowolska and Dobrowolski, *The Sultan's Fountain*.

While this volume focuses primarily on pre-modern water infrastructure, this contribution demonstrates that our understanding of how these historic *sabīls* have survived for centuries can also be enriched by comprehending how they were managed during the modern period, and why and how they eventually stopped functioning.

The Comité de conservation des monuments de l'art arabe: A Major Player in Managing *Sabīls*?

One of the many players involved in the restoration and safeguarding of *sabīls* between the end of the 19th and the middle of the 20th century was the Comité de conservation des monuments de l'art arabe/Lajnat ḥifẓ al-āthār al-ʿArabiyya/ The Commission for the Preservation of Monuments of Arab Art.[10] The Comité was established through an official decree/superior order, issued by Egypt's then Khedive, Méhmét Thewfik (Muḥammad Tawfīq) in December 1881. The Comité was placed under the Ministry of Awqāf (more on this below) and the foundation decree was signed by the Khedive, as well as the president of the Council of Ministers and the Minister of Awqāf, Mohamed Zéki (Muḥammad Zakī).[11] The issuing of this decree was strongly related on the one hand to the growing fascination with Cairo's historic monuments and urban fabric during the 19th century, and on the other, to the condition of many of these structures, which were deteriorating or at risk of being lost due to lack of maintenance, neglect, vandalism, and also modernization.

The overarching mandate of the Comité was to inventory Arab/Islamic monuments of an artistic and historic nature and to ensure that they were maintained, restored and conserved. Two commissions were created under the Comité; the first was in charge of preparing an inventory of monuments to be classified and conserved by the Comité, and the second was responsible for the technical aspects and was later renamed the Technical Commission/Technical Section. It remained

10 Several scholars have written about this committee, its foundation and its members including but not limited to Donald Malcolm Reid, Mercedes Volait, Alaa El-Habashi, Philipp Speiser, István Ormos and others, including the author of this chapter. For a summary of the historiography of this committee see Bakhoum, "The Restorations of the *Comité*," 19–28.

11 The foundation decree was published in the First Bulletin of the Comité. The Comité published 41 Bulletins. These will be referenced here as BC followed by the volume number, the period the Bulletin covers (which is sometimes different from the year when it was published) and the page numbers. The transliteration of names in the Comité Bulletins does not match the IJMES transliteration used for this volume; in such cases, the original transliteration is followed in brackets by the IJMES one. French to English translations of quotations from the Bulletins are by the author.

active till the 1950s, when laws were issued that led to its transformation (more on this below).

Among the common narratives of the work of the Comité (or the Committee) and its impact on *sabīls* is one given by the conservation architect Hossam Mahdy, as follows:[12]

> The Committee identified, documented, inventoried, studied, and restored the aesthetic, stylistic, and archaeological aspects of Islamic historic buildings as single monuments in isolation from their urban and environmental contexts with disregard to their functions, religious dimensions, intangible aspects, and their social interests and associations. Accordingly, listed *sabil*s, *kuttab*s (elementary schools for teaching the Qur'an, reading, and writing), madrasas (religious schools), *khanqah*s (convents for Sufis), and other Islamic buildings were divorced from their original meanings and functions. The restoration of a historic *sabil*, for example, included the discontinuation of the function of offering drinking water. The justification for that was that water could damage the historic fabric of the building. Notably, this argument was not considered in the conservation of historic fountains in Rome and all over Europe. These double standards were due to the dismissal of the Islamic meaning and function of *sabil*s by European conservators. Furthermore, the Committee's restoration of a *sabil* contradicted and ignored its *waqf* (charitable endowments) arrangements and conditions, which are mandatory under Islamic sharia law.

I have argued in my research on the work of the Comité and its members that their interventions with regard to Islamic monuments, especially mosques and madrasas, reveal that restoration decisions taken by the Comité members conserved not only the buildings' artistic and historic aspects but also their social and religious value. I have particularly explored this matter in response to scholarly notions that argued that the Comité's interventions were catering to the interests of foreign tourists and ignored the social and religious value of Islamic monuments. In fact, however, in many circumstances it was thanks to the Comité's work that the function of certain mosques, which were in a state of ruin prior to the Comité's establishment, was reinstated after they were restored through the Comité's intervention.[13] My earlier research did not however deal with *sabīls*, whose management system differed from that of mosques. Mosques were mainly charitable endowments founded by sultans, emirs, etc. Since the 19th century, they had predominantly fallen under

[12] Mahdy, "Decolonizing the Conservation," 374–75.

[13] For an overview of these notions about the Comité and my counterarguments, see Bakhoum, "Mamluk Minarets," 150–51, 194–95; "The Restorations of the *Comité*," 23–28 and 216–22; "The *Comité*... and the secularization".

the purview of the Administration/Ministry of Awqāf.[14] Many *sabīl*s, even if they provided a charitable service, were privately endowed and financed, and managed through a *nāẓir* (supervisor) or alternatively through the Awqāf Administration as will be further discussed below. Moreover, the mosque, unlike the *sabīl*, has a strictly religious function, and it can be argued that it would have been impossible for the Comité to put an end to a mosque's use. In this light, the above statement, as well as discussions I have had with its author, have prompted me to explore the following question, specifically regarding the Comité and *sabīl*s: To what extent did the Comité influence the fate of *sabīl*s as functioning water fountains?

To tackle this question, I will look at the conditions of *sabīl*s prior to the establishment of the Comité, how and by whom they were managed and funded prior to and during the time of the Comité, and what actions were taken by the Comité regarding these structures. While I am not claiming to provide a comprehensive answer to these questions in this chapter, I hope to contextualize this matter and to highlight the complex circumstances in which these *sabīl*s were managed and preserved during the modern period.

Cairo's *Sabīl*s during the 19th Century and Prior to the Establishment of the Comité

An important source on Cairo's and Egypt's architecture and urban environment during the 19th century is the twenty-volume publication *al-Khiṭaṭ al-Tawfīqiyya al-jadīda li-Miṣr al-Qāhira wa-mudunihā wa-bilādihā al-qadīma wa-l-shahīra* [New Guide to the Districts ruled by Tawfīq] written by ʿAlī Mubārak (c. 1823–93) and published between 1887 and 1889.[15] When discussing *sabīl*s, Mubārak provides the religious and social reasons for endowing such structures, indicating that the stipulations placed in the endowment deeds and other documents aim to retain the benefits of the endowment in perpetuity. He highlights, however, that these stipulations were not always respected, noting that in Cairo there are around 200 *sabīl*s, some of them *ʿāmir* (see more on this term below) and some *kharāb* (in a state of ruin and/or abandonment).[16]

[14] In 1835 many *awqāf* properties were centralized under Dīwān ʿUmūm al-Awqāf, which was re-established in 1851 as Dīwān al-Awqāf/The *Awqāf* Administration. This body was transformed into a Ministry in 1878, back into an Administration in 1884, then again into a Ministry in 1913. For a brief overview of these transformations, including in relation to the Comité, see Bakhoum, "The Restorations of the *Comité*," 41–42. Here the label Ministry/Administration of Awqāf will be used.

[15] For more on him see, Vollers, "ʿAlī Pa<u>sh</u>a Mubārak;" Agai, "ʿAlī Mubārak."

[16] Mubārak, *al-Khiṭaṭ al-Tawfīqiyya* 6, 58.

Figure 11.3: The *sabīl* of Khalīl Aghā Mustaḥfiẓān ($11^{th}/17^{th}$ century). Photograph by Matjaž Kačičnik.

The term *ʿāmir* is intentionally not translated above, as it and its derivatives have several meanings. Edward William Lane (1801–76) in his *Lexicon* explains—specifically with regard to places and buildings—that these are: "inhabited, peopled, well stocked with people and the like, in a flourishing state, in a state contrary of desolate or waste or ruin, or in a state of good repair."[17]

In his *Khiṭaṭ*, ʿAlī Mubārak described a number of significant *sabīl-kuttāb*s (*sabīl*s that have a *maktab* or *kuttāb*) from different periods. While in some cases he specifies that a building is *ʿāmir bi-l-aṭfāl* (*ʿāmir* with children), in other cases he only mentions that it is *ʿāmir*, without clarifying whether it is actively used only for teaching, or whether it is also *ʿāmir* for providing water. In one instance, for the *sabīl* of Khalīl Aghā Mustaḥfiẓān (11th/17th century) he notes that both the *sabīl* and the *maktab* on top of it are *ʿāmirayn* (the dual form of *ʿāmir*), which means that both were functioning at that time.[18] Today this *sabīl* is not registered as a monument and is in deplorable condition (Fig. 11.3).

For a handful of other *sabīl*s there is a clear mention of the place where water was stored or the location, receptacle or instrument from which people drew water,[19] indicating that these were still being used to provide water. In certain cases, Mubārak clearly mentions the ruined state of the *sabīl*; in others, he does not mention the *sabīl*'s state of conservation, but it becomes clear from his description whether it is functioning or not.

Mubārak also often mentions how expenditure on *sabīl*s was administered, with some using the revenue of their own endowments through the management of the *nāẓir* (supervisor), others using the revenue of their own endowments but through the Dīwān al-Awqāf (the Awqāf Administration), and still others receiving money directly from the Dīwān al-Awqāf. This demonstrates the variety of ways in which these *sabīl*s were managed, funded and maintained. This in turn influenced

17 Online version: http://lexicon.quranic-research.net (last accessed December 2024); https://lexicon.quranic-research.net/data/18_E/196_Emr.html. Partially quoted also in El-Habashi, "*Athar* to Monuments," 31. For more on why I prefer not to translate this term, as well as some others, see Bakhoum, "The Waqf," 29, 56–57.

18 Mubārak, *al-Khiṭaṭ al-Tawfīqiyya* 6, 57-65 and particularly 59 for the *sabīl* of Khalīl Aghā Mustaḥfiẓān. The location of this *sabīl* as mentioned by Mubārak (al-Magharbellīn Street) corresponds with a *sabīl* still standing on this street. See Raymond, "Les fontaines publiques," 245–46 and Abul Abū al-ʿAmayim, *Athār al-Qāhira* 2-1, 220–21. Warner, *The Monuments of Historic Cairo*, 178 gives the *sabīl* on this location the number U44 and lists its exact address; he dates it to possibly the 12th/18th century and does not provide additional information regarding the patron. See notes 26 and 57 for the registration number and what the letter "U" indicates.

19 For example, the *mazmala*, "[a] place where jars of water were stored to be cooled. They were usually found in corridors and covered by a carved wooden screen. Other terms used in its stead are *mazyara* or *bayt al-azyār*"; see letter M in the "Technical glossary", https://www.islamic-art.org/Glossary/; or the *bazbūz*, a nozzle or spout from which people took water.

whether individual *sabīl*s remained functioning or were abandoned or not in full use.

What is clear from Mubārak's description is that although by the end of the 19th century some *sabīl*s were still providing water, many (if not most) of the Mamluk and Ottoman *sabīl*s had already undergone deterioration, damage, and in some cases destruction. This was a result of financial and administrative difficulties linked to managing these structures. But it was not only the management of each individual *sabīl* that impacted its state of conservation. Changes in the provision of water in Cairo also played a major role in this regard. As is well known, the 19th century is considered to be a turning point in Egypt's (particularly Alexandria's and Cairo's) history because of the modernization of urban infrastructure. This was the time when the way drinking water was distributed and managed began to change in a substantial manner, especially following the establishment of the Water Company, first in Alexandria in 1856, then in Cairo in 1865 (though only beginning to operate in 1870).[20] Samir Saul, Shehab Ismail, Alexandra Schultz and Emilie Pasquier have all studied, from a variety of perspectives, the Water Company and the modern distribution of water in Alexandria and Cairo, providing us with valuable information on the transformations taking place in water supply in Egypt between the mid-19th and the early to mid-20th centuries.[21] Their research also sheds light on how the creation of the Cairo Water Company (Société des eaux du Caire/ Société anonyme des eaux du Caire) impacted the role of the water carriers, whose profession gradually shrank, before finally disappearing. This process can also be traced through the emergent Egyptian popular culture of the first half of the 20th century. In 1918, a song titled *Laḥn al-saqqā* (Song of the Water Carriers) was written by the Egyptian folk lyricist Badīʿ Khayrī (1893–1966) and composed by Sayed Darwish (1892–1923). Ismail highlights how this song "is a testament to a now forgotten controversy concerning the taste and purity of drinking water that engulfed British colonial Cairo during the first decades of the 20th century," when the company began to draw water from wells instead of the Nile. He also notes that the song "was part of a series of songs that expressed the grievances of craftsmen and service workers."[22] The song describes the difficult but valuable work of the water carriers and how the foreign water company (referred to by the locals at

[20] Saul, "La France et l'Égypte", 41; Ismail, "Engineering Metropolis," 147; Schultz, "Living and Dying in Water," 40; Pasquier, "La société des eaux," para. 15.

[21] Saul, "La France et l'Égypte", 31–62; Ismail, "Epicures and Experts" and "Engineering Metropolis"; Schultz; "Living and Dying in Water"; Pasquier, "La société des eaux du Caire" *and La société des eaux du Caire.*

[22] Ismail, "Engineering Metropolis," 140–42, where you can also find the lyrics of the song and more about it. See also Ismail, "Epicures and Experts", 23–24 and 40, n.37; Schultz, "Living and Dying in Water," 37.

the time as the *kūbāniyya*) with its foreign managers are trying to stop the work of the water carriers, referred to in the song as the "the fathers' craft". In addition, Yūsuf al-Sibāʿī's novel *al-Saqqā māt* ('The Water Carrier is Dead'), published in 1952, acknowledged the ultimate "death" of the profession.[23]

The trajectory of the water carrying profession is mirrored in the life and death of *sabīl*s and explains that the need for *sabīl*s as sources of water was gradually reduced until it disappeared. To date we do not have, however, a dedicated study that looks in detail at how the function of *sabīl*s came to an end and how the Comité, as a new player, dealt with these *sabīl*s as structures that were possibly either still in use, or already abandoned and no longer serving their original function of providing water. The following section provides an overview of the Comité, the scope of its work and its involvement in managing *sabīl*s at a time when Cairo and its water infrastructure was at a turning point, and when the role and function of *sabīl*s was already diminishing.

The Comité de conservation des monuments de l'art arabe (1881–1950s) and Cairo's *Sabīl*s

The Comité under the Awqāf and What This Meant for Sabīls

Before delving into the Comité's work on *sabīl*s, the reason for its establishment under the Ministry of Awqāf deserves attention, since this establishment demonstrates an acknowledgement of the responsibility of the Awqāf toward these endowed structures. Regular maintenance of these structures was central to the *waqf* system. Among the most important stipulations listed in endowment deeds was that the revenues of the endowment be spent primarily on the maintenance and repair of the endowed structure, in order for it to remain *ʿāmir*, and hence for the benefits it provided to continue in perpetuity.[24] Nevertheless, in practice this was not always successful, for a number of financial and administrative reasons.

The establishment of the Comité under the Ministry of Awqāf demonstrates that the Ministry of Awqāf did not want to lose control, even if only of the technical aspects, over its endowments.[25] The Comité being under the purview of the Ministry (later Administration) of Awqāf also meant that the Awqāf took on the responsibility of funding the Comité's restoration works, at least in part, as will be elaborated upon

23 Raymond, "Les porteurs d'eau," 201–202; Ismail, "Epicures and Experts", 23–24; Schultz, "Living and Dying in Water," 25.

24 El-Habashi, "*Athar* to Monuments," 11–58; Bakhoum, "The *Waqf*"; "The *Waqf* System".

25 Bakhoum, "The Restorations of the *Comité*," 53–54.

below. Moreover, this also impacted what kind of structures the Comité classified (or was able to classify) and registered as monuments worth conserving,[26] as well as how restoration projects were carried out and financed. For example, the buildings listed by the Comité for conservation were initially and predominantly mosques and madrasas, hence charitable endowments (*awqāf khayriyya*) by sultans or emirs that were directly under the responsibility of the Ministry/Administration of Awqāf. Analysis of the Comité's list of classified monuments based on the typology of buildings (religious structures; *sabīls*; secular buildings) clearly demonstrates the smaller share of registered *sabīls* and secular buildings vis-à-vis religious buildings, especially during the earlier decades of the Comité's work.[27]

In 1936, the Comité was moved under the purview of the Ministry of Public Instruction.[28] Prior to this move, the Ministry/Administration of Awqāf had funded a great part of the Comité's restoration and conservation works, though as already noted, mainly for buildings directly under its purview. This did not, however, mean that the Comité was not interested in protecting and conserving other types of monuments, but it had to find other sources of funds for works on such buildings. This financial constraint was brought up in the Comité's second meeting (on 16 December 1882) by ʿAlī Mubārak, who had been appointed as a Comité member shortly before (on 27 November 1882). The matter was further discussed by the members in 1885 (Mubārak was no longer a member at that time; more below). The members agreed that the expenses for works deemed necessary by the Comité for monuments belonging to the government or private individuals would be borne

[26] The Comité's initial list was based on a map issued in 1874 by Pierre Grand Bey for Khedive Ismāʿīl. Inventory lists continued to be produced by the Comité (see note 27), listing and also delisting monuments. There were also debates among the members regarding what to classify. For more on this, see El-Habashi (Al-Ḥabašī) and Warner, "Recording the Monuments"; Bakhoum, "The Restorations of the *Comité*," 57–58. Each Cairene monument registered on these lists received a monument number; however, monuments in the provinces did not receive a number. After law no. 215 of the year 1951 for the protection of Egyptian antiquities was issued, the final list during the Comité's lifetime was published as "The Index to Mohammedan Monuments", attached to a 1951 ministerial decree (no. 10357). For a summary of the antiquities laws as well as the lists, see Bakhoum, "Awqaf properties," 18–20; https://www.urhcproject.org/content/studies/20_bakhoum.pdf. It is important to mention that a) some monuments registered by the Comité were sometime later delisted; and, b) monuments registered after the appearance of this index did not always receive a registration number, or their registration number is not widely available and known (see note 57).

[27] Speiser, *Die Geschichte*, 58 and table 4. Speiser has four tables that analyse the classified monuments based on the Comité inventories of 1890, 1903, 1910, 1914 and 1922, and also the situation in 1950.

[28] I have argued that moving the Comité from the purview of the Ministry of Awqāf to the Ministry of Public Instruction was a major turning point in the Comité's history and in the way *awqāf* properties were managed. It eventually led to the secularization of the management of religious endowed heritage in Egypt, though not necessarily to the secularization of the buildings themselves. Bakhoum, "The Restorations of the *Comité*," 62–65, 217–19; "The *Comité*...and the secularization".

by the government or the private owners.[29] In 1896 the Comité also received special funds from the government (particularly from the *Caisse de la dette publique*, the Public Debt Commission) to cover works on historic Arab buildings that were: a) owned by the government; b) abandoned and whose owners were unknown; c) owned by individuals (or, for many of them, family endowments: *waqf ahlī*; pl. *awqāf ahliyya*); d) endowed structures belonging to the Awqāf but that did not generate revenue.[30] As many of the *sabīls* were endowed and managed privately, it meant that funding for their restoration had to be taken from this special budget, or from their own revenues if available, but not directly from the Ministry/ Administration of Awqāf. Accordingly, in some years, a share of this special fund was allocated to individuals and/or family endowments. Regulations were issued for those interested in receiving it, in such a way that they (owners and *nāẓirs* of the endowments) would have to contribute to the cost of interventions; the work would be carried out under the supervision of the Comité; and finally the owners or *nāẓirs* would ensure that no work would be done without consulting with the Comité and that they would do everything in their power to ensure the conservation of the monument in question.[31]

Who benefited from such an agreement and how the details of such an engagement were worked out between the *nāẓirs* of the *sabīls* and the Comité varied depending on the specifics of each *sabīl*. My initial survey of the Comité's involvement on *sabīls* shows that this depended on: the date of construction; the exact type of the endowment and whether the *sabīl* was free-standing or attached to a mosque, madrasa or other type of building.; and who oversaw it, its physical condition, and its revenue, among other factors. Moreover, the period during which the attention of the Comité was drawn to a specific *sabīl* and when interventions took place also impacted conservation decisions. A comprehensive understanding of the Comité's interventions regarding *sabīls* would however require a large-scale research project that would, on the one hand, look at each *sabīl* individually, and on the other trace common trends and procedures followed by the Comité. While, as noted above, it is beyond the scope of this chapter to discuss all Cairene *sabīls*, the case studies selected here highlight some fundamental decisions taken by the Comité regarding *sabīls*, shedding light on their fate from the late 19th to the middle of the 20th centuries.

29 BC 3, 1885, xx–xxi.

30 BC 13, 1896, 9–12. Coptic monuments were added to the Comité's mandate in 1896 but did not receive funding from the Ministry of Awqāf. Rather, they received funding from the Coptic Patriarchate and also from this special fund. For more on the Comité's budget, see El-Habashi, "*Athar* to Monuments," 245–68.

31 BC 15, 1898, 61–62.

When researching the Comité's interventions on mosques and madrasas, that is to say buildings directly under the responsibility of the Awqāf, one can often trace detailed information in the Comité's minutes of meetings and in the technical reports published in its Bulletins as well as in the archives. This is unfortunately not the case for most of the *sabīls*; for many of them the available information is relatively scarce and inconsistent. For some *sabīls*, the Bulletins contain information only on their listing as monuments, whilst in other cases one can find very limited information on physical restoration or repair works and/or administrative matters. There are several possible reasons for this scarcity of information, including but not limited to the fact that many of the *sabīls* were endowed and financed privately, and managed through a *nāẓir*, meaning the works were carried out through him, albeit under the supervision of the Comité (in many but not necessarily in all of the cases).

It was a complicated matter for the Comité to deal with any privately owned, managed and financed endowments when it came to their registration as monuments and their subsequent conservation. Often the only way to protect such privately endowed structures was to purchase them, mainly through exchange (*istibdāl*).[32]

The Comité and Sabīls Still Providing Water

ʿAlī Mubārak's 19th-century account of *sabīls* demonstrates that at that time some of the 200 still standing were also functioning. Reports issued during the 1890s indicate that sixty *sabīls* and also ten mosques subscribed to receive water from the Cairo Water Company and distributed it for free,[33] and that it was individuals or the Awqāf Administration who paid for this subscription.[34] As will be further

[32] The registration of private endowments was particularly a complicated issue. For more, see Bakhoum, "Le classement partiel et tardif d'un *waqf* privé." Today, the registration of any historic building (secular or religious; Islamic or Christian) as a monument requires a long administrative process. I presented on this topic in 2024 during the annual workshop organized by the research project "La fabrique du Caire modern," focusing on the registration of three mosques from the 19th and 20th centuries: https://www.ifao.egnet.net/recherche/manifestations/ma1622/. I have also presented frequently on the impact of registration (or not) on the protection of historic buildings in Egypt.

[33] Ismail, "Engineering Metropolis," 152–53. He notes that: "Sixty fountains out of the city's roughly 200 subscribed to the company's filtered water and provided water freely to the public." The 200 he refers to are most likely those noted by Mubārak, but the report he refers to does not mention the 200, but does mention sixty public fountains; see note 34.

[34] Volait, "Le «sébile» moderne." https://sites.duke.edu/cairemoderne/2019/10/26/le-sebile-moderne-السبيل-الحديث/ The report Ismail and Volait refer to is the "Rapport de la commission internationale de l'assainissement du Caire" published in 1892. Thanks are due to Mercedes Volait for sending me the relevant pages of this report.

Figure 11.4: The *sabīl-kuttāb* of Bashīr Aghā Dār Saʿāda (no. 309; 1131/1718). Photograph by Matjaž Kačičnik.

discussed below, even today the façades of many of Cairo's *sabīls* (attached to mosques, madrasas, *kuttābs*, dwellings or other structures) still display the marks placed by the Société/Companie des eaux (C^IE DES EAUX) with an individual number for each subscription (Fig. 11.8b).

To what extent was the Comité involved with those *sabīls* that still supplied water? Did its interventions with regards to—or restorations of—*sabīls* lead to "the discontinuation of the function of offering drinking water" as suggested by Mahdy (as quoted above)? My investigation of the Comité's documents did not come across any decisions taken by it to suspend the distribution of water from *sabīls*. As noted above, few *sabīls* were under the direct supervision of the Comité, and very few still maintained their original function as water fountains. I would like to however draw attention here to two *sabīls* where water was explicitly mentioned in the Comité Bulletins. The first is the *sabīl-kuttāb* of Bashīr Aghā Dār Saʿāda (no. 309; 1131/1718) (Fig. 11.4).[35] The Comité's attention was drawn to this *sabīl* in 1888, when contacted by its *nāẓir*, who wanted to carry out repair and consolidation works to the interior and exterior of the *sabīl* and needed the permission of the Awqāf Administration, as he could not get permission directly from the Tanẓīm Department of the Ministry of Public Works. Accordingly, the Comité's second commission went to visit the building to provide its opinion and recommendations. Among these was that the cistern of the *sabīl* could still be used if the column in the centre were repaired, noting that this could be done at low cost. It was also suggested that the *sabīl* be classified as a monument to be conserved. Moreover, the members granted the *nāẓir* the permit needed to ask the Tanẓīm to proceed with the execution of the work, under the supervision of the Awqāf's technical office, on the condition that the work would not alter the character of the *sabīl*.[36] While no further discussions in the Bulletins reveal whether the function of the cistern was reinstated or not, the Comité members' suggestions demonstrate that they were not against the functioning of the *sabīls* as water fountains, at least not during the Comité's early period.

The second case where water was mentioned in Comité reports is the *sabīl-kuttāb* of Qāytbāy near al-Azhar Mosque (no. 76; 882/1477) (Fig. 11.5).[37] The Comité was concerned by the poor condition of the building and had already started in the 1880s and 1890s to carry out some consolidation and cleaning works on it. During

[35] For the monuments discussed in this chapter, their registration number and date of construction are listed, with the exception of those monuments which did not receive a monument number (see notes 26 and 57).

[36] BC 5, 1887–1888, 52–53.

[37] Although the *sabīl-kuttāb* is physically attached to the *wikāla* of Qāytbāy (no. 75; 822/1477), each was classified separately (Warner, *The Monuments of Historic* Cairo, 103). In my opinion this might have been linked to the different ownerships of the two parts of the building, and the complex nature of the *wikāla* with multiple owners (for more see BC 3, 1886, VIII–IX).

Figure 11.5: The *sabīl-kuttāb* of Sultan Qāytbāy (no. 75; 882/1477) attached to his *wikāla* (no. 75) near al-Azhar Mosque. Photograph by Matjaž Kačičnik.

these interventions, the Awqāf Administration and the Comité were pressured by the Ministry of Public Works to complete the works as fast as possible, suggesting that otherwise the building should be demolished.[38] The building was in the end not demolished, and in 1900, as the members of the Comité's technical section were passing by the *sabīl*, they decided to visit it. In their report they noted their surprise on seeing that strong cast iron pipes had been installed in the *sabīl* without the Comité being consulted. Consequently, the Comité expressed its desire to the Awqāf that prior to future works on any monument the Comité should first be consulted.[39] The Comité's report did not include any further elaboration on this matter, but in a 1910 Comité report the incident was discussed in more detail, giving a brief technical explanation of how the pipes, installed to supply the *sabīl*'s reservoir with water from the cistern, functioned. The members noted that "from a purely aesthetic point of view" these exposed pipes, visible all around the walls of the north-east corner of the room, could not be left in place. They accordingly suggested that instead of having the pipes run through the interior of the *sabīl*, they might be installed on the exterior of the room without causing any disruption to the way the pipes connected to the *sabīl*'s reservoir, and that this should be communicated to the Awqāf Administration.[40]

This story reveals some important points regarding the work of the Comité when it came to *sabīl*s classified as monuments. Firstly, when work was carried out by the Awqāf Administration, the Comité was mainly interested in being consulted regarding the technical aspects of the work. Secondly, the Comité members did not object to having the *sabīl* function as a water fountain, but their concern, as they noted, was mainly regarding aesthetics and the visual impact the pipes had on this historical monument.

The Comité Relocating Sabīls

One major intervention that some *sabīl*s faced was their dismantling and subsequent reconstruction elsewhere, sometimes only a few metres away, sometimes a bit farther off. While today responsibility for relocating *sabīl*s is mainly attributed to the Comité, it was not the sole or even the main actor in this affair. Why would the Comité have been interested in moving *sabīl*s from their original location? As will be demonstrated through the case studies below, the main reason for such

[38] BC 1, 1882–1883, 28–29; BC 5, 1887–1888, 9; BC 6, 1889, 55–56; BC 7, 1890, 35 and 101.

[39] BC 18, 1901, 8.

[40] BC 27, 1910, 16. Original quote in French.

interventions was to safeguard *sabīls* from being demolished, and forever lost, due to street widening and other urban projects.[41]

Among the earliest cases raised regarding such an issue after the establishment of the Comité was for the *zāwiya* and *sabīl* of Faraj ibn Barqūq (no. 203; 811/1408) (Fig. 11.6a). The structure had very early on attracted the attention of the members of the Comité's second commission, who in 1882 were visiting one of Cairo's city gates, Bāb Zuwayla (also referred to as "Bab-el-Metoualli"), and the minarets of the mosque of al-Mu'ayyad Shaykh on top of the gate, which stand on the opposite side of the street from the *sabīl* of Faraj ibn Barqūq. The members commented on the poor state of conservation of the *sabīl* and suggested several restoration interventions. Nevertheless, during the Comité's second meeting (on 16 December 1882), ʿAlī Mubārak noted that in his opinion this *sabīl* should be removed (or demolished) because its location, very close to the gate, made it difficult for vehicles and carriages to pass. Several other members objected to this suggestion, noting that it was against the mandate of the Comité, whose responsibility was "the conservation not the demolition of ancient monuments".[42]

El-Habashi, discussing this matter, notes the following regarding Mubārak's views:[43]

> For Mubarak, insistence on preserving each historical fragment would constrain the State's advocates from modernizing the city of Cairo. Mubarak confirmed that a building had to be used in order to survive. In order to clarify his position, he made an analogy to the French people demolishing the Bastille to symbolize that such a building type had no use in the new climate of freedom. The historical importance of the building did not prevent the French from eradicating that jail, just as the Egyptians should not be prevented from demolishing the water fountain. For Mubarak, enlarging Cairo streets to facilitate circulation had preference over conserving every single historic building.

[41] In many circumstances relocating monuments is considered the only possible solution for saving them from being demolished or flooded under water (such as for example the Temples of Abu Simbel and other Nubian monuments). And while this intervention has become widely accepted, especially when it comes to urban expansion, I have suggested that we need to critically think about it and instead propose solutions that would ensure the conservation of monuments in their original contexts. I have presented on this subject during the International Conference: "Exploring Reuse. Egypt, Mediterranea and the East", organized at IFAO as part of the programme "Destins d'objets" and my presentation was titled: "Dismantle. Move. Reconstruct. A critical examination of relocation as a conservation solution". https://www.ifao.egnet.net/uploads/manifestations/2025/Exploring-reuse-programme_02.pdf

[42] BC 1, 1882–1883, 16. Original quote in French.

[43] El-Habashi, "*Athar* to Monuments," 149–50.

After attending a few more Comité meetings, ʿAlī Mubārak resigned, presumably because of this issue.[44] Shortly after, a decree was issued, on 27 June 1883, for "Arab monuments" inventoried and classified as per the decree issued on 18 December 1881, stating that they were not subject to the regulations regarding street alignment issued by the Tanẓīm Department of the Ministry of Public Works. This decree saved the building for a couple of decades, until the issue of widening the street in front of Bāb Zuwayla (also known as Taḥt al-Rabʿ) was brought up again by the Tanẓīm in the 1910s.[45] This time, in 1917, a compromise was reached between the Comité and the Tanẓīm, where the Comité would dismantle the structure and reconstruct it a few meters to the south of its original location, thus allowing for the street to be widened and additionally for the *sabīl* to be conserved for future generations. The work was carried out over a few months in 1922 and 1923. During the transfer, and as the foundations on which the structure would be reconstructed were being prepared, the Comité discovered the cistern of the *sabīl*. Arguing that it was not of artistic and historical value, the members decided to backfill it, after demolishing part of it as per the requirements needed for laying the new foundations.[46] The appendix to Comité Bulletin no. 33, covering works carried out between 1920 and 1924, includes an entry on the *zāwiya* and *sabīl* of Faraj ibn Barqūq written by Ahmad as-Sayyed bey (Aḥmad al-Sayyid), the Comité's chief architect and director of Arab Monuments at the time, and translated from Arabic into French by the Comité's translator (and later inspector of monuments) Mahmoud Akkoush.[47] It includes some historical notes on the building, a description of its elements, as well as a summary of the Comité's work. Moreover, the appendix contains eight illustrations of the building, including a map and seven photographs showing the location and views of the building before and after its transfer (Fig. 11.6b).[48]

More than a decade later, two Ottoman *sabīl*s, namely *sabīl* al-Bāzdār (no. 27;1050-51/1640-41) and *sabīl-kuttāb* Ismāʿīl ibn Aḥmad (al-Maghlawī/al-Manawī) (no. 57; 1068/1657) faced a similar fate. Both *sabīl*s, which were small in size, were

[44] El-Habashi, "*Athar* to Monuments," 150.

[45] BC 32, 1915–1919, 647–48.

[46] BC 33, 1920–1924, 231–32.

[47] In November 2024 at the conference *Bodies of Knowledge: Arabic Language, Egyptian Labor, and Communities of Practice in the History of Archaeology & Egyptology* I presented a paper on "The Languages of the Comité de conservation des monuments de l'art arabe and its Egyptian and European Members" in which I discuss the different translators of the Comité including Akkoush and his work. https://www.universiteitleiden.nl/en/nvic/events-nvic/symposia-and-workshops/international-conference-bodies-of-knowledge; https://www.universiteitleiden.nl/binaries/content/assets/customsites/nvic/bodies-of-knowledge_abstract-booklet_final.pdf, 36.

[48] BC 33, 1920–1924, 397–406; map and plates I–VII.

Figure 11.6a: The *zāwiya* and *sabīl* of Faraj ibn Barqūq in its new location (no. 203; 811/1408). Photograph by Matjaž Kačičnik.

classified by the Comité as monuments due to their artistic and historical value, despite their poor state of conservation. Around 1928 the Tanẓīm had started to expropriate buildings between al-Azhar and al-Ḥusayn mosques to create a square in the area; among these buildings were these two *sabīls*. As they were classified by the Comité as historic buildings to be conserved, the Tanẓīm asked the Comité for its opinion regarding the matter; the members agreed on the importance of safeguarding these structures by moving them to a new nearby location.[49] It took a couple of years for the *sabīls* to be transferred, due to discussions between the Comité and the Tanẓīm regarding the financing of this intervention. Finally in 1936, the members visited the *sabīls* in their new location, expressing "their satisfaction with the work carried out but regretting however that [the *sabīls*] are surrounded by unsightly constructions, which detract from their interest".[50] The area has since then changed and the *sabīls* have also been restored (Fig. 11.7).

[49] BC 35, 1927–1929, 53–54.

[50] BC 37, 1933–1935, 393. Original in French. The Comité Bulletins do not provide any information regarding discovering—and possibly relocating—the *sabīls*' cisterns, and Warner notes that "the original water-cisterns were abandoned in the move" (Warner, *The Monuments of Historic Cairo*, 66). He also notes that the numbers of both *sabīls* are sometimes confused; Warner, *The Monuments of Historic Cairo*, 92 and 99; 63 for the map (Fig. 41a) showing the old and new location.

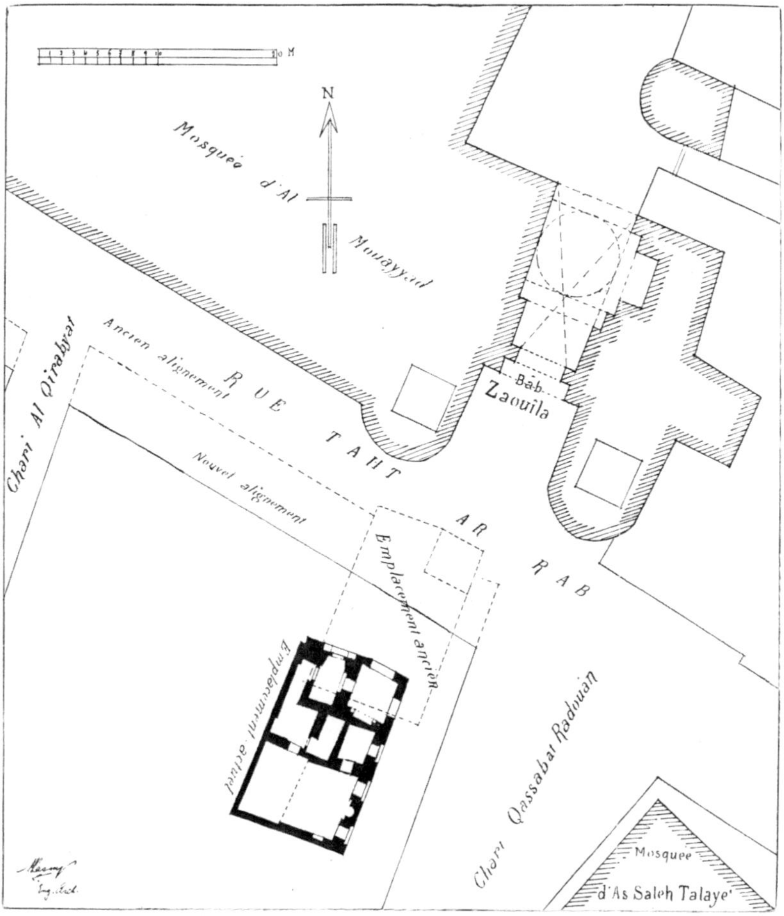

Figure 11.6b: The plan of the *zāwiya* and *sabīl* of Faraj ibn Barqūq showing in dotted lines its original location close to Bāb Zuwayla and its new location. (Bulletin of the Comité 33, 428.)

Also in 1928, the Tanẓīm sent a letter to the Comité regarding moving another Ottoman *sabīl-kuttāb*, namely that of Sitt Ṣālḥa (no. 313; 1154/1741) (sometimes referred to as Sitt Ṣāliḥa; or that of ʿĀʾisha Hanim). This was also to accommodate the widening of a street, this time al-Khalīj al-Miṣrī.[51] The *sabīl* was located on Darb al-Shamsī in al-Sayyida Zaynab.[52] Its relocation took place and works were paid for by the Tanẓīm (Fig. 11.8a).[53] By early 1938, the Comité members had visited the *sabīl*,

[51] BC 35, 1927–1929, 120–21.

[52] BC 26, 1909, 93.

[53] BC 40, 1946–1953, 240.

Figure 11.7: *Sabīl* al-Bāzdār (no. 27;1050-51/1640-41) and *sabīl-kuttāb* Ismāʿīl ibn Aḥmad (al-Maghlawī/al-Manawī) (no. 57; 1068/1657) in their new location. Photograph by Matjaž Kačičnik.

which had been pushed back to allow for the street's new alignment, and expressed their satisfaction at the work carried out.[54] It is important to note that the attention of the Comité members was drawn to this *sabīl* in 1908 when Mohamed Rostom bey (Muḥammad Rustum), the *nāẓir* of the *waqf* of ʿĀʾisha Hānim,[55] sent a letter to the Comité asking it to intervene on the building, which deserved to be listed. He noted that the *waqf* to which this *sabīl* belonged had revenues that were collected by the Awqāf Administration and that he could not cover the costs needed for the maintenance. He promised that if the Comité agreed to pay for the restoration costs, he would conserve the *sabīl* without making any changes to it and would also ensure that the neighbours would not encroach on it. The Comité members agreed to classify the building as a monument to be conserved and to pay for the necessary restoration works, while also checking the validity of Rostom's statement regarding the *waqf*'s revenue with the Awqāf Administration.[56] Nevertheless, no further information regarding this matter was published in the Bulletins until the

[54] BC 38, 1936–1940, 119.

[55] ʿAlī Mubārak mentions the *sabīl* of ʿĀʾisha Hānim, and that some argue that it has a different name. Aletihad News Centre, "Sabīl wa-Kuttāb ʿal-Sitt Ṣāliḥaʾ." According to the Bulletins it was her *waqf*, and the *nāẓir* is referred to as the *nāẓir* of this *waqf* [BC: el-Sett Aïcha Hanem].

[56] BC 26, 1909, 93.

Figure 11.8a: The *sabīl-kuttāb* of Sitt Ṣālḥa (no. 313; 1154/1741) in its new location. It is sometimes referred to as the sabīl of Sitt Ṣāliḥa or that of ʿĀʾisha Hānim). Photograph by Matjaž Kačičnik.

Figure 11.8b: The Cairo Water Company (La société des eaux) subscription number that was placed on the *sabīl-kuttāb* of Sitt Ṣālḥa. Photograph by Matjaž Kačičnik.

issue of the *sabīl*'s relocation was raised in 1928. The subscription number with the mark of the Société/Companie des eaux can be found on the building, which indicates that those managing the *sabīl* subscribed (or re-subscribed) to the Cairo Water Company after the relocation (Fig. 11.8b).

I did not find information regarding the Comité's agreement or objection to this subscription, but this mark, along with many others placed on Cairo's *sabīls* (attached to other endowed structures), indicates that the Comité was not against the continuation or the re-instatement of the use of *sabīls* as water fountains.

Two more *sabīls* were relocated to accommodate requests by the Tanẓīm due to urban development projects taking place in Cairo during the 1950s and 1960s. One is the *sabīl* of Umm Ḥusayn (U36;[57] 1270/1853), which like the *sabīl* of Sitt Ṣālḥa was relocated at the expense of the Tanẓīm. It was moved in the 1950s.[58] The *sabīl* was originally located on what used to be referred to as Gāmiʿ al-Banāt street,[59] a section of al-Khalīj al-Miṣrī street. It was close to the mosque of ʿAbd al-Ghanī al-Fakhrī (no. 184; 821/1418) also known as Gāmiʿ al-Banāt, hence the name of the street. This

57 "U" is a designation by Warner for monuments that are "unregistered" or others that "have been registered subsequent to 1950 but do not follow the sequence of the earlier published index." Warner, *The Monuments of Historic Cairo*, 172; 177 for "U36". See also note 26.

58 BC 40, 1946–1953, 240.

59 Mubārak, *al-Khiṭaṭ al-Tawfīqiyya* 6, 58.

sabīl had to be relocated due to the widening of al-Khalīj al-Miṣrī street.[60] The new location was further south on the same street, behind the mosque of Qāḍī Yaḥyā Zayn al-Dīn (bīn al-Nahḍīn) (no. 182; 848/1444). The Comité published photographs of the *sabīl* before and after its transfer (Fig. 11.9a and Fig. 11.9b).[61] According to ʿAlī Mubārak, the *sābil* was *ʿāmir* and the spending on it came from its own *waqf* revenues via the Dīwān al-Awqāf.[62] While it is not clear from his statement whether its water supply function was still operative, it is quite possible that the *sabīl* did still serve as a water fountain, given that it was only constructed around fifty years prior to the publication of Mubārak's *Khiṭaṭ*. Nevertheless, the move took place in the 1950s, more than sixty years after the publication of his *Khiṭaṭ*, and it is possible that at that time it was no longer functioning as a water fountain. It is also worth noting that the Comité only paid attention to this *sabīl* in the 1930s, because of its relatively young age, which meant that it was not eligible to be listed as a monument according to the law for the protection of Arab monuments issued in 1918. This law stated that the Arab monuments falling under this law were those dating from the time of the Arab conquest of Egypt (18/639) until the death of Mehmed Ali (1265/1848).[63] It was proposed, however, to extend this period to cover buildings constructed up until the end of the reign of Khedive Ismāʿīl (r. 1863–79), and it seems that the desire to conserve and register this *sabīl* was the reason (or one of the reasons) for this extension, as will be explained below.[64]

While during the Comité's early period a number of Ottoman buildings were listed as monuments, it was mainly during the 1930s that attention was paid to them, as attested by several studies prepared by the Comité member Edmond Pauty on this topic. These were published in the Bulletins of the Comité as well as in a combined article in the *Bulletin de l'Institut français d'archéologie orientale*, titled "L'architecture au Caire depuis la conquête ottomane". Pauty dedicated a section of his study to Ottoman *sabīl-kuttābs*.[65] In the same Bulletin where this section of the study was published,[66] Maḥmūd Aḥmad, another Comité member, suggested that six *sabīl*s constructed after the death of Mehmed Ali deserved to be listed as monuments and asked the members to examine these *sabīl*s and give their opinion regarding their protection; the list included the *sabīl* of Umm Ḥusayn as well as

[60] BC 40, 1946–1953, 233.

[61] BC 40, 1946–1953, pl. VIII and pl. IX.

[62] Mubārak, *al-Khiṭaṭ al-Tawfīqiyya* 6, 58–59.

[63] Law no. 8 of the year 1918.

[64] BC 40, 1946–1953, 240. In another Comité report it was noted that it had been suggested that the period be extended till the "death" of Khedive Ismāʿīl, but this is most likely a mistake.

[65] Pauty, "L'architecture au Caire," 22–30.

[66] BC 37, 1933–1935, 322–32.

Pl. VIII

٨ — سبيل أم حسين (قبل نقله)

8.—La fontaine publique (Sébil) d'Om Ḥussein (avant son transfert).

Pl. IX

٩ — سبيل أم حسين (بعد نقله خلف مسجد القاضي يحي)

9.—La fontaine publique (Sébil) d'Om Ḥussein (après son transfert derrière la mosquée d'Al Ḳadi Yéḥia)

Figures 11.9a and 11.9b: The *sabīl* of Umm Ḥusayn (U36; 1270/1853). Views published by the Comité de conservation des monuments de l'art arabe showing its condition before and after its relocation. (Bulletin of the Comité 40, plate VIII and plate IX.)

other *sabīl*s constructed after the death of Mehmed Ali.[67] Today, this *sabīl* is listed as a monument but has no number, which indicates that it was most likely inscribed as a monument during or after the 1950s. In 1951, Law No. 215 for the Protection of Egyptian Antiquities was issued, which did indeed extend the protection period to monuments constructed up until the end of the reign of Khedive Ismāʿīl. A ministerial decree issued the same year included a list of registered monuments, with each Cairene monument having a registration number based on earlier numbers.[68]

The other *sabīl* that was moved to accommodate for street widening, namely the Taḥt al-Rabʿ street, was that of Ḥasan Aghā Arzinjān (no. 420; 1246/1830) (Fig. 11.10). The relocation took place around 1961, and the new and old locations can be seen on maps that predate and postdate the move as well as on maps produced

[67] BC 37, 1933–1935, 333–34.

[68] See note 26. Other *sabīl*s on the list discussed by the members were eventually registered.

by the Egyptian Antiquities Organization, the successor of the Comité.[69] The Comité registered this *sabīl* in 1930.[70]

The relocation of these *sabīl*s required a lot of advanced engineering knowledge. Their dismantling and reconstruction, with all their structural and decorative elements, required the proper documentation of the structure prior to intervention as well as in-depth knowledge of the highly intricate technical work necessary.[71] Moreover, it also entailed a great deal of administrative work, especially for example the expropriation of properties constructed on the land where the *sabīl*s would be relocated. It is not known whether at the time these *sabīl*s were moved, they were working as water fountains or not. It seems that some were not, which might have in turn facilitated both administrative and technical tasks, especially for example the relocation of the underground cistern and the reinstatement of the use of the *sabīl*s as water fountains. Physical evidence, however, shows that in certain circumstances, the function of the *sabīl* was reinstated after relocation, as indicated by the Water Company subscription number of the *sabīl* of Sitt Ṣālḥa. While this does not necessarily mean that the cistern was relocated, it is another strong indication that the Comité was not against the use of the *sabīl*s as water fountains. Such evidence is often not taken into consideration by scholars who look at the Comité through the post-colonial lens to argue that its work created a medieval Cairo for foreign tourists while ignoring the religious and social value these buildings represented for the community.

Some would still wonder: why did the Comité not consider reinstating *sabīl*s that had fallen out of use as working water fountains? It might be argued that the Comité and its members only cared about the aesthetic, artistic and historical value of the *sabīl*s and that for them safeguarding their architecture and decoration was the priority, not protecting their function. As noted above, for mosques this was surely not the case, and the Comité managed to balance the different values embodied in these religious buildings. When it came to *sabīl*s, there is no doubt that the Comité—as for other types of Islamic architecture—cared about the artistic and historical significance of these buildings. Nevertheless, and as demonstrated by the case studies discussed above, this did not mean, however, that it was against

[69] Comité archive, Markaz Tasjīl al-Āthār al-Islāmiyya wa-l-Qibṭiyya, Supreme Council of Antiquities under MoTA. This archive houses the drawings produced by the Comité as well as later ones by the Egyptian Antiquities Organization. See also Warner, *The Monuments of Historic Cairo*, 79 (Fig. 46) for the location of the *sabīl* before and after its relocation.

[70] BC 36, 1930–1932, 11.

[71] The Comité and the contractors and craftsmen it collaborated with had a great deal of experience when it came to dismantling and reconstructing structures and individual elements of buildings such as minarets. For more on the Comité's interventions with regard to Mamluk minarets, see Bakhoum, "Mamluk Minarets".

Figure 11.10: The *sabīl* of Ḥasan Aghā Arzinjān (no. 420; 1246/1830) in its new location. Photograph by Matjaž Kačičnik.

the function of the *sabīls per se*. At a time when modernization was at the doorstep with the creation of the Cairo Water Company and more generally the promotion of clean and healthy water, I think it would have been totally unacceptable for the wider community if the Comité, with its European and Egyptian members, had suggested to, or imposed on the Awqāf or the *nāẓir*s of the *sabīls* the reinstatement of the historic *sabīls* as water fountains. There is no evidence that such discussions took place during the Comité's meetings. This most likely indicates that reinstating the functions of *sabīls* was completely beyond the scope, capacity and responsibility of the Comité. Likewise, and as demonstrated by the examples presented here, it was beyond the Comite's authority to put an end to the *sabīls*' functioning.

At a time when historic *sabīls* were suffering from lack of funds and discontinuity in their function, and also at risk of being demolished as part of urban modernization, the Comité, concerned with the conservation of such valuable buildings, involved itself in their protection. The Comité's listing of most of them as monuments and the other physical interventions it undertook or supported (such as restoration or relocation) were actions that saved many of them from being forever lost. And although the Comité considered relocation to be one solution for conservation, such decisions were not solely taken by the Comité, as is often thought. It was an intervention that demanded a great deal of financial and administrative work, in collaboration with the Tanẓīm and other interest groups.

A Short Note on *Sabīls* Since the 1950s

In this chapter, I have particularly focused on exploring to what extent the Comité was involved in managing and restoring *sabīls* between the 1880s and 1950s. The outcomes of the research conducted so far encourage me to argue the following: to have a more profound understanding of how *sabīls* reached their current condition, one must also look at their legal ownership status and the changes they have undergone since the 1950s. As noted, the 1950s saw the issuance of new laws allowing for the registration of more monuments. These laws eventually also led to the transformation of the Comité. Moreover, 1952 saw the abolition of any type of endowment other than those established for charitable purposes, which affected privately endowed, financed and managed properties.[72]

The registration of *sabīls* as monuments during the Comité's lifetime or later provided them with a certain degree of protection. Today, listed *sabīls* are state owned. Information regarding non-listed *sabīls* is not always readily available,

72 Decree-law (*marsūm bi-qānūn*) no. 180 of the year 1952; it was published in the official gazette. For more see Bakhoum, "*Awqaf* Properties," 14–16, 27–28.

which hinders setting up potential projects for their restoration and conservation. As noted, *sabīl*s are often attached to mosques, madrasas, *kuttāb*s or dwellings. Each of these types of buildings is under the jurisdiction of a different authority, which means that the ownership and management of a single building may be split between two or more entities. In turn, this also complicates the conservation process for such buildings.[73] It was also during the 1950s that the Water Company was nationalized.

It is beyond the scope of this chapter to further explore and discuss all these issues and their impact on *sabīl*s. However, the intent of this short section is firstly, to demonstrate the complexity of managing and protecting these structures whose function ceased to exist and whose management and ownership status changed and is not always known; and, secondly, to highlight the importance of looking beyond the Comité's lifetime to understand how certain endowed structures and other historic buildings reached their current status.

Concluding Remarks

Numerous *sabīl*s, including those endowed privately, were constructed in Cairo as charitable endowments (*awqāf)* to provide the city's inhabitants with free drinking water. However, beginning in the mid-19th century, modern water infrastructure was built in the city. This gradually diminished the need for these *sabīl*s and marked the end of the water carriers who had previously provided water to people's homes as well as to *sabīl*s. It is generally accepted that this in turn was the main reason why the initial function of the *sabīl*s as water fountains came to an end. Nonetheless, it has also been argued that the Comité de conservation des monuments de l'art arabe, the body established under the Ministry of Awqāf in 1881 in Egypt to conserve the country's Islamic monuments, was only concerned with the *sabīl*s' historical and artistic value and that their restorations led to these endowed structures ceasing to function.

[73] For example, during the recent conservation project for the *sabīl* of ʿUmar Aghā (no. 240; 1063/1652) it was discovered that some of the living units of the adjacent house (also attributed to ʿUmar Aghā; U59) and which overlook the *sabīl* are owned by individuals. While these units were initially part of an endowment, they were sold to other families in the 1950s. The *sabīl*, which is registered as a monument, is under the responsibility of MoTA, while the units are not. This meant that certain interventions that must be carried out on both the *sābil* and the living units must be coordinated and approved by different entities. My thanks to the project manager architect Nader Ali for this information. The project received funding from the Antiquities Endowment Fund of the American Research Center in Egypt and was carried out in collaboration with the Egyptian Ministry of Tourism and Antiquities. Acknowledgements: I would like to thank Anne-Marie Shaker for the initial copy-editing of the article and Matjaž Kačičnik for the photographs.

This chapter has readdressed this notion by tracing the conditions of *sabīls* during the 19th century and carefully examining the attention paid by the Comité to a number of them. The Comité, established under the Ministry of Awqāf and receiving funding from it for endowed structures, nonetheless faced financial constraints limiting its involvement in private endowments that had their own revenues, such as many *sabīls*. These were managed and funded in diverse ways, as per ʿAlī Mubārak's 19th-century account as well as Comité documents. As many of the *sabīls* were also historic structures of artistic and scientific significance, the Comité and its members did feel that they had a responsibility towards them and strove to protect them by attempting to register them as monuments when possible, securing special funding for their restoration when needed, and ensuring that the Comité was involved in supervising restoration works. Nonetheless, it must be also noted that in many circumstances, the Comité's attention was drawn to a certain *sabīl* because of its *nāẓir*, who for example needed technical or financial assistance to restore the structure.

Focusing on the Comité's work on *sabīls*, this chapter has demonstrated, drawing on archival and published sources as well as in-situ evidence, that although there were limits to the extent of Comité involvement with *sabīls*, some general trends regarding its philosophy and methodologies can be traced. Particularly for *sabīls* that were still functioning as water fountains, the two examples presented here clearly show that the Comité members did not show any objection regarding their function as water fountains and did not at any point in time propose that their functioning cease. Moreover, my research on the Comité and its work on several *sabīls* has not found evidence showing that the Comité members at any point discussed their desire to discontinue the original function of *sabīls* in order to ensure their protection, and thus to transform them into purely aesthetic and historic monuments with no practical use. There is evidence that many historic *sabīls* did subscribe to the Cairo Water Company and continued to provide water during the Comité's lifetime. In fact, as demonstrated with the example of the *sabīl* of Sitt Ṣālḥa, it seems that this took place even after the building's relocation. Relocating *sabīls* to a new location was seen by the Comité as a solution to conserve many of these *sabīls* for future generations, especially at a time in the 20th century when numerous urban projects in Cairo (such as the widening of al-Khalīj al-Miṣrī street) put many *sabīls* at risk of demolition. It was also an action that required much coordination and administrative cooperation with other entities and interest groups, including but not limited to the Tanẓīm.

As for reinstating the function of other *sabīls* that were no longer supplying water, I believe that such an endeavour would have been beyond the Comité's responsibility and capacity, especially at a time when so many other players had more power to impact the transformation of Cairo's urban fabric and its water supply infrastructure.

With time, the Comité was able to register more *sabīls* of historical and artistic value, especially after the Law for the Protection of Arab Monuments was issued in 1918, and also after 1936 when the Comité was transferred to the purview of the Ministry of Public Instruction. Nonetheless, the Comité could only register buildings constructed up until the death of Mehmed Ali, which meant that many *sabīl*s built during the 19th and early 20th centuries were not under the purview of the Comité. The law issued in 1951 for the protection of Egyptian antiquities allowed for the registration of buildings constructed up until the end of the reign of Khedive Ismāʿīl, which meant that modern *sabīl*s could be registered as monuments. It was however also in the 1950s that several changes regarding *awqāf* and water infrastructure took place in Egypt; moreover it was the time when the Comité was transformed. What happened, however, to *sabīl*s after the 1950s as a result of these changes is a topic beyond the scope of this chapter and deserves further research, as it would further contribute to contextualizing the work of the Comité vis-à-vis later interventions.

Finally, I would like to note that the protection of *sabīl*s requires a multi-disciplinary research project that explores from the art-historical, conservational and legal perspectives what has led to the current condition of each of these *sabīl*s. The outcomes of such a study would not only be a valuable scholarly contribution, but would also impact the future protection of these structures. Well-informed research allows for the formulation of creative solutions regarding how to deal technically, managerially, legally and financially with these buildings, in a way that would ensure their long-term conservation and (adaptive) reuse. The only way to prevent this heritage from deterioration and/or demolition is to keep it alive and maintain the building as *ʿāmir*, even if through a new form of use.

Bibliography

Primary Sources

Bulletins of the Comité de conservation des monuments de l'art arabe, 41 volumes.

Mubārak, ʿAlī. *al-Khiṭāṭ al-Tawfīqiyya al-jadīda li-Miṣr al-Qāhira wa-mudunihā wa-bilādihā al-qadīma wa-l-shahīra*. al-Maṭbaʿa al-Kubra al-Amīriyya, 1887–1889.

Pauty, Edmond. "L'architecture au Caire depuis la conquête ottoman (vue d'ensemble) [avec 20 planches]." *Bulletin de l'Institut français d'archéologie orientale* 36 (1936): 1–69.

al-Sibāʿī, Yūsuf. *al-Saqqā māt*, 1952.

Secondary Sources

Abū al-ʿAmayim, Muḥammad. *Athār al-Qāhira al-islāmiyya fī al-ʿaṣr al-ʿUthmānī. Al-mujallad al-thānī. Al-asbila wa-aḥwād al-ṣaqqiyya*. Research Center for Islamic History, Art and Culture, 2011.

Agai, Bekim. "ʿAlī Mubārak." *Encyclopaedia of Islam Three Online*, edited by K. Fleet, G. Krämer, D. Matringe. J. Nawas and D.J. Steward. Brill, 2007. https://doi.org/10.1163/1573-3912_ei3_SIM_0046.

Aletihad News Centre. 'Sabīl wa-Kuttāb "al-Sitt Ṣāliḥa" yujāwir maqām al-Sayyida Zaynab'. 27 August 2011. https://www.aletihad.ae/article/79774/2011/%D8%B3%D8%A8%D9%8A%D9%84-%D9%88%D9%83%D9%8F%D8%AA%D8%A7%D8%A8-%C2%AB%D8%A7%D9%84%D8%B3%D8%AA-%D8%B5%D8%A7%D9%84%D8%AD%D8%A9%C2%BB-%D9%8A%D8%AC%D8%A7%D9%88%D8%B1-%D9%85%D9%82%D8%A7%D9%85-%D8%A7%D9%84%D8%B3%D9%8A%D8%AF%D8%A9-%D8%B2%D9%8A%D9%86%D8%A8.

Alleaume, Ghislaine. "Hygiène publique et travaux publics: Les ingénieurs et l'assainissement du Cairo (1882-1907)." *Annales Islamologiques* 20 (1984): 151–82.

Bakhoum, Dina Ishak. "The *Waqf* in Relation to Maintenance and Repair: The Medieval Sources and Their Uses for Contemporary Practices." M.A. thesis, The American University in Cairo, 2004.

Bakhoum, Dina Ishak. "The Madrasa of Umm al-Sultan Shaʿban before and after Creswell." In *Creswell Photographs Re-examined, Perspectives on Islamic Architecture*, edited by Bernard O'Kane, 99-119. The American University in Cairo Press, 2009.

Bakhoum, Dina Ishak. "The *Waqf* System: Maintenance, Repair and Upkeep." In *Held in Trust: Waqf in the Islamic World*, edited by Pascale Ghazaleh, 179–96. The American University in Cairo Press, 2011.

Bakhoum, Dina Ishak. "*Awqaf* Properties. Maintenance and Management." Report Produced in the framework of the Urban Regeneration Project for Historic Cairo, UNESCO, World Heritage Centre, 2014. https://www.urhcproject.org/content/studies/20_bakhoum.pdf, accessed April 2026.

Bakhoum, Dina Ishak. "Mamluk Minarets in Modern Egypt: Tracing Restoration Interventions." *Annales Islamologiques* 50 (2016): 147–98.

Bakhoum, Dina Ishak. "The Restorations of the *Comité de conservation des Monuments de l'Art Arabe* in Egypt: 'Conservation' or 'Reinvention' of Monuments?/Les restaurations du Comité de conservation des monuments de l'art arabe en Égypte: 'conservation' ou 'réinvention' des monuments?" PhD diss., Université Paris 1 Panthéon Sorbonne & Universiteit Leiden, 2021.

Bakhoum, Dina Ishak. "The *Comite de conservation des monuments de l'art arabe* (1881-1950s) and the secularization of the management of the conservation of religious heritage in Egypt." In *Pyramids and Progress. Perspectives on the Entanglement of Imperialisms and Early Egyptology (1800-1950). Proceedings of the International Symposium Held at Leuven & Brussels, 8-10 November 2021*, edited by Marleen De Meyer, Vincent Oeters, Jan Vandersmissen, and Harco Willems. Peeters Publishers, Orientalia Lovaniensia Analecta, in press.

Bakhoum, Dina Ishak. "Le classement partiel et tardif d'un *waqf* privé." In *La maison du cheikh al-Sadat, deux siècles d'histoire*, edited by Dina Ishak Bakhoum and Mercedes Volait. Institut français d'archéologie orientale, forthcoming.

Bent, Josephine van den, and Angela Isoldi. "Pleasing God, Serving the Citizens: Charity and Water Supply in Cairo and Baghdad." In *City, Citizens, Citizenship, 400-1500*, edited by Els Rose, Robert Flierman, Merel de Bruin-van de Beek, 199–229. Palgrave Macmillan, 2024. https://doi.org/10.1007/978-3-031-48561-9.

Bosworth, Clifford E. and Doris Behrens-Abouseif. "Sabīl." *Encyclopaedia of Islam New Edition Online (EI-2 English)*, edited by P. J. Bearman et al. Brill, 2012. https://doi.org/10.1163/1573-3912_islam_COM_0954.

Dobrowolska, Agnieszka and Khaled Fahmy. *Muhammad ʿAli Pasha and His Sabil*. The American University in Cairo Press, 2005.

Dobrowolska, Agnieszka and Jarosław Dobrowolski. *The Sultan's Fountain. An Imperial Story of Cairo, Istanbul and Amsterdam*. A Netherlands-Flemish Institute in Cairo Edition. The American University in Cairo Press, 2011.

Ebeid, Sophie W. "Early *sabīls* and their standardization." M.A. thesis, The American University in Cairo, 1976.

Griffiths, J. Gwyn. "Hecataeus and Herodotus on 'A Gift of the River'." *Journal of Near Eastern Studies* 25, no. 1 (1966): 57–61.

El-Habashi, Alaa El-Din [al-Ḥabashī, ʿAlāʾ], and Nicholas Warner. "Recording the Monuments of Cairo: an Introduction and Overview." *Annales Islamologiques* 32 (1998): 81–99.

El-Habashi, Alaa El-Din. "*Athar* to Monuments: The Intervention of the *Comité de Conservation des Monuments de l'Art Arabe.*" PhD diss., University of Pennsylvania, 2001.

El-Habashi, Alaa El-Din, and Hanan Abd-Elrahman Nada. "Monuments or Heritage Buildings: The Historic Water Dispensaries (Sabils) of Cairo." In *Proceedings of the IX International Forum "Le Vie dei Mercanti"*, 1–12. La scuola di Pitagora, 2011.

al-Ḥusaynī, Maḥmūd Ḥamīd. *al-Asbila al-ʿUthmāniyya bi-madinat al-Qāhira 1517-1798*. Maktabat Madbūlī, 1977.

"Citadelle du Caire." *Institut français d'archéologie orientale*. Accessed February 15, 2026. https://www.ifao.egnet.net/recherche/archeologie/citadelle-du-caire/.

"Technical Glossary." *Islamic Art Network*. Accessed February 15, 2026. https://www.islamic-art.org/Glossary/.

Ismail, Shehab. "Engineering Metropolis: Contagion, Capital, and the Making of British Colonial Cairo, 1882-1922." PhD diss., Columbia University, 2017.

Ismail, Shehab. "Epicure and Experts: The Drinking Water Controversy in British Egypt." *The Arab Studies Journal* 26, no. 2 (2018): 8–43.

Lane, Edward William. "An Arabic-English Lexicon." Accessed February 15, 2026. https://lexicon.quranic-research.net/data/18_E/196_Emr.html

Mahdy, Hossam. "Decolonizing the Conservation of Islamic Built Heritage in Egypt." In *Islamic Architecture Today and Tomorrow. (Re)Defining the Field*, edited by Mohammad Gharipour and Daniel E. Coslett, 373–86. Intellect Books, 2022.

Mostafa, Heba. "The Nile as Nexus. The Nilometer at al-Rawda Island between Veneration and Mediation in Medieval Islamic Egypt." In *The Nile Delta: Histories from Antiquity to the Modern Period*, edited by Katherine Blouin, 421–52. Cambridge University Press, 2024.

Mostafa, Saleh Lamei. "The Cairene *Sabil*: Form and Meaning." *Muqarnas* 6 (1989): 33–42.

Pasquier, Émilie. "La société des eaux du Caire (1865-1919): entre bien commun et intérêts privés." *Histoire Politique* 48 (2022), https://doi.org/10.4000/histoirepolitique.8478.

Pasquier, Émilie. *La société des eaux du Caire: Une histoire de l'équipement en infrastructures hydrauliques de la capitale Égyptienne 1865-1954*. Éditions Ithaka, 2023.

Raymond, André. "Les porteurs d'eau du Caire." *Bulletin de l'Institut français d'archéologie orientale* 57 (1958): 183–202.

Raymond, André. "Les fontaines publiques (*sabīl*) du Caire à l'époque ottomane (1517-1798) (I) [avec 1 carte]." *Annales Islamologiques* 15 (1976): 235–91.

Saul, Samir. *La France et l'Égypte de 1882 à 1914: Intérêts économiques et implications politiques*. Institut de la gestion publique et du développement économique, 1997.

Speiser, Philipp. *Die Geschichte der Erhaltung arabischer Baudenkmäler in Ägypten. Die Restaurierung der Madrasa Tatar al-Ḥiǧāzīya und des Sabīl Kuttāb ʿAbd ar-Raḥmān Katḫudā im Rahmen des Darb-al-Qirmiz-Projektes in Kairo*. Heidelberger Orientverlag, 2001.

Schultz, Alexandra Camille. "Living and Dying in Water: Fluid Infrastructure Disruptions in Urban Egypt (1870-1935)." PhD diss., University of California, 2022.

Volait, Mercedes. "Le «sébile» moderne." Last accessed February 15, 2026. https://sites.duke.edu/caire-moderne/2019/10/26/le-sebile-moderne-السبيل-الحديث/, October 2019.

Vollers, K. "ʿAlī Pas̲h̲a Mubārak." *Encyclopaedia of Islam New Edition Online (EI-2 English)*, edited by P. J. Bearman et al. Brill, 2012. doi: https://doi.org/10.1163/1573-3912_islam_COM_0954.

Warner, Nicholas. *The Monuments of Historic Cairo: A Map and Descriptive Catalogue*. The American Research Center in Egypt (Conservation Series I). The American University in Cairo Press, 2005.

CHAPTER 12

Afterword

Alan Mikhail

Taken together, this book's chapters represent a significant advance in the study of the environmental history of the Middle East. With its creative use of different kinds of sources and original conceptualization, it offers us grounded histories that extend the environmental history of the Middle East in time and space.

The focus on water is the first reason for this volume's importance. It is, as the editors, Josephine van den Bent, Maaike van Berkel, and Edmund Hayes, say, the universal substance. Water is most of the world. It courses through us and surrounds us. We drink it, worship it, fear it, bathe and swim in it, point to it as evidence of life on other planets, waste it, contemplate it, fight over it, store it, and move on it. Most of human history happened in proximity to water. Without it, there would have been no irrigation for the beginning of agriculture, no fish protein, no fresh drinking water. Bound as we are to water, it is the substance most vital to our existence and thus the one that can most easily destroy us. It is indeed "urgent," as Ibn Khaldūn offers. Too much or too little of it kills us. Two hydrogen atoms bonded to an oxygen atom have shaped the earth and all living beings on it arguably more than anything else.

In the Middle East, these universalities have usually been studied as the story of large states and their massive projects of environmental manipulation—aqueducts, canals, dams, or the Nilometer, for example. Some scholars have, alternatively, examined small-scale, often informal communities built around the sharing of water resources, or how groups at some remove from state power have organized their reliance on water. We might also think here of stories of resistance against aqueous grand design—farmers cutting into canals to siphon off water, cultivators withholding foodstuffs, or people fleeing. These are the two ends of the spectrum captured in this book's title: forces from "above" and "below."

Part of the importance of this book's casting is to focus not on either extreme of human interaction with water, but on the spaces between them. The book dwells in the connective zones between massive state projects and the creativity of local communities. Sidestepping simple stories of oppression or resistance, we are instead in the middle realm, we might say, where arguably the majority of history took place, or, more modestly, where the necessary work of connecting state and society functioned. Such a focus serves as an important contribution to the environmental history of water, in the Middle East and beyond.

This book's mesohistory of water thinks through questions of temporality and geography. Most, but significantly not all, Middle East environmental history has tended to cluster around two broad chronological foci. There is scholarship—largely from an earlier generation, Middle East environmental history *avant la lettre*, we might say—interested in the ancient states of the region. It is from this work that well-known grand theories of state, stasis, and collapse have been derived. There are also newer studies generally focused on the early modern and modern periods that have tended toward more social historical work on the environmental histories of the Ottoman Empire and its successor states in the modern period. This book is one of the first to cover that massive period in between these two areas of scholarly concentration. In the non-environmental study of the Middle East, there has long been a focus on these earlier periods of Islamic history, the period before and after the rise of Islam and the so-called classical period. In offering a robust set of environmental analyses of these crucial centuries of Islamic and Middle Eastern history, this book then helps to set the agenda for what might come next, and will no doubt serve as a catalyst for more work on these periods.

Moreover, by covering much of the region, this volume differs from most work done in Middle East environmental history that tends to focus on only one area, or perhaps on relations between an imperial centre or city and one particular hinterland. As the editors say, only by studying the environmental history of the Middle East *as a region*, instead of as a collection of case studies, can we evaluate supposed differences, exceptions, and commonalities. Bringing together Iraq, Syria, Egypt, and, perhaps most importantly because it is understudied, the Hijaz, this book makes a better case than most for being a history of the Middle East, with all its ecological diversity. It is a collective effort, of course, suggesting a model for collaborative work in environmental history.

Several specific contributions derive from this volume's temporal and geographic focus on the waters in-between. The first are elaborations of notions of governance. Nathalie de Haan offers us the idea of "polycentric governance," one with precedents in the Roman past that helps us to conceive of how imperial rule functioned later in the Middle East. Cities and the hinterlands upon which they depended operated as units within the larger imperial realm. Water, and the infrastructure built to harness this vital substance to allow cities and countrysides to flourish, prove some of the most important lenses through which to think about these dynamics of rule. Cities were semi-autonomous but ultimately subservient to the larger polity. As the editors make clear in the introduction, an environmental analysis of *urban* governance necessarily becomes a history of connections between cities and the countryside, cities and the empires to which they belonged.

This helps us to break a conception of the Middle East as a hierarchically organized set of actors: state elites, provincial governors, and the mass of farmers. Rather, we have a more nuanced and complex picture in which power did not simply emanate out from Cairo or Baghdad; power existed in multiple places throughout an imperial realm, often independently from the imperial centre. The picture in our heads should be one of undulating three-dimensional plains of various thickness, instead of straight lines and discrete points.

To populate this more complicated scene, this volume keeps the focus on an interesting set of actors, people like water carriers, *sabīl* managers, pipefitters, caravan leaders, families with cisterns, small-scale infrastructural tinkerers, craftspeople, janitors, doormen, and even a lone female water merchant in Baghdad. In line with the book's themes, these are, we might say, middling peoples, not state elites and not farmers, but those within the city-countryside complex engaged in supporting all the connective work of water management. Echoing the volume's editors, these people *are* the system, not simply cogs in its machine. The picture is not only a livelier one than we usually get but also one that allows us to ask different questions about the economy, labour, and expertise—questions that revolve around the larger notion of governance.

Another major contribution of this book is that it presents a new analysis of a kind of infrastructure I have come to think of as rhizomatic infrastructure. This is infrastructure that is again neither of the monumental sort—say a huge canal or dam project—nor of the simple variety—say a dug basin or cut canal. It is the material world made up of waterpipes, wells, cisterns, gutters, waterskins, modest *sabīl*s, spouts, and the like. These are the figurative and literal conduits that allowed water to move, diffuse, and to be directed where human and other animal actors desired and needed it. Along with others, the middling individuals discussed above manufactured and maintained these necessary components of urban waterworks as part of the "diversified craft sector" identified by Marcus Milwright. This rhizomatic infrastructure represented, we might venture, the majority of the networked water infrastructure of Middle Eastern cities and also, as Josephine van den Bent says, proved a major reason for the system's overall resiliency. In a place like Abbasid Baghdad, a city of nearly half a million people, the enormity of the project of allowing water to flow to all these people in far-flung neighbourhoods throughout the city, the challenging final mile, was never the work of a single large canal or the river itself. It was rather the pipes, spouts, fountains, and spigots that made all of this possible. As this volume shows us, these waterworks deserve our attention.

The in-between artisans and the vital waterworks they made possible point to this book's important analytical argument about the dual nature of infrastructure as both tangible and intangible. Infrastructure is never simply the material objects that channel and harness water, but also the traditions, ethical imperatives, social institutions, and practices that govern and shape the uses of these physical things. Infrastructure is not a static object but a process. The infrastructures of water examined in this book created arenas of connection, shared reliance, and negotiation that had to be mediated among all those implicated in their use. The material and the discursive work together dialectically, of course, but this is never straightforward. This volume offers a theory of the alloyed nature of infrastructure and examples of the intricacies of its function.

The book's chapters offer other novel approaches for environmental history as well. One is struck, for example, in Peter Brown's chapter by the discussion of the environmental politics of shade and the necessity of understanding the seasonality of the hajj, or by Edmund Hayes's analysis of distance and pipe seepage or the ways ritual purity laws might reflect lived realities. What would it mean to understand the geography of the Middle East as, more than anything, a honeycomb of cisterns, as Bethany Walker suggests? How should we think about the cultural practice of naming irrigation works like cisterns? What does this say about their place in society? Thanks to this book, these are all new avenues to pursue, beyond of course the story of the management of urban water supply and the relationships it forged.

In all of this, the question of sources is a central one, and this too is an area in which this volume makes robust contributions. Judiciously and creatively combining archaeological and textual evidence—an echo of the dialectics of infrastructure—allows many of the book's authors to get at subjects that might not have been accessible through only one kind of source base. From the archaeological record, we learn, for example, of water harvesting techniques in Tall Ḥisbān that are invisible in the textual record, when it exists. Dina Ishak Bakhoum alerts us in her study of Cairo *sabīls* to how the material past can be manipulated over time for political and ideological purposes.

In some places, like the Hijaz, as Harry Munt notes, we have scant extant archaeological evidence and therefore must rely almost entirely on the written record. This is obviously not a given. As Vanessa Van Renterghem says, rare is the source dedicated to water or water management (*waqf* documents about *sabīls* or other relevant structures prove exceptional in this regard, as Angela Isoldi notes). Rather, we must read for the passing reference or incidental account of water. Often the sources that remain are normative and prescriptive and do not tell us about the actual social practices in question. To get at that lived reality of water usage, authors in his volume have employed unique techniques. For example, in reading rare and highly specific examples from the hadith and *fiqh*, some of the most normative of all texts in the Islamic tradition, Edmund Hayes argues that these "are less likely to be catechistic rehearsals of standard juristic positions, and more likely to be specific cases that arose in fact." In this way, one can read reality into these sources to understand the history of water management in the early Islamic period. This allows both for an understanding of the general relations between Islamic thought and the natural world and a kind of social history of environmental practice. Throughout the volume, authors employ written sources such as *ḥisba* and *kharāj* manuals, legal documents, historical chronicles, *waqfiyyāt* and *sijillāt*, poetry, and geographies—all to great effect and often, again, in conjunction with archaeological and other material evidence.

The archaeological and the textual record offer proof of the consistency and indispensability of the history of water and water management in the Middle East over millennia. This makes the Middle East both unique and essential for understanding the global environmental history of water. Unlike most regions of the world, perhaps excluding China and India, the Middle East possesses a material and written record of this history since antiquity. With this volume, we now have histories of the region that we did not possess before—those of the early Islamic and medieval periods that push into new geographies, questions, and sources. Like its subject, this book thus provides vital sustenance for many fields to grow.

About the Authors

Dina Ishak Bakhoum (PhD) is an engineer and art historian specializing in cultural heritage conservation and management (2000–present). Her research focuses on Egypt's Islamic and Coptic architecture, the *waqf* as a maintenance system, the history of architectural conservation and restoration in Egypt as well as contemporary conservation issues of living heritage sites.

Josephine van den Bent is assistant professor of medieval history at the University of Amsterdam. Her research interests include ethnicity and ethnic stereotyping, urban organisation, and in general the social and cultural history of the premodern Islamicate world. She previously worked as a postdoctoral researcher on the NWO Vici-project "Source of Life" at Radboud University.

Maaike van Berkel is professor of medieval history at Radboud University, Nijmegen. She is a social and cultural historian of the Middle East, with a particular interest in court culture and urban history. Her current research project investigates water management in the premodern Middle Eastern city.

Peter J. Brown is a landscape archaeologist of the medieval world. His research focuses on interactions between medieval societies and their environments, especially in relation to water management. He is currently a consultant for the Heritage Commission, KSA, and an honorary fellow in the Department of Archaeology, Durham University, UK.

Nathalie de Haan is an associate professor in ancient history at Radboud University, Nijmegen. Her research interests include private and public baths and Roman bathing culture, ancient water management, and Reception Studies.

Edmund Hayes is a lecturer at Leiden University. His research focus is the social history of early Shiʿa Islam, which gave rise to his 2022 monograph, *Agents of the Hidden Imam: Forging Twelver Shiʿism, 850-950 CE*, and a 2024–2028 ERC project "Embodied Imamate: Mapping the Development of the Early Shiʿi Community 700–900 CE."

Angela Isoldi specialises in the use of premodern Arabic manuscript sources. She earned her PhD from Radboud University in 2026. Her doctoral research, largely based on the analysis of Mamluk-era *waqf* documents, reconstructed the complex intersections of social agents, economic dynamics, and environmental factors that shaped urban water management in Mamluk Cairo.

Alan Mikhail is the Chace Family Professor of History at Yale University. He is the author of six books, including *Nature and Empire in Ottoman Egypt: An Environmental History*, *The Animal in Ottoman Egypt*, and *Under Osman's Tree: The Ottoman Empire, Egypt, and Environmental History*.

Marcus Milwright is British Academy Global Professor at the Department of History of Art, University of York. His research focuses on Islamic archaeology, craft practice in the Middle East, and cross-cultural interaction. His books include *Islamic Arts and Crafts: An Anthology* (2017) and *A Story of Islamic Art* (2024).

Harry Munt is Senior Lecturer in Medieval History at the University of York. He is the author of several articles on the history of the early Islamic Arabian Peninsula as well as the book *The Holy City of Medina: Sacred Space in Early Islamic Arabia* (Cambridge University Press, 2014).

Vanessa Van Renterghem is Professor of Arab Medieval History at Inalco (Paris). She specialises in social and urban history, particularly of Baghdad and Iraq (11th–13th centuries), and is also interested in Arab historiography, social categorisation, and gender issues in the medieval Middle East.

Bethany Walker, Professor of Islamic Archaeology (University of Bonn), is a historian and archaeologist, with thirty years of field experience. Her publications include *Life on the Farm in Late Medieval Jerusalem*, *Jordan in the Late Middle Ages: Transformation of the Mamluk Frontier*, and the *Oxford Handbook of Islamic Archaeology*. She edits the *Journal of Islamic Archaeology*.

Index

FSC
www.fsc.org
MIX
Papier | Fördert gute Waldnutzung
FSC® C083411

Kolibri 360 GmbH
Ferdinand-Jühlke-Straße 7
99095 Erfurt, Deutschland
produktsicherheit@kolibri360.de